TRUE DIGITAL CONTROL

TRUE DIGITAL CONTROL

STATISTICAL MODELLING AND NON-MINIMAL STATE SPACE DESIGN

C. James Taylor, Peter C. Young and Arun Chotai
Lancaster University, UK

WILEY

This edition first published 2013

Registered office
John Wiley & Sons Ltd, The Atrium, Southern Gate, Chichester, West Sussex, PO19 8SQ, United Kingdom

For details of our global editorial offices, for customer services and for information about how to apply for permission to reuse the copyright material in this book please see our website at www.wiley.com.

Library of Congress Cataloging-in-Publication Data

Taylor, C. James.
True digital control : statistical modelling and non-minimal state space design / C. James Taylor, Peter C. Young, Arun Chotai.
pages cm
Includes bibliographical references and index.
ISBN 978-1-118-52121-2 (cloth)
1. Digital control systems–Design. I. Young, Peter C., 1939- II. Chotai, Arun. III. Title.
TJ223.M53T38 2013
629.8′95–dc23

2013004574

A catalogue record for this book is available from the British Library

ISBN: 978-1-118-52121-2

Typeset in 10/12pt Times by Aptara Inc., New Delhi, India

Printed and bound in Singapore by Markono Print Media Pte Ltd

1 2013

To Ting-Li

To Wendy

In memory of Varsha

Contents

Preface

This book develops a *True Digital Control* (TDC) design philosophy that encompasses data-based (statistical) model identification, through to control algorithm design, robustness evaluation and implementation. Treatment of both stochastic system identification and control design under one cover highlights the important connections between these disciplines: for example, in quantifying the model uncertainty for use in closed-loop stochastic sensitivity analysis. More generally, the foundations of linear state space control theory that are laid down in early chapters, with *Non-Minimal State Space* (NMSS) design as the central worked example, are utilised subsequently to provide an introduction to other selected topics in modern control theory. MATLAB®[1] functions for TDC design and MATLAB® scripts for selected examples are being made available online, which is important in making the book accessible to readers from a range of academic backgrounds. Also, the CAPTAIN Toolbox for MATLAB®, which is used for the analysis of all the modelling examples in this book, is available for free download. Together, these contain computational routines for many aspects of model identification and estimation; for NMSS design based on these estimated models; and for offline signal processing. For more information visit: http://www.wiley.com/go/taylor.

The book and associated software are intended for students, researchers and engineers who would like to advance their knowledge of control theory and practice into the state space domain; and control experts who are interested to learn more about the NMSS approach promoted by the authors. Indeed, such non-minimal state feedback is utilised throughout this book as a unifying framework for generalised digital control system design. This includes the *Proportional-Integral-Plus* (PIP) control systems that are the most natural outcome of the NMSS design strategy. As such, the book can also be considered as a primer for potentially difficult topics in control, such as optimal, stochastic and multivariable control.

As indicated by the many articles on TDC that are cited in this book, numerous colleagues and collaborators have contributed to the development of the methods outlined. We would like to pay particular thanks to our good friend Dr Wlodek Tych of the Lancaster Environment Centre, Lancaster University, UK, who has contributed to much of the underlying research and in the development of the associated computer algorithms. The first author would also like to thank Philip Leigh, Matthew Stables, Essam Shaban, Vasileios Exadaktylos, Eleni Sidiropoulou, Kester Gunn, Philip Cross and David Robertson for their work on some of the practical examples highlighted in this book, among other contributions and useful discussions while they studied at Lancaster. Philip Leigh designed and constructed the Lancaster forced

[1] MATLAB®, The MathWorks Inc., Natick, MA, USA.

ventilation test chamber alluded to in the text. Vasileios Exadaktylos made insightful suggestions and corrections in relation to early draft chapters of the book. The second author is grateful to a number of colleagues over many years including: Charles Yancey and Larry Levsen, who worked with him on early research into NMSS control between 1968 and 1970; Jan Willems who helped with initial theoretical studies on NMSS control in the early 1970s; and Tony Jakeman who helped to develop the *Refined Instrumental Variable* (RIV) methods of model identification and estimation in the late 1970s. We are also grateful to the various research students at Lancaster who worked on PIP methods during the 1980s and 1990s, including M.A. Behzadi, Changli Wang, Matthew Lees, Laura Price, Roger Dixon, Paul McKenna and Andrew McCabe; to Zaid Chalabi, Bernard Bailey and Bill Day, who helped to investigate the initial PIP controllers for the control of climate in agricultural glasshouses at the Silsoe Research Institute; and to Daniel Berckmans and his colleagues at the University of Leuven, who collaborated so much in later research on the PIP regulation of fans for the control of temperature and humidity in their large experimental chambers at Leuven.

Finally, we would like to express our sincere gratitude to the UK Engineering and Physical Sciences, Biotechnology and Biological Sciences, and Natural Environmental Research Councils for their considerable financial support for our research and development studies at Lancaster University.

C. James Taylor, Peter C. Young and Arun Chotai
Lancaster, UK

List of Acronyms

ACF	AutoCorrelation Function
AIC	Akaike Information Criterion
AML	Approximate Maximum Likelihood
AR	Auto-Regressive
ARIMAX	Auto-Regressive Integrated Moving-Average eXogenous variables
ARMA	Auto-Regressive Moving-Average
ARMAX	Auto-Regressive Moving-Average eXogenous variables
ARX	Auto-Regressive eXogenous variables
BIC	Bayesian Information Criterion
BJ	Box–Jenkins
CAPTAIN	Computer-Aided Program for Time series Analysis and Identification of Noisy systems
CLTF	Closed-Loop Transfer Function
CT	Continuous-Time
DARX	Dynamic Auto-Regressive eXogenous variables
DBM	Data-Based Mechanistic
DC	Direct Current
DDC	Direct Digital Control
DF	Directional Forgetting
DT	Discrete-Time
DTF	Dynamic Transfer Function
EKF	Extended or generalised Kalman Filter
EWP	Exponential-Weighting-into-the-Past
FACE	Free-Air Carbon dioxide Enrichment
FIR	Finite Impulse Response
FIS	Fixed Interval Smoothing
FPE	Final Prediction Error
GBJ	Generalised Box–Jenkins
GPC	Generalised Predictive Control
GRIVBJ	Generalised RIVBJ or RIVCBJ
GRW	Generalised Random Walk
GSRIV	Generalised SRIV or SRIVC
IPM	Instrumental Product Matrix
IRW	Integrated Random Walk

IV	Instrumental Variable
IVARMA	Instrumental Variable Auto-Regressive Moving-Average
KF	Kalman Filter
LEQG	Linear Exponential-of-Quadratic Gaussian
LLS	Linear Least Squares
LLT	Local Linear Trend
LPV	Linear Parameter Varying
LQ	Linear Quadratic
LQG	Linear Quadratic Gaussian
LTR	Loop Transfer Recovery
MCS	Monte Carlo Simulation
MFD	Matrix Fraction Description
MIMO	Multi-Input, Multi-Output
MISO	Multi-Input, Single-Output
ML	Maximum Likelihood
MPC	Model Predictive Control
NEVN	Normalised Error Variance Norm
NLPV	Non-Linear Parameter Varying
NMSS	Non-Minimal State Space
NSR	Noise–Signal Ratio
NVR	Noise Variance Ratio
PACF	Partial AutoCorrelation Function
PBH	Popov, Belevitch and Hautus
PEM	Prediction Error Minimisation
PI	Proportional-Integral
PID	Proportional-Integral-Derivative
PIP	Proportional-Integral-Plus
PRBS	Pseudo Random Binary Signal
RBF	Radial Basis Function
RIV	Refined Instrumental Variable
RIVAR	Refined Instrumental Variable with Auto-Regressive noise
RIVBJ	Refined Instrumental Variable for Box–Jenkins models
RIVCBJ	Refined Instrumental Variable for hybrid Continuous-time Box–Jenkins models
RLS	Recursive Least Squares
RML	Recursive Maximum Likelihood
RW	Random Walk
RWP	Rectangular-Weighting-into-the-Past
SD	Standard Deviation
SDARX	State-Dependent Auto-Regressive eXogenous variables
SDP	State-Dependent Parameter
SE	Standard Error
SISO	Single-Input, Single-Output
SP	Smith Predictor
SRIV	Simplified Refined Instrumental Variable
SRIVC	Simplified Refined Instrumental Variable for hybrid Continuous-time models
SRW	Smoothed Random Walk

SVF	State Variable Feedback
TDC	True Digital Control
TF	Transfer Function
TFM	Transfer Function Matrix
TVP	Time Variable Parameter
YIC	Young Information Criterion

List of Examples, Theorems and Estimation Algorithms

Examples

Theorems

Estimation Algorithms

1

Introduction

Until the 1960s, most research on model identification and control system design was concentrated on continuous-time (or analogue) systems represented by a set of linear differential equations. Subsequently, major developments in discrete-time model identification, coupled with the extraordinary rise in importance of the digital computer, led to an explosion of research on discrete-time, sampled data systems. In this case, a 'real-world' continuous-time system is controlled or 'regulated' using a digital computer, by sampling the continuous-time output, normally at regular sampling intervals, in order to obtain a discrete-time signal for sampled data analysis, modelling and *Direct Digital Control* (DDC). While adaptive control systems, based directly on such discrete-time models, are now relatively common, many practical control systems still rely on the ubiquitous 'two-term', *Proportional-Integral* (PI) or 'three-term', *Proportional-Integral-Derivative* (PID) controllers, with their predominantly continuous-time heritage. And when such systems, or their more complex relatives, are designed offline, rather than 'tuned' online, the design procedure is often based on traditional continuous-time concepts. The resultant control algorithm is then, rather artificially, 'digitised' into an approximate digital form prior to implementation.

But does this 'hybrid' approach to control system design really make sense? Would it not be both more intellectually satisfying and practically advantageous to evolve a unified, truly digital approach, which would allow for the full exploitation of discrete-time theory and digital implementation? In this book, we promote such a philosophy, which we term *True Digital Control* (TDC), following from our initial development of the concept in the early 1990s (e.g. Young *et al.* 1991), as well as its further development and application (e.g. Taylor *et al.* 1996a) since then. TDC encompasses the entire design process, from data collection, data-based model identification and parameter estimation, through to control system design, robustness evaluation and implementation. The TDC approach rejects the idea that a digital control system should be initially designed in continuous-time terms. Rather it suggests that the control systems analyst should consider the design from a digital, sampled-data standpoint throughout. Of course this does not mean that a continuous-time model plays no part in TDC design. We believe that an underlying and often physically meaningful continuous-time model should still play a part in the TDC system synthesis. The designer needs to be assured that the

---

True Digital Control: Statistical Modelling and Non-Minimal State Space Design, First Edition.
C. James Taylor, Peter C. Young and Arun Chotai.

discrete-time model provides a relevant description of the continuous-time system dynamics and that the sampling interval is appropriate for control system design purposes. For this reason, the TDC design procedure includes the data-based identification and estimation of continuous-time models.

One of the key methodological tools for TDC system design is the idea of a *Non-Minimal State Space* (NMSS) form. Indeed, throughout this book, the NMSS concept is utilised as a unifying framework for generalised digital control system design, with the associated *Proportional-Integral-Plus* (PIP) control structure providing the basis for the implementation of the designs that emanate from NMSS models. The generic foundations of linear state space control theory that are laid down in early chapters, with NMSS design as the central worked example, are utilised subsequently to provide a wide ranging introduction to other selected topics in modern control theory.

We also consider the subject of stochastic system identification, i.e. the estimation of control models suitable for NMSS design from noisy measured input–output data. Although the coverage of both system identification and control design in this unified manner is rather unusual in a book such as this, we feel it is essential in order to fully satisfy the TDC design philosophy, as outlined later in this chapter. Furthermore, there are valuable connections between these disciplines: for example, in identifying a parametrically efficient (or parsimonious) 'dominant mode' model of the kind required for control system design; and in quantifying the uncertainty associated with the estimated model for use in closed-loop stochastic uncertainty and sensitivity analysis, based on procedures such as *Monte Carlo Simulation* (MCS) analysis.

This introductory chapter reviews some of the standard terminology and concepts in automatic control, as well as the historical context in which the TDC methodology described in the present book was developed. Naturally, subjects of particular importance to TDC design are considered in much more detail later and the main aim here is to provide the reader with a selective and necessarily brief overview of the control engineering discipline (sections 1.1 and 1.2), before introducing some of the basic concepts behind the NMSS form (section 1.3) and TDC design (section 1.4). This is followed by an outline of the book (section 1.5) and concluding remarks (section 1.6).

1.1 Control Engineering and Control Theory

Control engineering is the science of altering the dynamic behaviour of a physical process in some desired way (Franklin *et al.* 2006). The scale of the process (or system) in question may vary from a single component, such as a mass flow valve, through to an industrial plant or a power station. Modern examples include aircraft flight control systems, car engine management systems, autonomous robots and even the design of strategies to control carbon emissions into the atmosphere. The control systems shown in Figure 1.1 highlight essential terminology and will be referred to over the following few pages.

This book considers the development of digital systems that control the output variables of a system, denoted by a vector $\boldsymbol{y}$ in Figure 1.1, which are typically positions or levels, velocities, pressures, torques, temperatures, concentrations, flow rates and other measured variables. This is achieved by the design of an online control algorithm (i.e. a set of rules or mathematical equations) that updates the control input variables, denoted by a vector $\boldsymbol{u}$ in

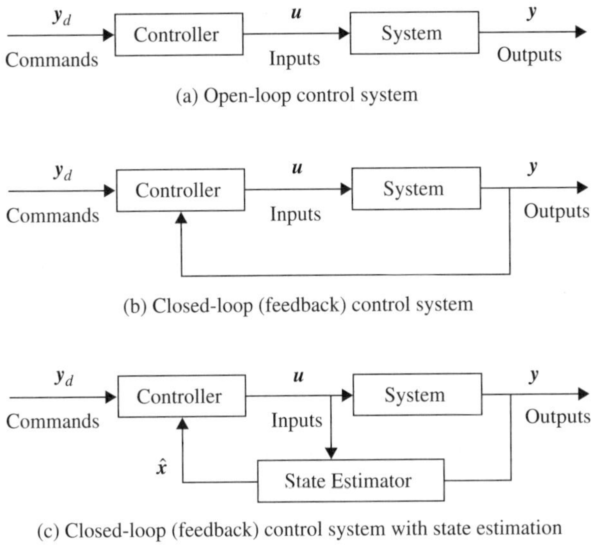

Figure 1.1 Three types of control system

Figure 1.1, automatically and without human intervention, in order to achieve some defined control objectives. These control inputs are so named because they can directly change the behaviour of the system. Indeed, for modelling purposes, the engineering system under study is *defined* by these input and output variables, and the assumed causal dynamic relationships between them. In practice, the control inputs usually represent a source of energy in the form of electric current, hydraulic fluid or pneumatic pressure, and so on. In the case of an aircraft, for example, the control inputs will lead to movement of the ailerons, elevators and fin, in order to manipulate the attitude of the aircraft during its flight mission. Finally, the command input variables, denoted by a vector $\boldsymbol{y}_d$ in Figure 1.1, define the problem dependent 'desired' behaviour of the system: namely, the nature of the short term pitch, roll and yaw of an aircraft in the local reference frame; and its longer term behaviour, such as the gradual descent of an aircraft onto the runway, represented by a time-varying altitude trajectory.

Control engineers design the 'Controller' in Figure 1.1 on the basis of control system design theory. This is normally concerned with the mathematical analysis of dynamical systems using various analytical techniques, often including some form of optimisation over time. In this latter context, there is a close connection between control theory and the mathematical discipline of optimisation. In general terms, the elements needed to define a control optimisation problem are knowledge of: (i) the dynamics of the process; (ii) the system variables that are observable at a given time; and (iii) an optimisation criterion of some type.

A well-known general approach to the optimal control of dynamic systems is 'dynamic programming' evolved by Richard Bellman (1957). The solution of the associated *Hamilton–Jacobi–Bellman* equation is often very difficult or impossible for nonlinear systems but it is feasible in the case of linear systems optimised in relation to quadratic cost functions with quadratic constraints (see later and Appendix A, section A.9), where the solution is a 'linear feedback control' law (see e.g. Bryson and Ho 1969). The best-known approaches of this type

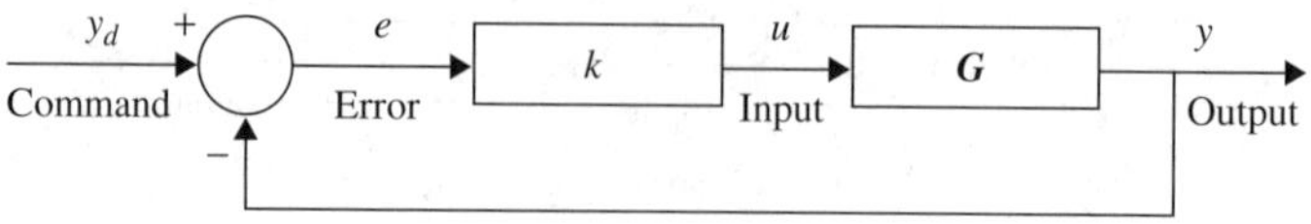

Figure 1.2 The archetypal negative feedback system

are the *Linear-Quadratic* (LQ) method for deterministic systems; and the *Linear-Quadratic-Gaussian* (LQG) method for uncertain stochastic systems affected by noise. Here the system relations are linear, the cost is quadratic and the noise affecting the system is assumed to have a Gaussian, 'normal' amplitude distribution. LQ and LQG optimal feedback control are particularly important because they have a complete and rigorous theoretical background, while at the same time introduce key concepts in control, such as 'feedback', 'controllability', 'observability' and 'stability' (see later).

Figure 1.1b and Figure 1.1c show two examples of such closed-loop feedback control. These are in contrast to the open-loop formulation of Figure 1.1a, where, given advanced knowledge of $\boldsymbol{y}_d$, a sequence of decisions $\boldsymbol{u}$ could be determined offline, i.e. there is no feedback. The potential advantages of closed-loop feedback control are revealed by Figure 1.2, where a *single-input, single-output* (SISO) system is denoted by $\boldsymbol{G}$ and k is a simple feedback control gain (that is adjusted by the control systems designer).

It is easy to see that $y = k\,\boldsymbol{G}\,e$, where $e = y_d - y$ is the error between the desired output and the actual output, so that after a little manipulation,

$$y = \frac{k\,\boldsymbol{G}}{1 + k\,\boldsymbol{G}} y_d \tag{1.1}$$

This is a fundamental relationship in feedback control theory and we see that if $k\,\boldsymbol{G} \gg 1$ *and provided the closed-loop system remains stable*, then the ratio $k\boldsymbol{G}/(1 + k\boldsymbol{G})$ approaches unity and the control objective $y = y_d$ is achieved, *regardless of the system* $\boldsymbol{G}$. Of course, this is a very simplistic way of looking at closed-loop control and ensuring that the gain k is selected so that stability is maintained and the objective is achieved, can be a far from simple problem. Nevertheless, this kind of thinking, followed in a more rigorous fashion, shows that the main advantages of a *well designed* closed-loop control system include: improved transient response; decreased sensitivity to uncertainty in the system (such as modelling errors); decreased sensitivity to disturbances that may affect the system and tend to drive it away from the desired state; and the stabilisation of an inherently unstable system. One property that the high gain control of equation (1.1) does not achieve is the *complete* elimination of steady-state errors between y and y_d, which only occurs when the gain k is infinite, unless the system $\boldsymbol{G}$ has special properties. But we will have more to say about this later in the chapter.

The disadvantage of a closed-loop system is that it may be more difficult to design because it has to maintain a good and stable response by taking into account the potentially complex manner in which the dynamic system and its normally dynamic controller interact within the closed-loop. And it may be more difficult (and hence more expensive) to implement because it requires sensors to accurately measure the output variables, as well as the construction of either analogue or digital controller mechanisms. Control system design theory may also have to account for the uncertain or 'stochastic' aspects of the system and its environment, so that key variables may be unknown or imperfectly observed. This aspect of the problem is implied by Figure 1.1c, in which an 'observer' or 'state estimator' is used to estimate the system state $\boldsymbol{x}$ from the available, and probably noisy, measurements $\boldsymbol{y}$. However, in order to introduce the concept of the system state, we need to return to the beginnings of control theory.

1.2 Classical and Modern Control

The mathematical study of feedback dynamics can be traced back to the nineteenth century, with the analysis of the stability of control systems using differential equations. James Clerk Maxwell's famous paper *On Governors*[1], for example, appeared in 1868 and the stability criteria introduced by Routh in 1877 (Routh 1877) are still commonly taught in undergraduate control courses. With the development of long distance telephony in the 1920s, the problem of signal distortion arose because of nonlinearities in the vacuum tubes used in electronic amplifiers. With a feedback amplifier, this distortion was reduced. Communications engineers at Bell Telephone Laboratories developed graphical methods for analysing the stability of such feedback amplifiers, based on their frequency response and the mathematics of complex variables. In particular, the approaches described by Nyquist (1932) and Bode (1940) are still in common usage.

The graphical 'Root Locus' method for computing the controller parameters was introduced a few years later by Evans (1948). While working on the control and guidance of aircraft, he developed rules for plotting the changing position of the roots of the closed-loop characteristic equation as a control gain is modified. As we shall see in Chapter 2, such roots, or closed-loop poles, define the stability of the control system and, to a large extent, the transient response characteristics. This root locus approach to control system design became extremely popular. In fact, the very first control systems analysis carried out by the second author in 1960, when he was an apprentice in the aircraft industry, was to use the device called a 'spirule' to manually plot root loci: how times have changed! Not surprisingly, numerous textbooks were published over this era: typical ones that provide a good background to the methods of analysis used at this time and since are James *et al.* (1947) and Brown and Campbell (1948).

However, new developments were on the way. Influenced by the earlier work of Hall (1943), Wiener (1949) and Bode and Shannon (1950), Truxal (1955) discusses an optimal 'least squares' approach to control system design in chapters 7 and 8 of his excellent book *Control System Synthesis*; while the influential book *Analytical Design of Linear Feedback Controls,* published in 1957 by Newton *et al.*, built a bridge to what is still known as 'Modern Control', despite its origins over half a century ago.

In contrast to classical control, modern control system design methods are usually derived from precise algorithmic computations based on a 'state space' description of a dynamic system, often involving optimisation of some kind. Here, the 'minimal' linear state space model of a dynamic system described by an nth order differential equation is a set of n, linked, first order differential equations that describe the dynamic evolution of n associated 'state variables' in response to the specified inputs. And the output of the system is defined as a linear combination of these state variables. For any nth order differential equation that describes the input–output behaviour of this system, the definition of these state variables is

[1] See http://rspl.royalsocietypublishing.org/content/16/270.full.pdf.

not unique and different sets of state variables in the state vector can be defined which yield the same input–output behaviour. Moreover, the state variables for any particular realisation often have some physical interpretation: for instance, in the case of mechanical systems, the states are often defined to represent physical characteristics, such as the positions, velocities and accelerations of a moving body.

Conveniently, the state equations can be represented in the following vector-matrix form, with the state variables forming an $n \times 1$ 'state vector' $\boldsymbol{x}$, in either continuous- or discrete-time, i.e.

Continuous-time:

$$\begin{aligned} \dot{\boldsymbol{x}}(t) &= \boldsymbol{A}\boldsymbol{x}(t) + \boldsymbol{B}\boldsymbol{u}(t) \qquad \left(\dot{\boldsymbol{x}}(t) = \frac{d\boldsymbol{x}(t)}{dt}\right) \\ \boldsymbol{y}(t) &= \boldsymbol{C}\boldsymbol{x}(t) \end{aligned} \tag{1.2}$$

Discrete-time:

$$\begin{aligned} \boldsymbol{x}(k+1) &= \boldsymbol{A}\boldsymbol{x}(k) + \boldsymbol{B}\boldsymbol{u}(k) \\ \boldsymbol{y}(k) &= \boldsymbol{C}\boldsymbol{x}(k) \end{aligned} \tag{1.3}$$

Referring to Appendix A for the vector-matrix nomenclature, $\boldsymbol{x} = [x_1\ x_2\ \ldots\ x_n]^T$ is the $n \times 1$ state vector; $\boldsymbol{y} = [y_1\ y_2\ \ldots\ y_p]^T$ is the $p \times 1$ output vector, where $p \leq n$; and $\boldsymbol{u} = [u_1\ u_2\ \ldots\ u_q]^T$ is the $q \times 1$ input vector (Figure 1.1).

Here, $\boldsymbol{A}$, $\boldsymbol{B}$ and $\boldsymbol{C}$ are constant matrices with appropriate dimensions. The argument t implies a continuous-time signal; whilst k indicates that the associated variable is sampled with a uniform sampling interval of Δt, so that at the kth sampling instant, the time $t = k\Delta t$. Using the state space model (1.2) or (1.3), control system synthesis is carried out based on the concept of *State Variable Feedback* (SVF). Here, the SVF control law is simply $\boldsymbol{u} = -\boldsymbol{K}\boldsymbol{x}$, in which $\boldsymbol{K}$ is a $q \times n$ matrix of control gains; and the roots of the closed-loop characteristic polynomial $\det(\lambda \boldsymbol{I} - \boldsymbol{A} + \boldsymbol{B}\boldsymbol{K})$, can be arbitrarily assigned *if and only if* the system (1.2) or (1.3) is 'controllable' (see Chapter 3 for details). In this case, it can be shown that the closed-loop system poles can be assigned to desirable locations on the complex plane. This is, of course, a very powerful result since, as pointed out previously, the poles determine the stability and transient response characteristics of the control system.

This elegant state space approach to control systems analysis and design was pioneered at the beginning of the 1960s, particularly by Rudolf Kalman who, in one year, wrote three seminal and enormously influential papers: 'Contributions to the theory of optimal control' (1960a[2]); 'A new approach to linear filtering and prediction problems' (1960b), which was concerned with discrete-time systems; and 'The theory of optimal control and the calculus of variations' (Report 1960c; published 1963). And then, in the next year, together with Richard Bucy, he published 'New results in linear filtering and prediction theory' (1961), which provided the continuous-time equivalent of the 1960b paper. The first paper deals with LQ feedback control, as mentioned above; while the second and fourth 'Kalman Filter' papers develop recursive

[2] This is not available from the source but can be accessed at: http://citeseerx.ist.psu.edu/viewdoc/summary?doi=10.1.1.26.4070.

(sequentially updated) algorithms for generating the optimal estimate $\hat{x}$ of the state vector x in equation (1.3) and equation (1.2), respectively. Here, the design of the optimal deterministic LQ control system and the optimal state estimation algorithm are carried out separately and then, invoking the 'separation' or 'certainty equivalence' principle (see e.g. Joseph and Tau 1961; Wonham 1968), they are combined to produce the optimal LQG stochastic feedback control system. In other words, the discrete or continuous-time Kalman Filter is used to estimate or 'reconstruct' the state variables from the available input and noisy output measurements; and then these estimated state variables are used in place of the unobserved state variables in the linear SVF control law: i.e. $\boldsymbol{u} = -\boldsymbol{K}\hat{\boldsymbol{x}}$ (Figure 1.1c).

Despite their undoubted power and elegance, these modern control system design methods, as initially conceived by Kalman and others, were not a panacea and there were critics, even at the theoretical level. For instance Rosenbrock and McMorran (1971) in their paper 'Good, bad or optimal?' drew attention to the excessive bandwidth and poor robustness of the LQ-type controller. And in 1972, the second author of the present book and Jan Willems, noted that the standard LQ controller was only a 'regulator': it lacked the integral action required for the kind of Type 1 'servomechanism' performance demanded by most practical control system designers, namely zero steady-state error to constant command inputs. As we shall see later, the solution to this limitation prompted the development, in the 1980s, of the NMSS-PIP design methodology discussed in this book.

As a result of these and other limitations of LQG control, not the least being its relative complexity, it was the much simpler, often manually tuned PI and PID algorithms that became established as the standard approaches to control system design. Indeed, all of the classical methods mentioned above still have their place in control systems analysis today and involve the analysis of either continuous or discrete-time *Transfer Function* (TF) models (Chapter 2), often exploiting graphical methods of some type (see recent texts such as Franklin *et al.* 2006, pp. 230–313 and Dorf and Bishop 2008, pp. 407–492). But 'modern' control systems theory has marched on regardless. For instance, the lack of robustness of LQG designs has led to other, related approaches, as when the H_2 norm that characterises LQ and LQG optimisation is replaced by the H_∞ norm (see e.g. Zames 1981; Green and Limebeer 1995; Grimble 2006). These robust design methods can be considered in the time or frequency domains and have the virtue that classical design intuition can be used in the design process. An equivalent, 'risk sensitive' approach is 'exponential-of-quadratic' optimisation, where the cost function involves the expectation of the exponential of the standard LQ and LQG criteria (Whittle 1990). This latter approach has been considered within a NMSS context by the present authors: see Chapter 6.

The NMSS-PIP control system design methodology has been developed specifically to avoid the limitations of standard LQ and LQG design. Rather than considering robust control in analytical terms by modifying the criterion function, the minimal state space model on which the LQ and LQG design is conventionally based is modified so that state estimation, as such, is eliminated by making all of the state variables available for direct measurement. As we shall see in section 1.3, this is achieved by considering a particular NMSS model form that is linked directly with the discrete-time TF model of the system, with the state variables defined as the present and stored past variables that appear in the discrete-time equation associated with this TF model. In this manner, the SVF control law $\boldsymbol{u} = -\boldsymbol{K}\boldsymbol{x}$ can be implemented directly using these measured input and output measurements as state variables, rather than indirectly using $\boldsymbol{u} = -\boldsymbol{K}\hat{\boldsymbol{x}}$, with its requirement for estimating the state variables and invoking the separation principle. As this book will demonstrate, the resulting PIP control system is not only easier to implement, it is also inherently more robust. Moreover, it can be interpreted in terms of feedback and forward path control filters that resemble those used in previous classical designs. In a very real sense, therefore, its heritage is in both classical and modern control system synthesis.

Finally, any model-based control system design requires a suitable mathematical representation of the system. Sometimes this may be in the form of a simulation model that has been based on a mechanistic analysis of the system and has been 'calibrated' in some manner. However, in the present TDC context, it seems more appropriate if the model has been obtained by statistical identification and estimation, on the basis of either experimental or monitored sampled data obtained directly from measurements made on the system. In the present book, this task is considered in a manner that can be linked with the Kalman Filter (Young 2010), but where it is the parameters in the model, rather than the state variables, that are estimated recursively. Indeed, early publications (see e.g. Young 1984 and the prior references therein) pointed out that the Kalman Filter represents a rediscovery, albeit in more sophisticated form, of the *Recursive Least Squares* (RLS) estimation algorithm developed by Karl Friedrich Gauss, sometime before 1826 (see Appendix A in Young 1984 or Young 2011, where the original Gauss analysis is interpreted in vector-matrix terms).

Chapter 8 in the present book utilises RLS as the starting point for an introduction to the optimal *Refined Instrumental Variable* (RIV) method for statistically identifying the structure of a discrete or continuous-time TF model and estimating the parameters that characterise this structure. Here, an optimal instrumental variable approach is used because it is relatively robust to the contravention of the assumptions about the noise contamination that affects any real system. In particular, while RIV estimation yields statistically consistent and efficient (minimum variance) parameter estimates if the additive noise has the required rational spectral density and a Gaussian amplitude distribution, it produces consistent, asymptotically unbiased and often relatively efficient estimates, even if these assumptions are not satisfied. This RIV method can also be used to obtain reduced order 'dominant mode' control models from large computer simulation models; and it can identify and estimate continuous-time models that provide a useful link with classical methodology and help in defining an appropriate sampling strategy for the digital control system.

1.3 The Evolution of the NMSS Model Form

The NMSS representation of the model is central to TDC system design. In more general terms, the state space formulation of control system design is the most natural and convenient approach for use with computers. It allows for a unified treatment of both SISO and multivariable systems, as well as for the implementation of the state variable feedback control designs mentioned above, which can include pole assignment, as well as optimal and robust control. Unfortunately, the standard minimal state space approach has three major difficulties. First, as pointed out in section 1.2, the required state vector is not normally available for direct measurement. Secondly, the parameterisation is not unique: for any input–output TF model, there are an infinite number of possible state-space forms, depending on the definition of the state variables. Thirdly, the number of parameters in an n-dimensional state space model is much higher than that in an equivalent TF model: e.g. a SISO system with an nth order denominator,

mth order numerator and a one sample time delay has $n + m$ parameters; while the equivalent state space model can have up to $n^2 + n$ parameters. The first problem has motivated the use of state variable estimation and reconstruction algorithms, notably the Kalman Filter (Kalman 1960b, 1961) for stochastic systems, as mentioned in section 1.2 and the Luenberger observer (Luenberger 1967, 1971) for deterministic systems. The second and third problems have led to a reconsideration of the state-space formulation to see if the uniqueness and parametric efficiency of the TF can be reproduced in some manner within a state space context.

The first movements in this direction came in the 1960s, when it was realised that it was useful to extend the standard minimal state space form to a non-minimal form that contained additional state variables; in particular, state variables that could prove advantageous in control system design. For example, as pointed out previously, Young and Willems (1972) showed how an 'integral-of-error' state variable, defined as the integral of the error between a defined reference or command input $\boldsymbol{y}_d$ and the measured output of the system $\boldsymbol{y}$, could be appended to an otherwise standard minimal state vector. The advantage of this simple modification is that a state variable feedback control law then incorporates this additional state, so adding 'integral-of-error' feedback and ensuring inherent Type 1 servomechanism performance, i.e. zero steady-state error to constant inputs, provided only that the closed-loop system remains stable. Indeed, in the 1960s, the present second author utilised this approach in the design of an adaptive autostabilisation system for airborne vehicles (as published later in Young 1981) and showed how Type 1 performance was maintained even when the system was undergoing considerable changes in its dynamic behaviour.

In the early 1980s, the realisation that NMSS models could form the basis for SVF control system design raised questions about whether there was a NMSS form that had a more transparent link with the TF model than the 'canonical', minimal state space forms that had been suggested up to this time. More particularly, was it possible to formulate a discrete-time state space model whose state variables were the present and past sampled inputs and outputs of the system that are associated directly with the discrete-time dynamic equation on which the TF model is based? In other words, the NMSS state vector $\boldsymbol{x}(k)$ for a SISO system would be of the form:

$$\boldsymbol{x}(k) = [y(k)\ y(k-1)\ \ldots\ y(k-n+1)\ u(k-1)\ u(k-2)\ \ldots\ u(k-m+1)]^T \qquad (1.4)$$

where $y(k)$ is the output at sampling instant k; $u(k)$ is the input at the same sampling instant; n and m are integers representing the order of the TF model polynomials (see Chapter 2). Here, the order $n + m - 1$ of the associated state space model is significantly greater than the order n of a conventional *minimal* state space model (see Chapters 3 and 4 for details). Such a NMSS model would allow the control system to be implemented straightforwardly as a full state feedback controller, without resort to state reconstruction, thus simplifying the design process and making it more robust to the inevitable model uncertainty.

A NMSS model based on the state vector (1.4) was suggested by Hesketh (1982) within a pole assignment control context. This 'regulator' form of the NMSS model is discussed in Chapter 4, where the term 'regulator' is used because the model is only really appropriate to the situation where the command inputs are zero (or fixed at constant values, in which case the model describes the perturbations about these constant levels). The purpose of the control system is then to 'regulate' the system behaviour by restoring it to the desired state, as defined by the command inputs, following any disturbance. In particular, it does not include any inherent integral action that will ensure Type 1 servomechanism performance and the ability to 'track' command input changes, with zero steady-state error, when the command input remains constant for some time period greater that the settling time of the closed-loop system.

Fortunately, following the ideas in Young and Willems (1972) mentioned above, an alternative 'servomechanism' NMSS model can be defined straightforwardly by extending the state vector in (1.4) to include an integral-of-error state, i.e. again in the SISO case,

$$\boldsymbol{x}(k) = [y(k)\; y(k-1)\; \ldots\; y(k-n+1)\; u(k-1)\; u(k-2)\; \ldots\; u(k-m+1)\; z(k)]^T \quad (1.5)$$

where, in this discrete-time setting, the integral-of-error state variable is defined by the following discrete-time integrator (summer),

$$z(k) = z(k-1) + (y_d(k) - y(k)) \quad (1.6)$$

and $y_d(k)$ is the control system command input. This NMSS form, which was first introduced and used for control system design by Young *et al.* (1987) and Wang and Young (1988), is discussed in Chapter 5 and it constitutes the main NMSS description used in the present book. Indeed, as this book shows, the NMSS model provides the most 'natural' and transparent state space description of the discrete-time TF model as required for control system design (Taylor *et al.* 2000a). State variable feedback control system design based on this NMSS model yields what we have termed *Proportional-Integral-Plus* (PIP) control algorithms since they provide logical successors to the PI and PID controllers that, as mentioned previously in section 1.2, have dominated control applications for so long. Here, the 'plus' refers to the situation with systems of second and higher order, or systems with pure time delays, where the additional delayed output and input variables appearing in the non-minimal state vector (1.5), lead to additional feedback and forward path control filters that can be interpreted in various ways, including the implicit introduction of first and higher order derivative action (see Chapter 5 and Appendix D for details).

As this book explains, the servomechanism NMSS model provides a very flexible basis for control system design. For example, the state vector is readily extended to account for the availability of additional measured or estimated information, and these additional states can be utilised to develop more sophisticated NMSS model structures and related control algorithms (see Chapter 6 and Chapter 7). Also, the PIP algorithm is quite general in form and resembles various other digital control systems developed previously, as well as various novel forms, including: feedback and forward path structures, incremental forms and the Smith Predictor for time-delay systems (Taylor *et al.* 1998a); stochastic optimal and risk sensitive control (Taylor *et al.* 1996b); feed-forward control (Young *et al.* 1994); generalised predictive control (Taylor *et al.* 1994, 2000a); multivariable control (Taylor *et al.* 2000b); and state-dependent parameter, nonlinear control (see e.g. Taylor *et al.* 2009, 2011 and the references therein). The multivariable extensions are particularly valuable because they allow for the design of full multi-input, multi-output control structures, where control system requirements, such as channel decoupling and multi-objective optimisation, can be realised.

Of course, a control system is intended for practical application and this book will only succeed if it persuades the reader to utilise PIP control systems in such applications. In this regard, its application record and potential is high: in the 25 years that have passed since the seminal papers on the servomechanism NMSS model and PIP control were published, the methodology has been extended and applied to numerous systems from the nutrient-film (hydroponic) control of plant growth (e.g. Behzadi *et al.* 1985); through various other agricultural (e.g. Lees *et al.* 1996, 1998; Taylor *et al.* 2000c, 2004a, 2004b; Stables and Taylor 2006), benchmark simulation, laboratory and industrial applications (e.g. Chotai *et al.* 1991; Fletcher *et al.* 1994; Taylor *et al.* 1996a, 1998b, 2004c, 2007; Seward *et al.* 1997; Ghavipanjeh *et al.* 2001; Quanten *et al.* 2003; Gu *et al.* 2004; Taylor and Shaban 2006; Taylor and Seward 2010; Shaban *et al.* 2008); and even to the design of emission strategies for the control of atmospheric carbon dioxide in connection with climate change (Young 2006; Jarvis *et al.* 2008, 2009). And all such applications have been guided by the underlying TDC design philosophy outlined in the next subsection.

1.4 True Digital Control

In brief, the TDC design procedure consists of the following three steps:

1. Identification and recursive estimation of discrete and continuous-time models based on the analysis of either planned or monitored experimental data, or via model reduction from data generated by a physically based (mechanistic) simulation model.
2. Offline TDC system design and initial evaluation based on the models from step 1, using an iterative application of an appropriate discrete-time design methodology, coupled with closed-loop sensitivity analysis based on MCS.
3. Implementation, testing and evaluation of the control system on the real process: in the case of self-tuning or self-adaptive control, employing online versions of the recursive estimation algorithms from step 1.

Step 1 above is concerned with stochastic system identification and estimation, i.e. the identification of the control model structure and estimation of the parameters that characterise this structure from the measured input–output data. In the present book, this will normally involve initial identification and estimation of TF models on which the NMSS model form, as required for PIP control system design, is based. It may also involve the identification and estimation of continuous-time models that allow for the direct evaluation of the model in physically meaningful terms, as well as the evaluation of different sampling strategies. Here, the continuous-time model, whose parameter estimates are not dependent on the sampling interval unless this is very coarse, can be converted easily (for example using the c2d conversion routine available in MATLAB®[3]) to discrete-time models defined at different sampling intervals, which can then be evaluated in control system terms in order to define the most suitable sampling frequency.

In step 2, the stochastic models also provide a useful means of assessing the robustness of the TDC designs to the uncertainty in the model parameters, as estimated in step 1 (see e.g. Taylor *et al.* 2001). However, if experimental data are unavailable, the models are instead obtained from a conventional, usually continuous-time and physically based, simulation model of the system (assuming that such a model is available or can be built). In this case, the identification

[3] MATLAB®, The MathWorks Inc., Natick, MA, USA.

step addresses the combined problem of model reduction and linearisation, and forms the connection between the TDC approach and classical mechanistic control engineering methods.

Step 3 is, of course, very problem dependent and it is not possible to generalise the approach that will be used. Implementation will depend on the prevailing conditions and technical aspects associated with specific case studies. We hope, therefore, that the various examples discussed in later chapters, together with other examples available in the cited references, will be sufficient to provide some insight into the practical implementation of TDC systems.

1.5 Book Outline

Starting with the ubiquitous PI control structure as a worked example, and briefly introducing essential concepts such as the backward shift operator z^{-1}, we try to make as few prerequisite assumptions about the reader as possible. Over the first few chapters, generic concepts of state variable feedback are introduced, in what we believe is a particularly intuitive manner (although clearly the reader will be the judge of this), largely based on block diagram analysis and straightforward algebraic manipulation. Conventional minimal state space models, based on selected canonical forms, are considered first, before the text moves onto the non-minimal approach.

More specifically, the book is organised as follows:

- In **Chapter 2**, we introduce the general discrete-time TF model, define the poles of the system and consider its stability properties. Here, as in Chapters 3–6, the analysis is based on this SISO model. Some useful rules of block diagram analysis are reviewed and these are the utilised to develop three basic, discrete-time control algorithms. The limitations of these simple control structures are then discussed, thereby providing motivation for subsequent chapters.
- **Chapter 3** considers *minimal* state space representations of the TF model and shows how these may be employed in the design of state variable feedback control systems. Two particularly well-known representations of this minimal type are considered, namely the controllable canonical form and the observable canonical form. These are then used to illustrate the important concepts of controllability and observability.
- We start **Chapter 4** by defining the regulator NMSS form, i.e. for a control system in which the command input $y_d = 0$ (Figure 1.1). Once the controllability conditions have been established, the non-minimal controller can be implemented straightforwardly. The final sections of the chapter elaborate on the relationship between non-minimal and minimal state variable feedback, while the theoretical and practical advantages of the non-minimal approach are illustrated by worked examples.
- **Chapter 5** develops the complete version of the NMSS-PIP control algorithm for SISO systems. Most importantly and in contrast to Chapter 4, an integral-of-error state variable (1.6) is introduced into the NMSS form to ensure Type 1 servomechanism performance, i.e. if the closed-loop system is stable, the output will converge asymptotically to a constant scalar command input y_d specified by the user. Two main design approaches are considered: pole assignment and optimal LQ control.
- In **Chapter 6**, we extend the NMSS vector in various ways to develop generalised linear PIP controllers. The robustness and disturbance response characteristics of the main control

structures that emerge from this analysis are considered, including incremental forms for practical implementation, the Smith Predictor for time-delay systems, stochastic optimal design, feed-forward control and predictive control.

- **Chapter 7** is important in practical terms because it considers the full NMSS-PIP control system design approach for more complex multivariable systems of the kind that are likely to be encountered in practical applications. Here, the system is characterised by multiple command and control inputs that affect the state and output variables in a potentially complicated and cross-coupled manner. Two design approaches are discussed: optimal LQ with multi-objective optimisation; and a combined multivariable decoupling and pole assignment algorithm. These can be contrasted with the classical 'multichannel' approach to design in which each channel of the system, between a command input and its associated output, is often designed separately and 'tuned' to achieve reasonable multivariable control and cross-coupling.
- **Chapter 8** provides a review of data-based modelling methods and illustrates, by means of simulation and practical examples, how the optimal RIV methods of statistical identification and estimation are able to provide the TF models utilised in previous chapters (including both SISO and multivariable). We also demonstrate how these stochastic models provide a useful means of assessing the robustness of the NMSS designs to uncertainty in the model parameters. In order to help connect with classical methods for modelling in engineering and also allow for the appropriate selection of the sampling interval Δt, both discrete-time and continuous-time estimation algorithms are considered.
- **Chapter 9** considers several additional topics that are not central to TDC design but could have increased relevance in future design studies. These include control system design for rapidly sampled systems using δ-operator models (Middleton and Goodwin 1990); and nonlinear NMSS control system design using *Time-Variable* (TVP) and *State-Dependent* (SDP) parameter models.
- Finally, **Appendix A** revises matrices and the essentials of matrix algebra, as used at various points in the text. The other appendices cover supplementary topics, such as selected theorem proofs, and are cited in the main text where appropriate.

These chapters blend together, in a systematic and we hope readable manner, the various theoretical and applied research contributions made by the authors and others into all aspects of TDC system design over the past half century. This allows for greater integration of the methodology, as well as providing substantially more background detail and examples than the associated academic articles have been able to do.

1.6 Concluding Remarks

The present book, taken as a whole, aims to provide a generalised introduction to TDC methods, including both NMSS design of PIP control systems, and procedures for the data-based identification and estimation of the dynamic models required to define the NMSS form. In this initial chapter, we have both introduced the TDC design philosophy, drawing a distinction between it and *Direct Digital Control*, and reviewed the historical context in which TDC and its associated methodological underpinning have been developed. In TDC design, all aspects of the control system design procedure are overtly digital by nature, with continuous-time concepts introduced only where they are essential for the purposes of describing, simulating and understanding the process in physically meaningful terms; or for making decisions about digitally related aspects, such as the most appropriate sampling strategies. We believe that the approach we have used for the development of the NMSS form and the associated PIP class of control systems provides a relatively gentle learning curve for the reader, from which potentially difficult topics, such as stochastic and multivariable control, can be introduced and assimilated in an interesting and straightforward manner.

References

Behzadi, M.A., Young, P.C., Chotai, A. and Davies, P. (1985) The modelling and control of nutrient film systems. In J.A. Clark (Ed.), *Computer Applications in Agricultural Environments*, Butterworth, London, Chapter 2.

Bellman, R. (1957) *Dynamic Programming*, Princeton University Press, Princeton, NJ.

Bode, H.W. (1940) Feedback amplifier design, *Bell Systems Technical Journal*, **19**, p. 42.

Bode, H.W. and Shannon, C.E. (1950) A Simplified derivation of linear least–squares smoothing and prediction theory, *Proceedinqs IRE*, **38**, pp. 417–425.

Brown, G.S. and Campbell, D.P. (1948) *Principles of Servomechanisms*, John Wiley & Sons, Ltd, New York; Chapman and Hall, London.

Bryson, A.E. and Ho, Y.C. (1969) *Applied Optimal Control*, Blaisdell, Waltham, MA.

Chotai, A., Young, P.C. and Behzadi, M.A. (1991) Self-adaptive design of a non-linear temperature control system, special issue on self-tuning control, *IEE Proceedings: Control Theory and Applications*, **38**, pp. 41–49.

Dorf, R.C. and Bishop, R.H. (2008) *Modern Control Systems*, Eleventh Edition, Pearson Prentice Hall, Upper Saddle River, NJ.

Evans, W.R. (1948) Analysis of Control Systems, *AIEE Transactions*, **67**, pp. 547–551.

Fletcher, I., Wilson, I. and Cox, C.S. (1994) A comparison of some multivariable design techniques using a coupled tanks experiment. In R. Whalley (Ed.), *Application of Multivariable System Techniques*, Mechanical Engineering Publications Limited, London, pp. 49–60.

Franklin, G.F., Powell, J.D. and Emami-Naeini, A. (2006) *Feedback Control of Dynamic Systems*, Fifth Edition, Pearson Prentice Hall, Upper Saddle River, NJ.

Ghavipanjeh, F., Taylor, C.J., Young, P.C. and Chotai, A. (2001) Data-based modelling and proportional integral plus (PIP) control of nitrate in an activated sludge benchmark. *Water Science and Technology*, **44**, pp. 87–94.

Green, M. and Limebeer, D. (1995), *Linear Robust Control*, Prentice Hall, London.

Grimble, M. (2006) *Robust Industrial Control Systems: Optimal Design Approach for Polynomial Systems*, John Wiley & Sons, Ltd, Chichester.

Gu, J., Taylor, C.J. and Seward, D. (2004) Proportional-Integral-Plus control of an intelligent excavator, *Journal of Computer-Aided Civil and Infrastructure Engineering*, **19**, pp. 16–27.

Hall, A.C. (1943) *The Analysis and Synthesis of Linear Servomechanisms*, The Technology Press, MIT, Cambridge, MA.

Hesketh, T. (1982) State-space pole-placing self-tuning regulator using input–output values, *IEE Proceedings: Control Theory and Applications*, **129**, pp. 123–128.

James, H.M., Nichols, N.B. and Phillips, R.S. (1947) *Theory of Servomechanisms*, volume 25 of MIT Radiation Laboratory Series, McGraw-Hill, New York.

Jarvis, A., Leedal, D., Taylor, C.J. and Young, P. (2009) Stabilizing global mean surface temperature: a feedback control perspective, *Environmental Modelling and Software*, **24**, pp. 665–674.

Jarvis, A.J., Young, P.C., Leedal, D.T. and Chotai, A. (2008) A robust sequential emissions strategy based on optimal control of atmospheric concentrations, *Climate Change*, **86**, pp. 357–373.

Joseph, P.D. and Tau, J.T. (1961) On linear control theory, *Transactions of the American Institute of Electrical Engineers*, **80**, pp. 193–196.

Kalman, R.E. (1960a) Contributions to the theory of optimal control, *Boletin de la Sociedad Matematica Mexicana*, **5**, pp. 102–119 (http://garfield.library.upenn.edu/classics1979/A1979HE37100001.pdf).

Kalman, R.E. (1960b) A new approach to linear filtering and prediction problems, *ASME Journal Basic Engineering*, **82**, pp. 34–45.

Kalman, R.E. (1960c) *The Theory of Optimal Control and the Calculus of Variations*. RIAS Technical Report, Defense Technical Information Center, Baltimore Research Institute for Advanced Studies.

Kalman, R.E. (1963) The theory of optimal control and the calculus of variations. In R.E. Bellman (Ed.), *Mathematical Optimization Techniques*, University of California Press, Berkley, Chapter 16.

Kalman, R.E. and Bucy, R.S. (1961) New results in linear filtering and prediction theory. *Transactions of the American Society of Mechanical Engineers, Journal of Basic Engineering*, **83**, pp. 95–108.

Lees, M.J., Taylor, C.J., Chotai, A., Young, P.C., and Chalabi, Z.S. (1996) Design and implementation of a Proportional-Integral-Plus (PIP) control system for temperature, humidity, and carbon dioxide in a glasshouse, *Acta Horticulturae (ISHS)*, **406**, pp. 115–124.

Lees, M.J., Taylor, C.J., Young, P.C. and Chotai, A. (1998) Modelling and PIP control design for open top chambers, *Control Engineering Practice*, **6**, pp. 1209–1216.

Luenberger, D.G. (1967) Canonical forms for linear multivariable systems, *IEEE Transactions on Automatic Control*, **12**, pp. 290–293.

Luenberger, D.G. (1971) An introduction to observers, *IEEE Transactions on Automatic Control*, **16**, pp. 596–603.

Newton, G.C., Gould, L.C. and Kaiser, J.F. (1957) *Analytical Design of Linear Feedback Controls*, John Wiley & Sons, Ltd, New York.

Middleton, R.H. and Goodwin, G.C. (1990) *Digital Control and Estimation – A Unified Approach*, Prentice Hall, Englewood Cliffs, NJ.

Nyquist, H. (1932) Regeneration theory, *Bell Systems Technical Journal*, **11**, pp. 126–147.

Quanten, S., McKenna, P., Van Brecht, A., Van Hirtum, A., Young, P.C., Janssens, K. and Berckmans, D. (2003) Model-based PIP control of the spatial temperature distribution in cars, *International Journal of Control*, **76**, pp. 1628–1634.

Rosenbrock, H. and McMorran, P. (1971) Good, bad, or optimal? *IEEE Transactions on Automatic Control*, **16**, pp. 552–554.

Routh, E.J. (1877) *A Treatise on the Stability of a Given State of Motion*, Macmillan, London.

Seward, D.W., Scott, J.N., Dixon, R., Findlay, J.D. and Kinniburgh, H. (1997) The automation of piling rig positioning using satellite GPS, *Automation in Construction*, **6**, pp. 229–240.

Shaban, E.M., Ako, S., Taylor, C.J. and Seward, D.W. (2008) Development of an automated verticality alignment system for a vibro-lance, *Automation in Construction*, **17**, pp. 645–655.

Stables, M.A. and Taylor, C.J. (2006) Nonlinear control of ventilation rate using state dependent parameter models, *Biosystems Engineering*, **95**, pp. 7–18.

Taylor, C.J., Chotai, A. and Burnham, K.J. (2011) Controllable forms for stabilising pole assignment design of generalised bilinear systems, *Electronics Letters*, **47**, pp. 437–439.

Taylor, C.J., Chotai, A. and Young, P.C. (1998a) Proportional-Integral-Plus (PIP) control of time delay systems, *IMECHE Proceedings: Journal of Systems and Control Engineering*, **212**, pp. 37–48.

Taylor, C.J., Chotai, A. and Young, P.C. (2000a) State space control system design based on non-minimal state-variable feedback: further generalisation and unification results, *International Journal of Control*, **73**, pp. 1329–1345.

Taylor, C.J., Chotai, A. and Young, P.C. (2001) Design and application of PIP controllers: robust control of the IFAC93 benchmark, *Transactions of the Institute of Measurement and Control*, **23**, pp. 183–200.

Taylor, C.J., Chotai, A. and Young, P.C. (2009) Nonlinear control by input–output state variable feedback pole assignment, *International Journal of Control*, **82**, pp. 1029–1044.

Taylor, C.J., Lees, M.J., Young, P.C. and Minchin, P.E.H. (1996a) True digital control of carbon dioxide in agricultural crop growth experiments, *International Federation of Automatic Control 13th Triennial World Congress* (IFAC-96), 30 June–5 July, San Francisco, USA, Elsevier, Vol. B, pp. 405–410.

Taylor, C.J., Leigh, P.A., Chotai, A., Young, P.C., Vranken, E. and Berckmans, D. (2004a) Cost effective combined axial fan and throttling valve control of ventilation rate, *IEE Proceedings: Control Theory and Applications*, **151**, pp. 577–584.

Taylor, C.J., Leigh, P., Price, L., Young, P.C., Berckmans, D. and Vranken, E. (2004b) Proportional-Integral-Plus (PIP) control of ventilation rate in agricultural buildings, *Control Engineering Practice*, **12**, pp. 225–233.

Taylor, C.J., McCabe, A.P., Young, P.C. and Chotai, A. (2000b) Proportional-Integral-Plus (PIP) control of the ALSTOM gasifier problem, *IMECHE Proceedings: Journal of Systems and Control Engineering*, **214**, pp. 469–480.

Taylor, C.J., McKenna, P.G., Young, P.C., Chotai, A. and Mackinnon, M. (2004c) Macroscopic traffic flow modelling and ramp metering control using Matlab/Simulink, *Environmental Modelling and Software*, **19**, pp. 975–988.

Taylor, J. and Seward, D. (2010) Control of a dual-arm robotic manipulator, *Nuclear Engineering International*, **55**, pp. 24–26.

Taylor, C.J. and Shaban, E.M. (2006) Multivariable Proportional-Integral-Plus (PIP) control of the ALSTOM nonlinear gasifier simulation, *IEE Proceedings: Control Theory and Applications*, **153**, pp. 277–285.

Taylor, C.J., Shaban, E.M., Stables, M.A. and Ako, S. (2007) Proportional-Integral-Plus (PIP) control applications of state dependent parameter models, *IMECHE Proceedings: Journal of Systems and Control Engineering*, **221**, pp. 1019–1032.

Taylor, C.J., Young, P.C. and Chotai, A. (1994) On the relationship between GPC and PIP control. In D.W. Clarke (Ed.), *Advances in Model-Based Predictive Control*, Oxford University Press, Oxford, pp. 53–68.

Taylor, C.J., Young, P.C. and Chotai, A. (1996b) PIP optimal control with a risk sensitive criterion, *UKACC International Conference* (Control–96), 2–5 September, University of Exeter, UK, Institution of Electrical Engineers Conference Publication No. 427, Vol. 2, pp. 959–964.

Taylor, C.J., Young, P.C., Chotai, A., McLeod, A.R. and Glasock, A.R. (2000c) Modelling and proportional-integral-plus control design for free air carbon dioxide enrichment systems, *Journal of Agricultural Engineering Research*, **75**, pp. 365–374.

Taylor, C.J., Young, P.C., Chotai, A. and Whittaker, J. (1998b) Non-minimal state space approach to multivariable ramp metering control of motorway bottlenecks, *IEE Proceedings: Control Theory and Applications*, **145**, pp. 568–574.

Truxal, T.G. (1955) *Control System Synthesis*, McGraw-Hill, New York.

Wang, C. and Young, P.C. (1988) Direct digital control by input–output, state variable feedback: theoretical background, *International Journal of Control*, **47**, pp. 97–109.

Whittle, P. (1990) *Risk-Sensitive Optimal Control*, John Wiley & Sons, Ltd, New York.

Wiener, N. (1949) *Extrapolation, Interpolation and Smoothing of Stationary Time Series with Engineering Applications*, MIT Press and John Wiley & Sons, Ltd, New York.

Wonham, W.M. (1968) On the separation theorem of stochastic control, *SIAM Journal on Control and Optimization*, **6**, pp. 312–326.

Young, P.C. (1981) A second generation adaptive autostabilization system for airborne vehicles, *Automatica*, **17**, pp. 459–470.

Young, P.C. (1984) *Recursive Estimation and Time-Series Analysis: An Introduction*, Springer-Verlag, Berlin.

Young, P.C. (2006) The data-based mechanistic approach to the modelling, forecasting and control of environmental systems. *Annual Reviews in Control*, **30**, pp. 169–182.

Young, P.C. (2010) Gauss, Kalman and advances in recursive parameter estimation. *Journal of Forecasting* (special issue celebrating 50 years of the Kalman Filter), **30**, pp. 104–146.

Young, P.C. (2011) *Recursive Estimation and Time-Series Analysis: An Introduction for the Student and Practitioner*, Springer-Verlag, Berlin.

Young, P.C., Behzadi, M.A., Wang, C.L. and Chotai, A. (1987) Direct digital and adaptive control by input–output, state variable feedback, *International Journal of Control*, **46**, pp. 1861–1881.

Young, P.C., Chotai, A. and Tych, W. (1991) True Digital Control: a unified design procedure for linear sampled data systems. In K. Warwick, M. Karny and A. Halouskova (Eds), *Advanced Methods in Adaptive Control for Industrial Applications*, volume 158 of Lecture Notes in Control and Information Sciences, Springer-Verlag, Berlin, pp. 71–109.

Young, P.C., Lees, M., Chotai, A., Tych, W. and Chalabi, Z.S. (1994) Modelling and PIP control of a glasshouse micro-climate, *Control Engineering Practice*, **2**, pp. 591–604.

Young, P.C. and Willems, J.C. (1972) An approach to the linear multivariable servomechanism problem, *International Journal of Control*, **15**, pp. 961–975.

Zames, G. (1981) Feedback and optimal sensitivity: model reference transformations, multiplicative seminorms, and approximate inverses, *IEEE Transactions on Automatic Control*, **26**, pp. 301–320.

2

Discrete-Time Transfer Functions

Throughout this book, discrete-time *Transfer Function* (TF) models will be utilised as a basis for control system design. In this initial, tutorial chapter, the models and control systems are represented conveniently in block diagram form, with TF models representing both the control model and control algorithm in various feedback arrangements. Later chapters will focus on *state space* methods. Although these methods, when implemented in vector-matrix form, yield a concise algorithm that is ideal for computational purposes, the authors feel that block diagram analysis provides a more transparent solution and, hence, can offer valuable insight into the algorithmic approach, particularly for tutorial purposes. Indeed, the development of a non-minimal state space approach to control system design, as discussed in later chapters, has been strongly motivated by the analysis of TF models at the block diagram level.

Classically, control systems have been analysed by means of continuous-time TF models. Here, the TF is the ratio of the Laplace transform of the output to the Laplace transform of the input, with the assumption that the initial conditions on the system are zero (see e.g. Franklin *et al.* 2006, pp. 72–165; Dorf and Bishop 2008, pp. 41–143). The numerator and denominator of this ratio are polynomials in the Laplace operator, s. A similar approach for discrete-time systems is based on the z-transform (Kuo 1980, pp. 106; Åström and Wittenmark 1984, pp. 52–55; Franklin *et al.* 2006, pp. 598–605; Dorf and Bishop 2008, pp. 901–912) and the TF is a ratio of polynomials in the z-transform operator z. In both the continuous and discrete-time cases, the denominator polynomial of the TF model plays a key role and is called the *characteristic polynomial*. The roots of the denominator polynomial, or the *poles*, define the stability and transient dynamic behaviour of the system.

In this book, we will normally exploit the operator notation in its simplest form, where the s operator is used to denote the differential operator, i.e. $s^i y(t) = d^i y(t)/dt^i$; while the z operator denotes the forward shift operation, i.e. $z^i y(k) = y(k+i)$. The inverse of the latter operation, i.e. $z^{-i} y(k) = y(k-i)$, is particularly important since we are dealing here with digital control based on sampled data, so that the controller is normally referring to present and past values of the input and output variables. In the present chapter, we introduce the general discrete-time TF model represented in z^{-1} terms (section 2.1), define the poles of the system and then use standard results (without derivation) to consider its stability properties (section 2.2). We review some useful rules of block diagram analysis in section 2.3 and utilise these in section

---

True Digital Control: Statistical Modelling and Non-Minimal State Space Design, First Edition.
C. James Taylor, Peter C. Young and Arun Chotai.

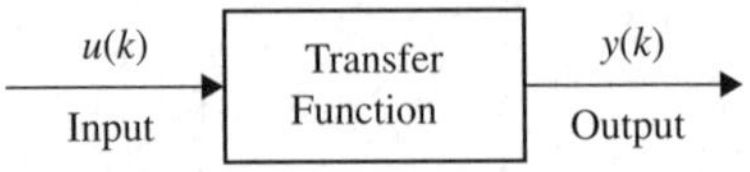

Figure 2.1 *Transfer Function* (TF) in block diagram form

2.4 to introduce three basic, discrete-time control algorithms: namely, proportional, integral and combined proportional-integral control. The limitations of these simple control structures are then discussed briefly, thereby providing motivation for subsequent chapters.

Finally, it is noted that most discrete-time TF models in the z^{-1} operator relate to an underlying continuous-time TF model in the s operator (equivalent to a differential equation) that normally exposes more clearly the physical nature of the system (section 2.5). It also raises the question of how the sampling interval should be selected, so that the discrete-time model is well identified and suitable for control system design.

2.1 Discrete-Time TF Models

Linear systems theory assumes a cause-and-effect relationship between the input and output variables. In this book, we aim to regulate the behaviour of a controlled output variable $y(k)$, typically position or level, velocity, pressure, torque, concentration, flow rate or some other measured variable, where the argument k indicates that the associated variable is sampled in time, normally uniformly with a constant sampling interval of Δt time units. At each sampling instant k, the control algorithm updates the control input variable $u(k)$, which usually represents an actuator of some kind (see Chapter 1).

In essence, the TF is a mathematical object that describes the transfer of an input signal to an output signal. This is illustrated by Figure 2.1, in which $u(k)$ and $y(k)$ denote the *sampled* input and output signals, respectively. For example, in the simplest case, the TF in Figure 2.1 represents a *scalar gain element* or *multiplier* K, and so $y(k) = K\,u(k)$; or it may describe a time delay (Figure 2.2). In all other cases, however, it describes the dynamic behaviour of the system in response to the input.

2.1.1 The Backward Shift Operator

As pointed out above, the basic discrete-time operator is the backward shift operator z^{-i}, i.e.

$$z^{-i}y(k) = y(k-i) \tag{2.1}$$

It is clear, therefore, that z^{-i} introduces a time delay of i samples. In Figure 2.2, for example, z^{-1} denotes a time delay of one sample between the input and output signals.

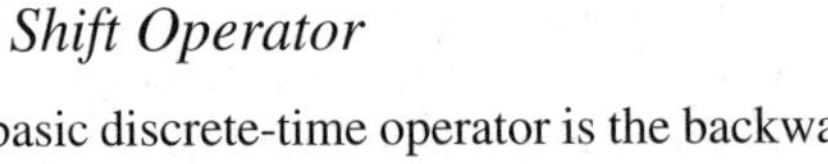

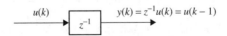

Figure 2.2 The backward shift operator

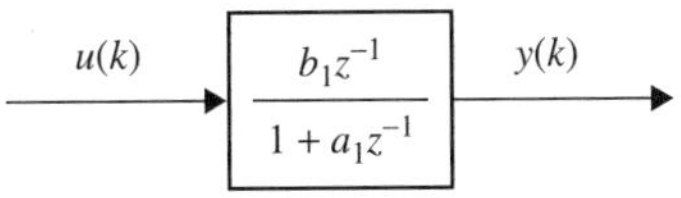

Figure 2.3 Block diagram form of the TF model (2.4)

Example 2.1 Transfer Function Representation of a First Order System Consider the following first order, scalar difference equation, in which a_1 and b_1 are constant coefficients:

$$y(k) + a_1 y(k-1) = b_1 u(k-1) \tag{2.2}$$

Substituting for the backward shift operator yields:

$$y(k) + a_1 z^{-1} y(k) = b_1 z^{-1} u(k) \tag{2.3}$$

and the equivalent TF representation is:

$$y(k) = \frac{b_1 z^{-1}}{1 + a_1 z^{-1}} u(k) \tag{2.4}$$

The block diagram of this TF model is illustrated in Figure 2.3.

Now consider the following numerical example, based on the TF model (2.4), with arbitrarily chosen $a_1 = -0.8$ and $b_1 = 1.6$,

$$y(k) = \frac{1.6z^{-1}}{1 - 0.8z^{-1}} u(k) \tag{2.5}$$

Figure 2.4 shows the unit step response of this system, i.e. the response of the output variable $y(k)$ to a unit step in the control input variable $u(k)$, with $u(k) = 0$ for $k < 5$ and unity thereafter; whilst the initial condition for $y(0)$ is zero.

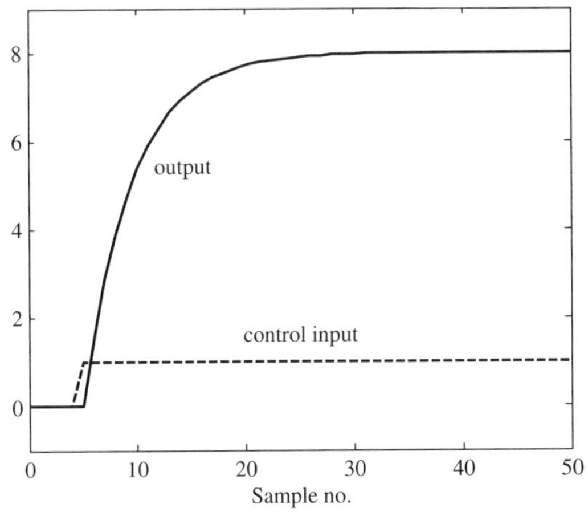

Figure 2.4 Unit step response of the first order TF model in Example 2.1

Here, $y(k)$ for $k = 1, 2, \ldots, 50$, as plotted in Figure 2.4, is generated recursively from this initial condition and $u(k)$, $k = 1, 2, \ldots, 50$, using the difference equation form of the model,

$$y(k) = 0.8y(k-1) + 1.6u(k-1) \tag{2.6}$$

This system has a *time delay* of one sample, i.e. there is a delay of one sample before a change in $u(k)$ starts to affect $y(k)$, as shown by equation (2.6) and Figure 2.4.

In order to provide the context for the simulation examples later in this chapter, we will briefly illustrate the practical utility of the TF model (2.4) with four practical examples:

- Permanent magnet *Direct Current* (DC) motors are commonly used to provide motion for a wide variety of electromechanical devices, including robotic manipulators, disk drives and machine tools. They convert DC electrical energy into rotational mechanical energy. Assuming linearity and ignoring various second order effects, TF models for DC motors can be developed from first principles (Dorf and Bishop 2008, pp. 62–66). Furthermore, if the electrical time constant is at least an order of magnitude faster than the mechanical time constant, then the behaviour of a DC motor can be approximated by the first order TF model (2.4). In this case, $y(k)$ represents the output shaft velocity and $u(k)$ the control input voltage, with a typical $\Delta t = 0.01$ s (100 Hz).
- The civil and construction industries use hydraulic actuators for heavy lifting. The control problem is generally made difficult by a range of factors that include highly varying loads, speeds and geometries, as well as the soil–tool interaction in the case of excavators. A commercial mini-tracked excavator and a 1/5th scale laboratory model have been used to investigate such issues at Lancaster University (e.g. Bradley and Seward 1998; Gu *et al.* 2004; Taylor *et al.* 2007b). In this regard, the laboratory excavator bucket dynamics are well approximated by the TF model (2.4), where $y(k)$ represents the joint angle and $u(k)$ the control input voltage. Here, the control input is scaled to lie in the range ± 1000, with the sign indicating the direction of movement. Using the statistical identification methods described in Chapter 8, together with experimental data collected at $\Delta t = 0.1$ s, yields $a_1 = -1$ and $b_1 = 0.0498$. Here, the normalised voltage has been calibrated so that there is no movement when $u(k) = 0$, which becomes clear when the model is expressed as: $y(k) = y(k-1) + 0.0498u(k-1)$. Similar models have been identified for other hydraulically actuated systems (e.g. Shaban *et al.* 2008; Taylor and Seward 2010), and an example of their control is considered in Chapter 5.
- *Free Air Carbon dioxide Enrichment* (FACE) systems enable the effects of elevated CO_2 on vegetation and other ecosystem components to be studied in the open air (Hendrey *et al.* 1999). They provide an alternative to studies inside glasshouses, by releasing a gas mixture from a distribution system of pipes surrounding an experimental plot. The absence of any enclosure enables research to be performed with *in situ* crops, natural vegetation or even mature forest trees. Using experimental data collected at $\Delta t = 10$s, the TF model (2.4) was identified with $a_1 = -0.713$ and $b_1 = 2.25$ for a FACE system installed in an uncut arable meadow adjacent to the Monks Wood National Nature Reserve, near Huntingdon in the United Kingdom. In this case, $y(k)$ is the CO_2 concentration (parts per million) and $u(k)$ is the voltage regulating a mass flow valve delivering the CO_2 gas (Taylor *et al.* 2000). The control of this system is considered in Chapter 5.

- Rainfall–flow and flow–flow (or 'flow routing') models are used in modelling water flow in rivers for the purposes of either forecasting flow at locations along the river ('flood forecasting': see e.g. Young 2010 and the previous references therein) or flow control and management (see e.g. Evans *et al.* 2011; Foo *et al.* 2011 and the previous references therein). A typical TF model (discrete or continuous-time) is of second order, between the rainfall (mm) and flow (m^3s^{-1}); and it is normally characterised by one short and one long time constant which, respectively, represent the surface and ground water flow pathways in the system (this is sometimes referred to as a stiff dynamic system). In such cases, each flow pathway is alternatively represented by a first order TF model such as (2.4), with the total flow obtained by the sum of these components, as shown later in this chapter (see Example 2.9).

Example 2.2 Transfer Function Representation of a Third Order System Consider the following third order, scalar difference equation:

$$y(k) + a_1 y(k-1) + a_2 y(k-2) + a_3 y(k-3) = b_1 u(k-1) + b_2 u(k-2) + b_3 u(k-3) \tag{2.7}$$

where a_i and b_i $(i = 1, 2, 3)$ are constant coefficients. Substituting for the backward shift operator yields:

$$y(k) + a_1 z^{-1} y(k) + a_2 z^{-2} y(k) + a_3 z^{-3} y(k) = b_1 z^{-1} u(k) + b_2 z^{-2} u(k) + b_3 z^{-3} u(k) \tag{2.8}$$

and the equivalent TF representation is:

$$y(k) = \frac{b_1 z^{-1} + b_2 z^{-2} + b_3 z^{-3}}{1 + a_1 z^{-1} + a_2 z^{-2} + a_3 z^{-3}} u(k) \tag{2.9}$$

The block diagram of this TF model is illustrated in Figure 2.5.

Now consider a numerical example based on the TF model (2.9):

$$y(k) = \frac{27.4671 z^{-1} + 65.6418 z^{-2} - 91.1006 z^{-3}}{1 - 2.4425 z^{-1} + 2.2794 z^{-2} - 0.8274 z^{-3}} u(k) \tag{2.10}$$

Equation (2.10) has been obtained from a model reduction (or 'nominal emulation') exercise (see later, Chapter 8) conducted on a high order, continuous-time simulation model of a horizontal axis, grid-connected wind turbine. The original continuous-time simulation represents a medium- to large-scale, constant speed, wind turbine, comprising a three-blade rotor with rigid hub, gearbox and induction generator (Leithead *et al.* 1991). The output $y(k)$ and input

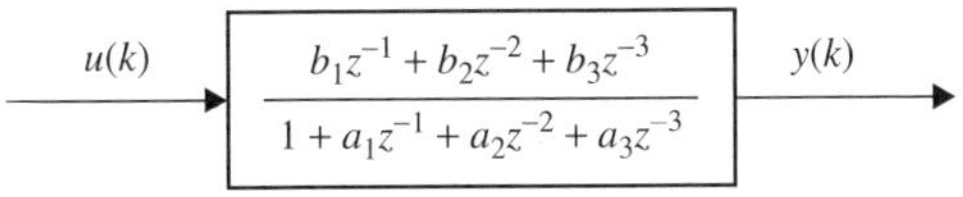

Figure 2.5 Block diagram form of the TF model (2.9)

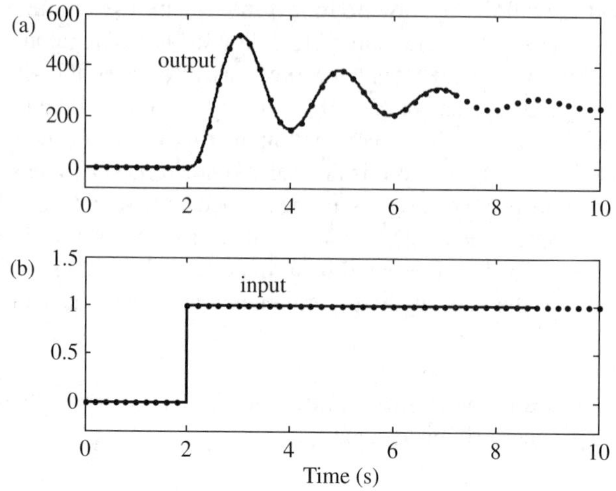

Figure 2.6 Unit step response of the wind turbine system in Example 2.2, comparing the discrete-time TF model (2.10) with the continuous-time simulation (solid trace)

$u(k)$ are the generator reaction torque and pitch angle of the wind turbine blades, respectively, with $\Delta t = 0.2$s.

Figure 2.6 shows the response of the output variable $y(k)$ to a unit step in the control input variable $u(k)$, highlighting the oscillatory behaviour typical of this poorly damped system (before the introduction of feedback control). Figure 2.6a compares simulated data obtained from the continuous-time model (solid trace), with the response of the reduced, third order TF model (points), based on the numerical values shown in equation (2.10). Figure 2.6b shows the input, chosen here as a unit step change in the pitch angle.

Note that the representation of a continuous-time model by a discrete-time TF model in this manner is discussed later in section 2.5, as well as in Chapter 8. For the present purposes, it is sufficient to note that equation (2.10) is a relatively straightforward mathematical model representing the practical wind turbine system. Since it is a linear discrete-time TF model and can be used subsequently for digital control system design, it will be termed the digital *control model*.

2.1.2 *General Discrete-Time TF Model*

Let us consider a general, nth order, deterministic model, with $(m + 1)$ coefficients associated with the input variable and no time delay. The system difference equation is given by:

$$y(k) + a_1 y(k-1) + \cdots + a_n y(k-n) = b_0 u(k) + b_1 u(k-1) + \cdots + b_m u(k-m) \quad (2.11)$$

where a_i $(i = 1 \cdots n)$ and b_i $(i = 0 \cdots m)$ are constant coefficients. The $b_0\, u(k)$ component on the right-hand side of equation (2.11) allows for an instantaneous response to an input signal. By contrast, a time delay of $\tau \geq 1$ samples can be accounted for by setting $b_0 = b_1 = \ldots = b_{\tau-1} = 0$. Representing the TF models (2.4) and (2.9) in this generalised framework,

$n = m = 1$ and $n = m = 3$, respectively. The time delay $\tau = 1$ and $b_0 = 0$ in both cases. Later, Example 2.8 considers a system with $\tau = 2$, i.e. $b_0 = b_1 = 0$.

In fact, discrete-time, model-based control system design generally requires at least one sample time delay. If there is no delay in the model of the system estimated from data, then it is necessary to introduce a 'false' delay of one sample on the control input $u(k)$. Hence, the generalised digital control model most commonly used in this book, is defined as follows:

$$y(k) + a_1 y(k-1) + \cdots + a_n y(k-n) = b_1 u(k-1) + \cdots + b_m u(k-m) \qquad (2.12)$$

Following a similar approach to Example 2.1 and Example 2.2, the equivalent TF model is:

$$y(k) = \frac{b_1 z^{-1} + b_2 z^{-2} + \cdots + b_m z^{-m}}{1 + a_1 z^{-1} + a_2 z^{-2} + \cdots + a_n z^{-n}} u(k) = \frac{B(z^{-1})}{A(z^{-1})} u(k) \qquad (2.13)$$

where $A(z^{-1})$ and $B(z^{-1})$ are the appropriately defined polynomials in z^{-1}.

Chapter 8 considers statistical methods for estimating a TF model such as (2.13), in which the objective is twofold: first, to 'identify' from an input–output data set $(u(k), y(k))$ of N samples (i.e. $k = 1, 2, \cdots, N$) the best model structure, as defined by the orders of the TF numerator and denominator polynomials, m and n, respectively, together with any pure time delay τ [see equation (2.4) and equation (2.9) for examples]; and secondly, to then 'estimate' numerical values for the coefficients a_i $(i = 1 \cdots n)$ and b_i $(i = 1 \cdots m)$ that characterise this identified structure. The aim of such an identification–estimation procedure is to ensure that the TF model adequately represents the system under consideration in some statistically defined sense (e.g. 'least-squares' where the sum-of-squares of the model error is minimised).

Note that the two-stage identification–estimation procedure follows the convention in the statistical literature. We believe this is a useful dichotomy because it explicitly raises the need to find the best 'identifiable' model prior to the final model parameter estimation, thus heightening awareness of the need for good model identifiability in model-based control system design. In the systems and control literature, however, it is more normal to refer to the whole procedure as simply 'identification'.

Given its importance in control design and other modelling applications (e.g. forecasting), it is not surprising that numerous software packages are available for such identification–estimation analysis. One example is the CAPTAIN Toolbox discussed in Chapter 8 (Appendix G; Taylor *et al.* 2007a), which runs within the MATLAB®[1] software environment and is used to obtain many of the models used in the present book. For now, however, it will be assumed that the control model (2.13) has already been estimated in this manner and is available to the designer.

2.1.3 Steady-State Gain

If the system is stable and the input signal takes a time invariant value, denoted by $\bar{u}$, then the output will reach an equilibrium level in which $y(k) \to y(k-1)$ as $k \to \infty$. Hence, the *steady-state gain* G of a discrete-time TF is determined straightforwardly by noting that

[1] MATLAB®, The MathWorks Inc., Natick, MA, USA.

$y(k) - y(k-1) = 0$ or $y(k)(1 - z^{-1}) = 0$, in the steady state, so that $z^{-1} = 1$. Making this substitution in the case of the general TF model (2.13) yields the following:

$$y(k \to \infty) = \frac{b_1 + b_2 + \cdots + b_m}{1 + a_1 + a_2 + \cdots + a_n}\bar{u} = G\bar{u} \tag{2.14}$$

where G is the steady-state gain. For example, the steady-state gain of the first order TF model (2.5) is given by,

$$G = \frac{1.6}{1 - 0.8} = 8.0 \tag{2.15}$$

as would be expected from Figure 2.4. The response of the system to a time invariant input $\bar{u}$ will approach $G\bar{u}$ in a time that is dependent on the dynamics of the system and will be insignificantly different from $y(\infty)$ thereafter.

2.2 Stability and the Unit Circle

There are numerous definitions of *stability*, particularly for nonlinear systems in which the concept of an equilibrium state becomes important (Zak 2003, pp. 149–153). For the present purposes, it is sufficient to say that a linear dynamic system is stable if its output is bounded for any bounded input. It is well known that a continuous-time TF is stable if the roots of the characteristic equation (i.e. the poles) lie in the left-hand side of the complex s-plane, where s is the Laplace transform operator (Franklin *et al.* 2006, pp. 130–131; Dorf and Bishop 2008, pp. 355–360). The equivalent condition for discrete-time systems is based on the z-operator (Franklin *et al.* 2006, pp. 602–604; Dorf and Bishop 2008, pp. 915–916).

In Chapter 3, the characteristic equation is derived directly from the state space model of the system. For now, however, it is sufficient to note that the TF model (2.13) is converted to the z-domain by multiplying by z^n (see Example 2.3, Example 2.4 and Example 2.7). In this regard, $A(z)$ obtained from the denominator of equation (2.13) is called the *characteristic polynomial*, while $A(z) = 0$ is the associated *characteristic equation*. The roots of $A(z) = 0$ are called the *poles* of the system and these are plotted on the complex z-plane, as illustrated in Figure 2.7.

For stability, all the poles p_i must have a magnitude less than unity, i.e.

$$|p_i| < 1 \qquad i = 1,\ 2,\ \ldots,\ n \tag{2.16}$$

where n is the order of the system: so that, for stability, the poles must all lie inside the unit circle on the complex z-plane, as shown in Figure 2.7, where it will be noted that, for $n > 1$, the poles may occur in complex conjugate pairs.

The roots of the numerator polynomial in equation (2.13), i.e. the solutions of $B(z) = 0$, are called the *zeros*. Although the stability of the system is not dependent on these zeros, they do shape the response of the system to an input signal, as illustrated below.

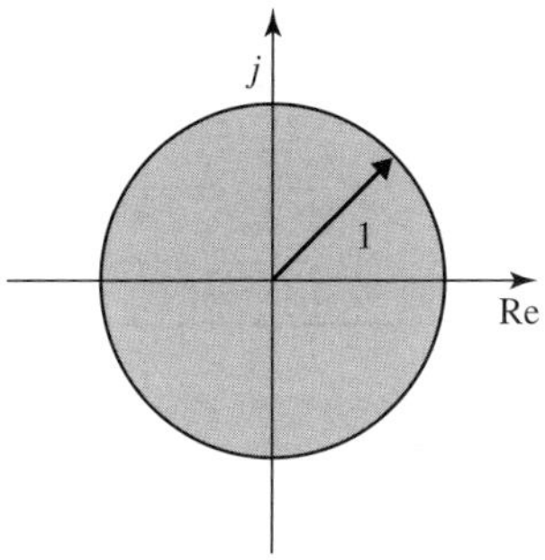

Figure 2.7 Unit circle on the complex z-plane with complex axis j and real axis Re. The shaded area shows the stable region with the magnitude of 1.0 highlighted by the arrow

Example 2.3 Poles, Zeros and Stability Consider a second order TF model based on equation (2.13) with $n = m = 2$,

$$y(k) = \frac{b_1 z^{-1} + b_2 z^{-2}}{1 + a_1 z^{-1} + a_2 z^{-2}} u(k) \tag{2.17}$$

With arbitrarily chosen numerical values for the coefficients, equation (2.17) becomes

$$y(k) = \frac{0.5z^{-1} - 0.4z^{-2}}{1 - 0.8z^{-1} + 0.15z^{-2}} u(k) \tag{2.18}$$

TF models (2.17) and (2.18) will be considered later in Chapter 3 and Chapter 4, where state space models and various algorithms for their control are considered. For now, however, we will examine their open-loop behaviour, i.e. before the introduction of feedback control. In this regard, the unit step response of the TF model (2.18) is illustrated by Figure 2.8a, which also highlights the steady-state gain (2.14) of the system, i.e. in this case $G = (0.5 - 0.4)/(1 - 0.8 + 0.15) = 0.29$.

Multiplying the denominator polynomial by z^2 and setting this equal to zero yields the following characteristic equation:

$$z^2 - 0.8z + 0.15 = (z - 0.3)(z - 0.5) = 0 \tag{2.19}$$

Hence, the system is defined by two poles, i.e. $p_1 = 0.3$ and $p_2 = 0.5$, both lying on the real axis of the complex z-plane. In this case, $|p_i| < 1$ $(i = 1,\ 2)$ and the system is stable, which is quite obvious from Figure 2.8a.

Next, consider a model with the same denominator polynomial, but revised numerator coefficients, as follows:

$$y(k) = \frac{-0.2z^{-1} + 0.3z^{-2}}{1 - 0.8z^{-1} + 0.15z^{-2}} u(k) \tag{2.20}$$

The stability and steady-state gain are unchanged, as illustrated by Figure 2.8b. However, the new numerator $B(z) = -0.2z + 0.3 = 0$ has one zero, which lies outside the unit circle, i.e. a

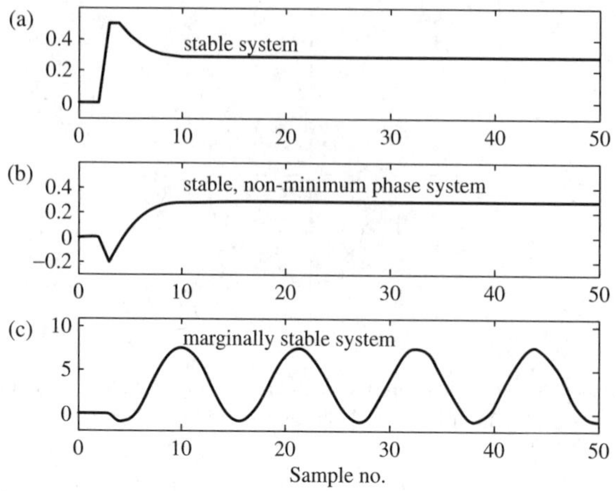

Figure 2.8 Unit step response of (a) stable (2.18), (b) stable non-minimum phase (2.20) and (c) marginally stable (2.21) TF models in Example 2.3

value of $0.3/0.2 = 1.5$ on the real axis. Consequently, the system is said to be *non-minimum phase* (Åström and Wittenmark 1984, pp. 59–60) and, in the case of the present example, this yields an initial negative response to the positive step input (examine the first few samples of Figure 2.8b).

Finally, consider another system based on equation (2.17) but here with $a_1 = -1.7, a_2 = 1.0$, $b_1 = -1$ and $b_2 = 2.0$. The characteristic equation

$$z^2 - 1.7z + 1 = 0 \tag{2.21}$$

in this case yields a complex conjugate pair of poles, i.e. $p_{1,2} = 0.85 \pm 0.5268j$, where the magnitude $|p_i| = \sqrt{0.85^2 + 0.5268^2)} = 1$ $(i = 1,\ 2)$ and the system is *marginally stable*. As a result, the system is characterised by an undamped, limit cycle response, as shown in Figure 2.8c. Furthermore, the numerator $B(z) = -z + 2 = 0$ has one zero, which again lies outside the unit circle, i.e. a value of 2 on the real axis. Consequently, this system is also non-minimum phase.

2.3 Block Diagram Analysis

When considering feedback control structures in later chapters, three key rules of block diagram manipulation will prove particularly useful, namely those dealing with series, parallel and feedback connections, as illustrated by Figure 2.9, Figure 2.10 and Figure 2.11, respectively. Here, $G_1(z^{-1})$ and $G_2(z^{-1})$ represent two connected (arbitrary) TF models.

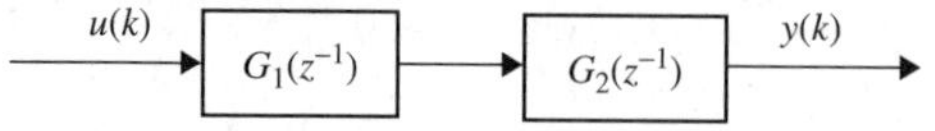

Figure 2.9 Two TF models connected in series

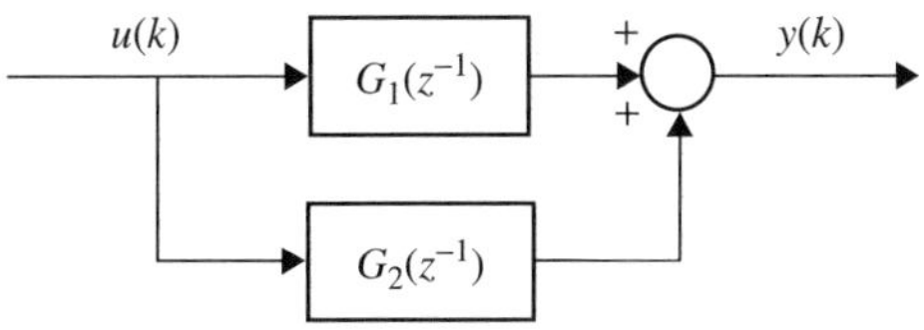

Figure 2.10 Two TF models connected in parallel

Since we are dealing with linear systems, the order of the blocks in the serial connection does not matter. Therefore, the essential input–output relationship defined by Figure 2.9 is obtained straightforwardly by multiplying the TF models, as follows:

$$y(k) = G_1(z^{-1})G_2(z^{-1})u(k) = G_2(z^{-1})G_1(z^{-1})u(k) \tag{2.22}$$

In Figure 2.10, $u(k)$ represents an input to both TF models, while the summation yields:

$$y(k) = G_1(z^{-1})u(k) + G_2(z^{-1})u(k) = \left(G_1(z^{-1}) + G_2(z^{-1})\right)u(k) \tag{2.23}$$

With regard to the feedback connection, inspection of Figure 2.11 shows that:

$$y(k) = G_1(z^{-1})\left(u(k) - G_2(z^{-1})y(k)\right) \tag{2.24}$$

and straightforward algebra yields the ubiquitous negative feedback relationship:

$$y(k) = \frac{G_1(z^{-1})}{1 + G_1(z^{-1})G_2(z^{-1})}u(k) \tag{2.25}$$

For a unity feedback system with $G_2(z^{-1}) = 1$, equation (2.25) reduces to [cf. equation (1.1)]:

$$y(k) = \frac{G_1(z^{-1})}{1 + G_1(z^{-1})}u(k) \tag{2.26}$$

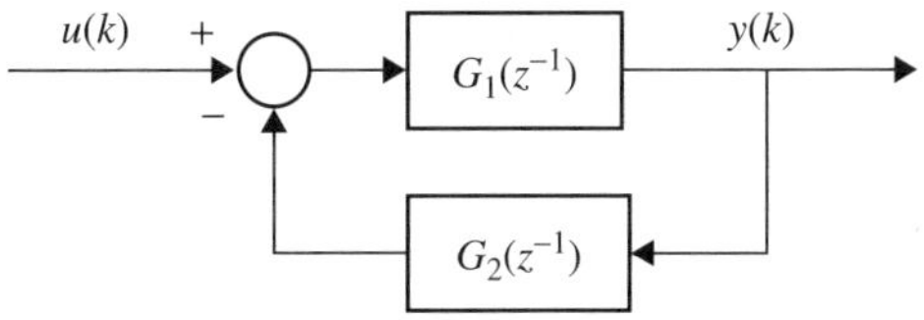

Figure 2.11 Two TF models connected in a negative feedback arrangement

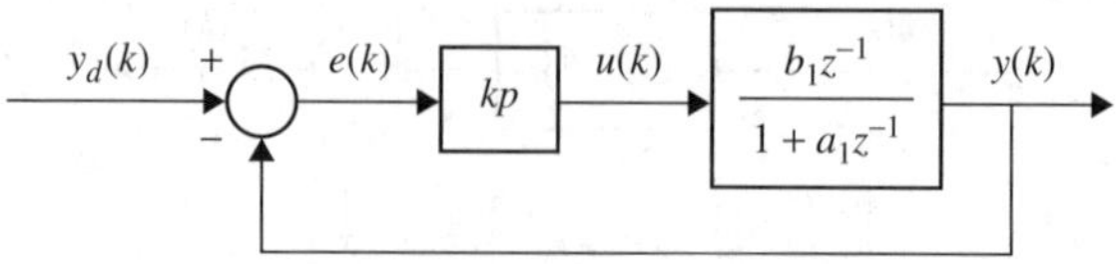

Figure 2.12 Proportional control of a first order TF model

To illustrate these rules of block diagram manipulation and to review some elementary concepts in discrete-time control, three basic control algorithms are introduced in section 2.4.

2.4 Discrete-Time Control

The following examples are concerned with developing discrete-time feedback control systems for the illustrative TF model (2.4), with parameter values (2.5), where $y(k)$ is the output variable and $u(k)$ is the control input. Note that the simplest first order control model has been chosen deliberately in these examples, for the reasons discussed at the end of the section.

Example 2.4 Proportional Control of a First Order TF Model One of the simplest closed-loop controllers is based on a negative feedback of the output variable and scalar proportional gain k_p, as illustrated in Figure 2.12.

In Figure 2.12, $y_d(k)$ is the *command input*, also commonly called the *set point* or *reference level*. Generally, this is dependent on the current requirements of the system: referring to the practical control systems introduced in Example 2.1, for example, it represents the *desired value* of the DC motor shaft velocity, bucket joint angle or elevated CO_2 concentration.

Examination of Figure 2.12 shows that:

$$u(k) = k_p e(k) = k_p \left(y_d(k) - y(k) \right) \tag{2.27}$$

Equation (2.27) represents the *control algorithm* for automatically generating the control input signal $u(k)$ at each sampling instant. Here, $e(k) = y_d(k) - y(k)$ is the difference between the desired and measured behaviour of the system. Most typically, the control objective is to minimise some measure of this error signal over time.

The simulated behaviour of this control system is illustrated by its response to a unit step in the command level $y_d(k)$, as shown by Figure 2.13. Here, the response is based on the proportional control structure illustrated in Figure 2.12, with $a_1 = -0.8$, $b_1 = 1.6$ and arbitrarily chosen $k_p = 0.2$. It is clear that the output variable $y(k)$ does *not* converge to the command input $y_d(k)$, as would be required for most practical control systems.

Utilising Figure 2.12, together with equation (2.22) and equation (2.26) yields:

$$y(k) = \frac{k_p \left(\dfrac{b_1 z^{-1}}{1 + a_1 z^{-1}} \right)}{1 + k_p \left(\dfrac{b_1 z^{-1}}{1 + a_1 z^{-1}} \right)} y_d(k) = \frac{k_p b_1 z^{-1}}{1 + a_1 z^{-1} + k_p b_1 z^{-1}} y_d(k) \tag{2.28}$$

Equation (2.28) is called the *Closed-Loop Transfer Function* (CLTF) and this is of particular interest to control designers. It represents the aggregated relationship between the controlled output variable $y(k)$ and the associated command input $y_d(k)$.

Recall that the TF model (2.4) has constant coefficients a_1 and b_1 defined by the physical behaviour of the system. Hence, a_1 and b_1 cannot be modified except by redesigning the device. Therefore, the control designer aims to achieve a satisfactory closed-loop response by selecting a suitable value for the proportional gain k_p.

The closed-loop characteristic equation:

$$z + a_1 + k_p b_1 = 0 \tag{2.29}$$

has one root, i.e. the pole of the CLTF is a real number, $p_1 = -a_1 - k_p b_1$.

Utilising (2.16), the system is stable if $|p_1| < 1$. Note that the gain must be a positive real number to ensure a positive response towards $y_d(k)$: hence, for stability,

$$k_p > 0 \quad \text{and} \quad k_p < \frac{1 - a_1}{b_1} = 1.125 \tag{2.30}$$

The value of $k_p = 0.2$ utilised in Figure 2.13 clearly lies within this stability range.

The steady-state gain of the CLTF, found by setting $z^{-1} = 1$ in equation (2.28) is:

$$G = \frac{k_p b_1}{1 + a_1 + k_p b_1} \tag{2.31}$$

Assuming that the closed-loop system is stable, $y(k \to \infty) = G\,\bar{y}_d = 0.615\bar{y}_d$ where $\bar{y}_d$ is a time invariant command input [e.g. in Figure 2.13, $y_d(k) = \bar{y}_d = 1.0$ for $k > 2$].

Equation (2.31) shows that the closed-loop steady-state gain is normally less than unity (except for in the special case that $a_1 = -1$), so that simple proportional control action of

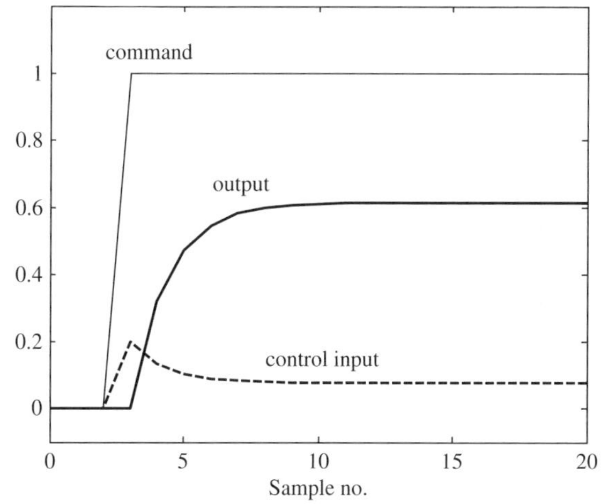

Figure 2.13 Closed-loop unit step response of the proportional control system in Example 2.4

this type yields a *steady-state error*, i.e. the output variable $y(k)$ does *not* converge to a time-invariant command input $y_d(k)$ for $k \to \infty$, as noted above. In fact, in Figure 2.13, the output asymptotically approaches a value of 0.615, yielding a steady-state error of $e(k \to \infty) = 1 - 0.615 = 0.385$.

Although equation (2.31) implies that this steady-state error can be reduced by simply increasing the value of the control gain, k_p, equation (2.30) shows that this will be limited by the stability requirement. For example, utilising equation (2.31) with $a_1 = -0.8$ and $b_1 = 1.6$, as before, together with $k_p = 1.1$, close to the stability limit given by equation (2.30), yields:

$$y(k \to \infty) = 0.9 y_d(k \to \infty) \tag{2.32}$$

and the steady-state error $e(k \to \infty) = 0.1$.

This example demonstrates that, when implemented in isolation, proportional control action does not have an inherent unity steady-state gain. In the control literature this is referred to as a *Type 0* control system. The disadvantages of such Type 0 control systems are considered further in Chapter 4.

Example 2.5 Integral Control of a First Order TF Model An alternative basic control action is illustrated by Figure 2.14, which shows an integral control algorithm applied to the first order TF model (2.4).

Examination of Figure 2.14 shows that the integral control algorithm is written in TF form as follows:

$$u(k) = \frac{k_I}{1 - z^{-1}} (y_d(k) - y(k)) \tag{2.33}$$

Using equation (2.1) and rearranging, this yields the following difference equation form of the algorithm:

$$u(k) = u(k-1) + k_I (y_d(k) - y(k)) \tag{2.34}$$

showing that the control input signal $u(k)$ is equal to its value at the previous sampling instant $u(k-1)$, together with a correction term based on the error $e(k) = y_d(k) - y(k)$.

Although outside the remit of the present book, it is straightforward to implement the control algorithm (2.34) in electronic hardware, using a micro-controller where, at each sampling instant k, the control input is determined using the current measured output variable $y(k)$, the stored value of the input variable $u(k-1)$ and the user specified command input $y_d(k)$.

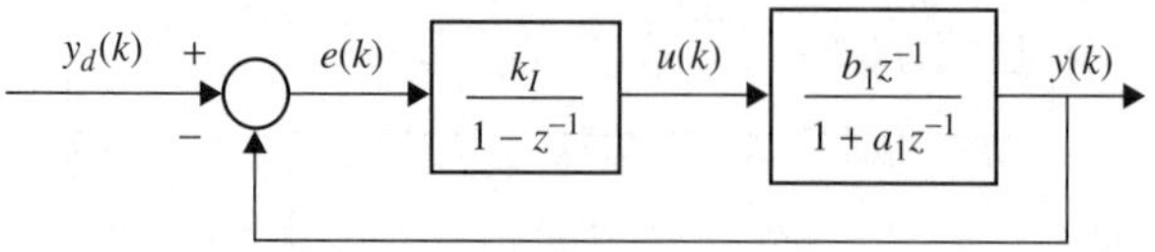

Figure 2.14 Integral control of a first order TF model

Utilising equation (2.22) and equation (2.26), the CLTF is determined from Figure 2.14 in the following form:

$$y(k) = \frac{\left(\frac{k_I}{1-z^{-1}}\right)\left(\frac{b_1 z^{-1}}{1+a_1 z^{-1}}\right)}{1+\left(\frac{k_I}{1-z^{-1}}\right)\left(\frac{b_1 z^{-1}}{1+a_1 z^{-1}}\right)} y_d(k) \tag{2.35}$$

Noting that $z^{-1} \cdot z^{-1} = z^{-2}$, straightforward algebra yields:

$$y(k) = \frac{k_I b_1 z^{-1}}{1+(k_I b_1 + a_1 - 1)z^{-1} - a_1 z^{-2}} y_d(k) \tag{2.36}$$

The steady-state gain of the CLTF, found by setting $z^{-1} = 1$ in equation (2.36), is now unity, i.e.

$$y(k \to \infty) = \frac{k_I b_1}{1 + k_I b_1 + a_1 - 1 - a_1} \bar{y}_d = \bar{y}_d \tag{2.37}$$

where $\bar{y}_d$ is the time-invariant command input. In other words, if the closed-loop system is stable, the output will converge asymptotically to the constant command input specified by the user with dynamics defined by the value of the integral gain k_I; see Example 2.7 for illustrations of such *steady-state tracking*. However, as we shall see, control over the nature of these closed-loop dynamics is limited by the fact that there is only a single control gain k_I. Nevertheless, the inherent unity steady-state gain introduced by the integral action is a very useful property of integral control. In the control literature, this is referred to as *Type 1* servomechanism performance. The subject is discussed more thoroughly in Chapter 5.

Example 2.6 Proportional-Integral Control of a First Order TF Model An important feedback algorithm widely used in industrial control systems is the *Proportional-Integral-Derivative* (PID) controller. An early citation is Challender *et al.* (1936) but the algorithm remains the focus of much practical and theoretical development into the current decade (e.g. Tan *et al.* 2006; Tavakoli *et al.* 2007). The term PID refers only to the *structure* of the algorithm, whilst a very wide range of design approaches may be utilised to tune the gains (Zhuang and Atherton 1993; Åström and Hagglund 1995). Let us consider, however, a special case of the PID controller: the 'two-term' *Proportional-Integral* (PI) control algorithm.

The previously cited introductory control engineering textbooks include numerous examples of both continuous-time (Franklin *et al.* 2006, p. 187; Dorf and Bishop 2008, p. 445) and discrete-time (Åström and Wittenmark 1984, p. 373) PI control algorithms. One particular PI control structure that proves particularly significant in the present book is illustrated in Figure 2.15. Here, the controller is implemented in a discrete-time form and again applied to the TF model (2.4).

Figure 2.15 exemplifies the way in which elements of a control algorithm, together with the control model, can be conveniently represented in block diagram form using Transfer Functions, scalar gains and summations.

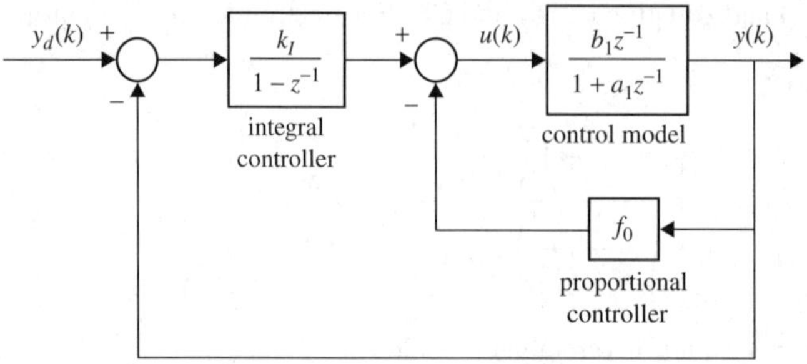

Figure 2.15 Proportional-Integral control of a first order TF model

Note that, in contrast to Figure 2.12, the proportional action is implemented in the feedback path. For this reason, and for consistency with later chapters, the proportional gain is denoted by f_0 rather than k_p, whilst k_I represents the integral gain, as before.

Examination of Figure 2.15 shows that the PI control algorithm is written in TF form:

$$u(k) = \frac{k_I}{1 - z^{-1}}(y_d(k) - y(k)) - f_0 y(k) \tag{2.38}$$

Using equation (2.1) and rearranging yields the difference equation form of the algorithm, suitable for implementation on a digital computer:

$$u(k) = u(k-1) + k_I(y_d(k) - y(k)) - f_0(y(k) - y(k-1)) \tag{2.39}$$

In a similar manner to the previous examples, we now use the rules of block diagram analysis to determine the closed-loop behaviour of the PI control system. In the first instance, the inner feedback loop of Figure 2.15, consisting of the control model together with a negative feedback of the output variable and the scalar proportional gain f_0, is simplified using equation (2.25), so that the control system is equivalently represented as shown in Figure 2.16.

Utilising equation (2.22) and equation (2.26), the CLTF determined from Figure 2.16 is:

$$y(k) = \frac{k_I b_1 z^{-1}}{1 + (f_0 b_1 + a_1 - 1 + k_I b_1)z^{-1} + (-a_1 - f_0 b_1)z^{-2}} y_d(k) \tag{2.40}$$

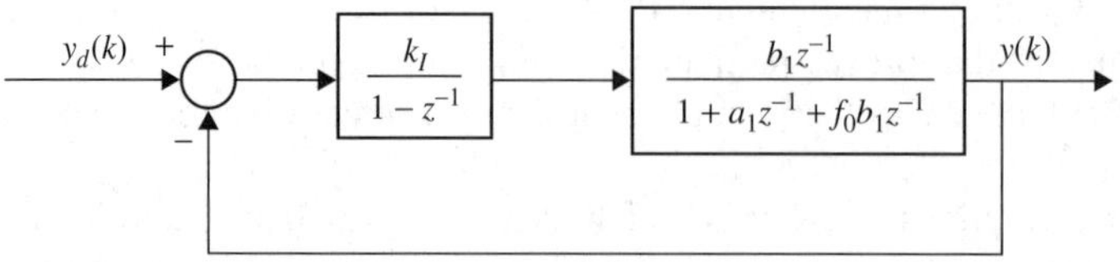

Figure 2.16 Reduced form of the control system in Figure 2.15

The steady-state gain of the closed-loop system, found by setting $z^{-1} = 1$ in equation (2.40) is again unity, i.e.

$$y(k \to \infty) = \frac{k_I b_1}{1 + f_0 b_1 + a_1 - 1 + k_I b_1 - a_1 - f_0 b_1} \bar{y}_d = \bar{y}_d \tag{2.41}$$

where $\bar{y}_d$ is the time-invariant command input.

Hence, if the closed-loop system is stable, the output will converge asymptotically to the constant command input, i.e. Type 1 servomechanism performance.

Note that the closed-loop dynamics are now defined by the two control gains f_0 and k_I so that, as we see in Example 2.7, the control systems designer has much more control over the nature of the closed-loop dynamics than in the pure integral control situation.

Example 2.7 Pole Assignment Design Based on PI Control Structure Continuing directly from the previous example, the poles p_i $(i = 1, 2)$ of the CLTF (2.40) are the roots of the characteristic equation:

$$z^2 + (f_0 b_1 + a - 1 + k_I b_1) z + (-a - f_0 b_1) = 0 \tag{2.42}$$

Now, if k_I and f_0 are selected so that both these poles lie at the origin of the complex z-plane, i.e. $p_1 = p_2 = 0$, then we obtain the *deadbeat* response illustrated by Figure 2.17. Here, the output signal $y(k)$ follows a step change in the command input $y_d(k)$ after just one sampling instant, the fastest theoretical response of a discrete-time control system. By contrast, the response shown in Figure 2.18 is obtained by selecting k_I and f_0 so that the closed-loop poles are complex with $p_{1,2} = 0.6 \pm 0.3j$.

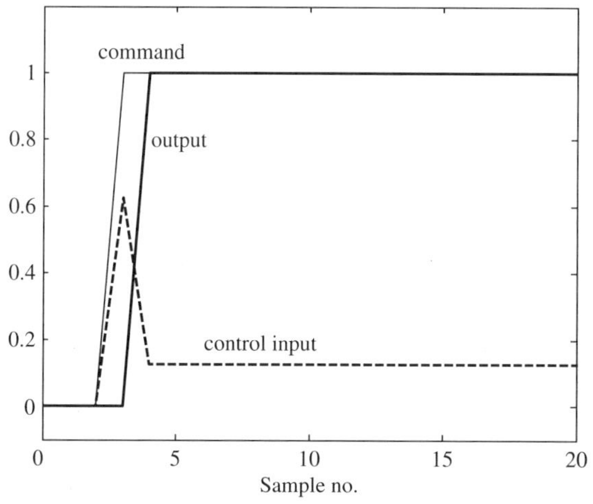

Figure 2.17 Closed-loop unit step response of the deadbeat PI control system using the characteristic equation (2.45) in Example 2.7

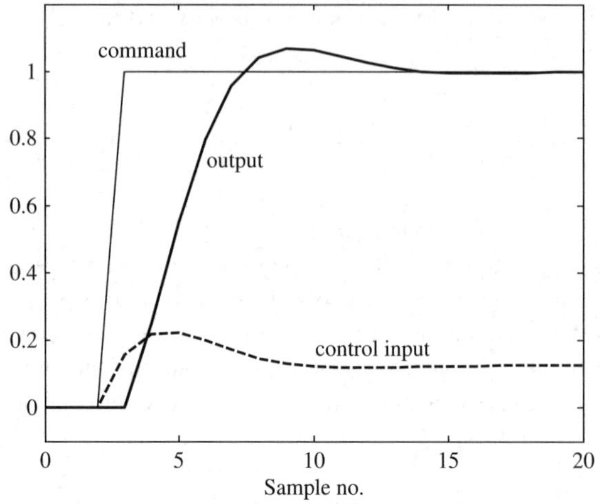

Figure 2.18 Closed-loop unit step response of the PI control system based on conjugate complex poles using the characteristic equation (2.43) in Example 2.7

The latter of these two controllers yields a slower speed of response and deliberately incorporates a small temporary overshoot of the command level, which is sometimes desirable in practice since it is more likely that the desired steady-state level is achieved despite practical limitations in the system, such as mechanical friction effects. In practical applications, it is also more robust to uncertainty in the model parameters and generates a less severe control input signal than the deadbeat design (see later chapters for discussions on robustness and other practical matters).

The characteristic equation of the closed-loop system associated with $p_{1,2} = 0.6 \pm 0.3j$ and Figure 2.18 is given by:

$$D(z) = (z - 0.6 + 0.3j)(z - 0.6 - 0.3j) = z^2 - 1.2z + 0.45 = 0 \tag{2.43}$$

where $D(z)$ denotes the 'desired' characteristic polynomial. In this case, the control gains are determined by equating (2.42) and (2.43), as follows:

$$(f_0 b_1 + a_1 - 1 + k_I b_1) = -1.2; \quad (-a_1 - f_0 b_1) = 0.45 \tag{2.44}$$

For the TF model (2.5) with $a_1 = -0.8$ and $b_1 = 1.6$, the simultaneous equations (2.44) are solved straightforwardly to obtain $f_0 = 0.2188$ and $k_I = 0.1563$. Using these control gains, the PI controller is implemented in the form of Figure 2.15 or, in practical applications, typically equation (2.39). The simulated response is illustrated in Figure 2.18.

The characteristic equation of the closed-loop system associated with $p_1 = p_2 = 0$ and the deadbeat response shown by Figure 2.17, is given by:

$$D(z) = (z - 0)(z - 0) = z^2 = 0 \tag{2.45}$$

and the control gains are obtained by equating (2.42) and (2.45) as follows:

$$(f_0 b_1 + a_1 - 1 + k_I b_1) = 0; \quad (-a_1 - f_0 b_1) = 0 \tag{2.46}$$

Solving these simultaneous equations yields: $f_0 = -a_1/b_1$ and $k_I = 1/b_1$. For the TF model (2.5) with $a_1 = -0.8$ and $b_1 = 1.6$, these relationships yield $f_0 = 0.5$ and $k_I = 0.625$.

In the above manner, the closed-loop behaviour is determined by the pole positions chosen by the designer. Such an approach to control system design is called *pole placement* or *pole assignment*. The important question of how to choose suitable pole positions, in order to satisfy a range of control design objectives, is addressed in later chapters.

Example 2.8 Limitation of PI Control Structure Taylor (2004) and Leigh (2002) describe a 1.0 m^2 by 2.0 m forced ventilation test chamber at Lancaster University. A computer-controlled axial fan is positioned at the outlet in order to draw air through the chamber, whilst an air velocity transducer measures the associated ventilation rate. In this regard, the following first order difference equation with $\Delta t = 2$ s, and two such sampling intervals defining a pure time delay of 4 s, has been estimated from experimental data collected from the chamber, using the statistical identification methods described in Chapter 8:

$$y(k) = 0.743y(k-1) + 0.027u(k-2) \tag{2.47}$$

where $y(k)$ is the air velocity (m s^{-1}) and $u(k)$ is the applied voltage to the fan expressed as a percentage. The TF form of the model is:

$$y(k) = \frac{0.027}{1 - 0.743z^{-1}} u(k-2) = \frac{0.027z^{-2}}{1 - 0.743z^{-1}} u(k) \tag{2.48}$$

where it is clear that $y(k)$ is expressed as a function of the delayed input $u(k-2)$. Comparing (2.48) with the generalised control model (2.13), it is clear that $n = 1$, $m = \tau = 2$ and $b_1 = 0$. Similar first order models with time delays have been identified and estimated for other ventilation systems (Taylor *et al.* 2004a, b).

Following a similar approach to Example 2.6, application of the PI controller (2.38) to this TF model yields the following CLTF:

$$y(k) = \frac{0.027k_I z^{-2}}{1 - 1.743z^{-1} + (0.743 + 0.027f_0 + 0.027k_I)z^{-2} - 0.027f_0 z^{-3}} y_d(k) \tag{2.49}$$

The CLTF is now third order, so that there are three closed-loop poles but still only two control gains, f_0 and k_I, that can be selected to achieve pole assignment. Consequently, the pole assignment problem cannot be solved, i.e. the designer cannot *arbitrarily* assign the closed-loop poles to any positions in the complex z-plane. Similar results emerge for PI control of second or higher order systems, such as the TF models (2.9) and (2.17). Hence, alternative approaches are required in order to handle the general nth order control model (2.13) and/or TF models with pure time delays. These alternative approaches are discussed in subsequent chapters.

2.5 Continuous to Discrete-Time TF Model Conversion

In the present control systems context, a discrete-time TF model is simply a convenient representation of a dynamic system. It normally relates to a real-world, physical system that we wish to control in some manner. However, models of physical systems are most often derived in continuous-time, differential equation terms on the basis of natural laws, such as Newton's Laws of Motion or conservation laws (mass, energy, momentum, etc.); and they are characterised by parameters that often have a prescribed physical meaning.

For instance, the parameters of a continuous-time TF model for a second order system with oscillatory dynamics reveal the natural frequency and damping of the oscillations (Franklin *et al.* 2006, p. 111). These are particularly useful because, together with the steady-state gain, they provide an immediate indication of the system's dynamic response characteristics.

In order to move from such a continuous-time differential equation model to its discrete-time equivalent, it is necessary to utilise some convenient method of transformation: e.g. in the case of a continuous-time TF model, we might use the MATLAB® c2d function. This transformation requires the user to make some assumptions about the inter-sample behaviour. For example, in the case of the c2d function, it is necessary to specify both the sampling interval Δt and the method of discretisation from amongst the following: zero order hold on the inputs; linear interpolation of the inputs (triangle approximation); impulse-invariant discretisation; bilinear (Tustin) approximation; and the Tustin approximation with frequency pre-warping. All of these are approximations because the actual inter-sample behaviour is normally not known. Moreover, since the resultant discrete-time TF is a function of the specified sampling interval Δt, its parameter values will change as Δt is changed. Finally, they will have no clear physical interpretation: in other words, they constitute a nominally infinite set of completely 'black-box' models.

It is clear from the above that, when compared with continuous-time models, discrete-time models have a number of disadvantages. Consequently, as an alternative to obtaining the model in discrete-time form by statistical estimation, it is possible to obtain the model directly in continuous-time form. This is achieved either by analysis based on the physical nature of the system, as is often the case when using simulation modelling software such as SIMULINK[2]; or by statistical estimation of the model in continuous-time form, based on input–output data. Example 2.9 helps to illustrate these possibilities but the topic receives a much more detailed examination in Chapter 8.

Example 2.9 Continuous- and Discrete-Time Rainfall–Flow Models A continuous-time model of flow in a river (Littlewood *et al.* 2010), obtained by statistical identification and estimation from the hourly measured effective rainfall $u(t)$ and flow $y(t)$, using the *Refined Instrumental Variable method for Continuous-time Box–Jenkins models* (RIVCBJ) algorithm developed in Chapter 8, has the following form:

$$y(t) = \frac{0.01706s^2 + 0.1578s + 0.001680}{s^2 + 0.3299s + 0.001768}u(t) \tag{2.50}$$

[2] Simulink™, The MathWorks Inc., Natick, MA, USA.

where s is the differential operator, i.e. $s^i y(t) = d^i y(t)/dt^i$. Applying partial fraction expansion to this TF, the model can be written in the following parallel pathway form:

$$y(t) = \left\{0.0171 + \frac{0.1496}{s + 0.3244} + \frac{0.002573}{s + 0.00545}\right\} u(t) \tag{2.51}$$

This has the following practical interpretation:

1. An instantaneous effect $0.0171\, u(t)$ accounting for the small amount of effective rainfall that affects the river flow within each hourly sampling period.
2. A 'quick-flow' effect with steady-state gain $G_1 = 0.1496/0.3244 = 0.461$ and residence time (or *time constant*: see Appendix B) $T_1 = 1/0.3244 = 3.1$ h that is associated with surface and near surface processes.
3. A 'slow-flow' effect with steady-state gain $G_2 = 0.002573/0.00545 = 0.472$ and residence time $T_2 = 1/0.00545 = 183.5$ h associated with groundwater processes.
4. The 'partitioning' of the flow, defined by $P_1 = 100(0.0171/0.950) = 1.8\%$, $P_2 = 0.461/0.950 = 48.5\%$ and $P_3 = 0.472/0.950 = 49.7\%$, where 0.950 is the steady-state gain of the composite TF model (2.50).

In other words, the parameters of this continuous-time model have immediate practical significance. Also, note that the model has 'stiff' dynamics: i.e. a combination of fairly widely spaced 'quick' and 'slow' modes. This has practical implications on model estimation, as discussed in Chapter 8.

Now, suppose we want to design a digital control system involving this model, initially considering implementation at a sampling interval of $\Delta t = 6$ h. MATLAB® c2d conversion at this sampling interval, using the zero order hold option, yields the following discrete-time model:

$$y(k) = \frac{0.01706 + 0.3915z^{-1} - 0.3824z^{-2}}{1 - 1.1106z^{-1} + 0.1382z^{-2}} u(k) \tag{2.52}$$

Naturally, this model has the same dynamic behaviour as the continuous-time model from which it is derived. If we only had this description, however, and wished to compute physical attributes, such as the residence times, it is clear that we would need to convert back to continuous time to do this. In other words, unlike the continuous-time model, the parameter values in this discrete-time model have no immediate physical interpretation. Moreover, it is found that, if this discrete-time model is actually estimated directly from the data decimated to the 6 h sampling interval (or, since it is possible in this particular hydrological application, the data are accumulated to yield samples every 6 h), then the estimates are biased away from the continuous-time estimates and so are misleading.

This biasing problem is a consequence of the rather coarse sampling at this decimation level. The natural period of the quick mode is: $2\pi/0.3244 = 19.4$ h. Therefore, in order to identify and control this mode, the data should be sampled faster than the Nyquist period (Franklin *et al.* 2006, p. 617) 19.4/2=9.7 h: i.e. a sampling interval less than about 10 h. Although $\Delta t = 6$ h does not contravene this rule, it is quite close to it and so the estimation is affected by this, as well as the reduced sample size. Clearly, this has implications as regards control system design and so this important question of choosing a suitable sampling interval for identification, estimation and control is considered more fully in Chapter 8.

2.6 Concluding Remarks

This chapter has introduced the nth order discrete-time TF model and examined both the steady-state gain and stability properties of systems via several simple, worked examples. It has also reviewed basic techniques in block diagram analysis, as demonstrated by the consideration of three well known control structures: namely, proportional control, integral control and PI control. This has led to the concept of pole assignment, using the PI control structure as a worked example. It has been noted that the PI pole assignment approach is limited to the simplest first order model with unity time delay. Finally, it has been emphasised that the discrete-time model is a function of the sampling interval and that it is often related to, and can be obtained from, a continuous-time model that is uniquely parameterised and normally physically meaningful. So, although this book is concerned with digital control, continuous-time concepts and models cannot be ignored, both in modelling and control system design.

Subsequent chapters of the book will expand on these simple beginnings and consider the development of control systems to handle the general TF model, with pure time delays and higher order model polynomials. This will set the scene for the extension of the concepts to the complexities of optimal, stochastic and multivariable systems.

References

Åström, K.J. and Hagglund, T. (1995) *PID Controllers: Theory, Design and Tuning*, Instrument Society of America, Research Triangle Park, NC.

Åström, K.J. and Wittenmark, B. (1984) *Computer-Controlled Systems Theory and Design*, Prentice Hall, Englewood Cliffs, NJ.

Bradley, D.A. and Seward, D.W. (1998) The development, control and operation of an autonomous robotic excavator, *Journal of Intelligent Robotic Systems*, **21**, pp. 73–97.

Challender, A., Hartree, D.R. and Porter, A. (1936) Time lag in a control system, *Philosophical Transactions of the Royal Society of London Series A Mathematical and Physical Sciences*, **235**, pp. 415–444.

Dorf, R.C. and Bishop, R.H. (2008) *Modern Control Systems*, Eleventh Edition, Pearson Prentice Hall, Upper Saddle River, NJ.

Evans, R., Li, L., Mareels, I., Okello, N., Pham, M., Qiu, W. and Saleem, S.K. (2011) Real–time optimal control of river basin networks. In L. Wang and H. Garnier (Eds), *System Identification, Environmetric Modelling and Control*, Springer-Verlag, Berlin, pp. 403–422.

Foo, M., Ooi, S.K. and Weyer, E. (2011) Modelling of rivers for control design. In L. Wang and H. Garnier (Eds), *System Identification, Environmetric Modelling and Control*, Springer-Verlag, Berlin, pp. 423–448.

Franklin, G.F., Powell, J.D. and Emami-Naeini, A. (2006) *Feedback Control of Dynamic Systems*, Fifth Edition, Pearson Prentice Hall, Upper Saddle River, NJ.

Gu, J., Taylor, J. and Seward, D. (2004) Proportional-Integral-Plus control of an intelligent excavator, *Journal of Computer-Aided Civil and Infrastructure Engineering*, **19**, pp. 16–27.

Hendrey, G.R., Ellsworth, D.S., Lewin, K.F. and Nagy, J. (1999) A free-air enrichment system for exposing tall forest vegetation to elevated atmospheric CO_2, *Global Change Biology*, **5**, pp. 293–309.

Kuo, B.C. (1980) *Digital Control Systems*, Holt, Rinehart and Winston, New York.

Leigh, P.A. (2002) *Modelling and Control of Micro-Environmental Systems*, PhD thesis, Environmental Science Department, Lancaster University.

Leithead, W.E., de la Salle, S.A. and Grimble, M.J. (1991) Wind turbine modelling and control, *IEE Conference Publication No. 332*.

Littlewood, I.G., Young, P.C. and Croke, B.F.W. (2010) Preliminary comparison of two methods for identifying rainfall–streamflow model parameters insensitive to data time–step: the Wye at Cefn Brwyn, Plynlimon, Wales. In C. Walsh (Ed.), *Role of Hydrology in Managing Consequences of a Changing Global Environment*, University of Newcastle, British Hydrological Society, Newcastle, UK, pp. 539–543.

Shaban, E.M., Ako, S., Taylor, C.J. and Seward, D.W. (2008) Development of an automated verticality alignment system for a vibro-lance, *Automation in Construction*, **17**, pp. 645–655.

Tan, N., Kaya, I, Yeroglu, C. and Atherton, D.P. (2006) Computation of stabilizing PI and PID controllers using the stability boundary locus, *Energy Conversion and Management*, **47**, pp. 3045–3058.

Tavakoli, S., Griffin, I. and Fleming, P.J. (2007) Robust PI control design: a genetic algorithm approach, *International Journal of Soft Computing*, **2**, pp. 401–407.

Taylor, C.J. (2004) Environmental test chamber for the support of learning and teaching in intelligent control, *International Journal of Electrical Engineering Education*, **41**, pp. 375–387.

Taylor, C.J., Leigh, P.A., Chotai, A., Young, P.C., Vranken, E. and Berckmans, D. (2004a) Cost effective combined axial fan and throttling valve control of ventilation rate, *IEE Proceedings: Control Theory and Applications*, **151**, pp. 577–584.

Taylor, C.J., Leigh, P., Price, L., Young, P.C., Berckmans, D. and Vranken, E. (2004b) Proportional-Integral-Plus (PIP) control of ventilation rate in agricultural buildings, *Control Engineering Practice*, **12**, pp. 225–233.

Taylor, C.J., Pedregal, D.J., Young, P.C. and Tych, W. (2007a) Environmental time series analysis and forecasting with the Captain toolbox, *Environmental Modelling and Software*, **22**, pp. 797–814.

Taylor, C.J. and Seward, D. (2010) Control of a dual-arm robotic manipulator, *Nuclear Engineering International*, **55**, pp. 24–26.

Taylor, C.J., Shaban, E.M., Stables, M.A. and Ako, S. (2007b) Proportional-Integral-Plus (PIP) control applications of state dependent parameter models, *IMECHE Proceedings: Systems and Control Engineering*, **221**, pp. 1019–1032.

Taylor, C.J., Young, P.C., Chotai, A., McLeod, A.R. and Glasock, A.R. (2000) Modelling and proportional-integral-plus control design for free air carbon dioxide enrichment systems, *Journal of Agricultural Engineering Research*, **75**, pp. 365–374.

Young, P.C. (2010) Gauss, Kalman and advances in recursive parameter estimation, *Journal of Forecasting*, **30**, pp. 104–146.

Zak, S.H. (2003) *Systems and Control*, The Oxford Series in Electrical and Computer Engineering, Oxford University Press, New York.

Zhuang, M. and Atherton, D.P. (1993) Automatic tuning of optimum PID controllers, *IEE Proceedings: Control Theory and Applications*, **140**, pp. 216–224.

3

Minimal State Variable Feedback

The classical approach to control systems analysis for *Single-Input, Single-Output* (SISO) systems involves continuous or discrete-time *Transfer Function* (TF) models, represented in terms of either the Laplace transform (s-operator) or the backward shift operator (z^{-1}), respectively. Here, closed-loop stability and satisfactory transient responses are obtained through techniques such as the Evans Root Locus method (Evans 1948, 1950; Ogata 1970) or Guillimin's synthesis method (Truxal 1955), both of which allow for the specification of closed-loop pole-zero patterns. Alternatively, a frequency domain approach is utilised involving Nyquist and Bode diagrams (Nyquist 1932; Bode 1940; Truxal 1955). One prominent feature of such classical approaches is that they are generally based on graphical analysis of some sort (see e.g. Kuo 1980, pp. 393–420; Franklin *et al.* 2006, pp. 230–313; Dorf and Bishop 2008, pp. 407–492).

By contrast, modern control system design methods are usually derived from precise algorithmic computations, often involving numerical optimisation. As mentioned in Chapter 1, one very important concept is the idea of the *state space*, which originates from the state variable method of describing differential equations (Kalman 1960, 1961, 1963; Zadeh and Descoer, 1963; Rosenbrock 1970, 1974; Kailath 1980; Franklin *et al.* 2006, pp. 438–593; Dorf and Bishop 2008, pp. 144–211). While the TF approach discussed in Chapter 2 is concerned only with input–output characteristics, the state space approach also provides a description of the non-unique, internal behaviour of the system. For mechanical systems, the states are often defined to represent physical characteristics, such as the positions and velocities of a moving body. However, in the more general case considered here, the state space formulation is derived directly from the previously estimated TF model using special *canonical* forms.

The present chapter considers *minimal* state space representations of a SISO-TF model and shows how these may be employed in the design of *State Variable Feedback* (SVF) control systems. In particular, a pole assignment SVF control law is developed for the general linear nth order system. Such a minimal dimension SVF always involves n states, where n is the highest power of z in the denominator polynomial of the discrete-time TF model. Two particularly well-known black-box (i.e. they do not relate obviously to any particular physical system) state space representations of this minimal type are considered: namely the *controllable canonical form* (section 3.1) and the *observable canonical form* (section 3.2). Some useful mathematical background is introduced in section 3.3, which discusses the transformation between state

True Digital Control: Statistical Modelling and Non-Minimal State Space Design, First Edition.
C. James Taylor, Peter C. Young and Arun Chotai.

space and TF models, so defining the *characteristic equation*, *eigenvalues* and *eigenvectors* of a system.

As we shall see, for a particular state variable representation of the system, the TF model is completely and uniquely specified. There are, however, an infinite number of possible state space representations for any given TF model. For example, certain state space representations can include states that are completely decoupled from the control input and so are uncontrollable by the control input signal. And again, there may be unobservable states that are entirely decoupled from the output and thus cannot influence the observed behaviour of the system. Section 3.4 considers these important concepts of *controllability* and *observability*. However, let us start with a straightforward example that demonstrates the non-uniqueness of a state space model.

Example 3.1 State Space Forms for a Third Order TF Model Consider the following third order, scalar difference equation:

$$y(k) + a_1 y(k-1) + a_2 y(k-2) + a_3 y(k-3) = u(k-1) \tag{3.1}$$

where $y(k)$ and $u(k)$ are the sampled output and input signals, respectively, whilst a_1, a_2 and a_3 are constant coefficients. These coefficients or parameters would normally be estimated from the data $\{y(k), u(k)\}$ using the identification and estimation methods discussed later in Chapter 8. When the difference between the true system and the estimated model is significant to the analysis, the *estimates* will be denoted by $\hat{a}_i$ $(i = 1, 2, 3)$. For simplicity here, however, it will be assumed initially that the parameters are known without any uncertainty. Consequently, the equivalent TF representation of equation (3.1) is:

$$y(k) = \frac{z^{-1}}{1 + a_1 z^{-1} + a_2 z^{-2} + a_3 z^{-3}} u(k) \tag{3.2}$$

in which z^{-1} denotes the backward shift operator (cf. Example 2.2). The block diagram of this TF model is illustrated in Figure 3.1.

An illustrative minimal state vector $\boldsymbol{x}(k)$ is defined as follows:

$$\boldsymbol{x}(k) = \begin{bmatrix} x_1(k) \\ x_2(k) \\ x_3(k) \end{bmatrix} = \begin{bmatrix} y(k) \\ y(k-1) \\ y(k-2) \end{bmatrix} \tag{3.3}$$

where $x_1(k)$, $x_2(k)$ and $x_3(k)$ are the state variables.

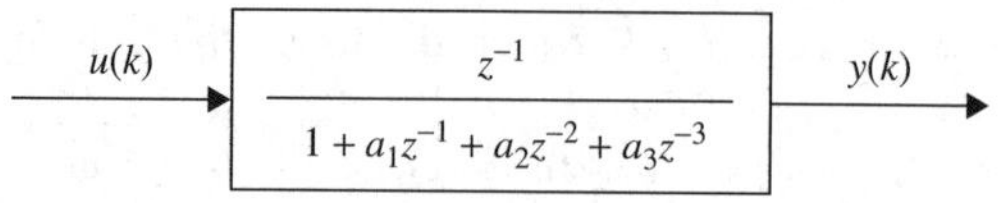

Figure 3.1 Block diagram form of the TF model (3.2)

Note that the model (3.1) or (3.2) is third order and there are three state variables. In this example, these state variables have been defined straightforwardly as the output signal $y(k)$ and its past values.

Equation (3.1) or its equivalent (3.2) can be written in the following state space form:

$$\mathbf{x}(k) = \begin{bmatrix} x_1(k) \\ x_2(k) \\ x_3(k) \end{bmatrix} = \begin{bmatrix} -a_1 & -a_2 & -a_3 \\ 1 & 0 & 0 \\ 0 & 1 & 0 \end{bmatrix} \begin{bmatrix} x_1(k-1) \\ x_2(k-1) \\ x_3(k-1) \end{bmatrix} + \begin{bmatrix} 1 \\ 0 \\ 0 \end{bmatrix} u(k-1) \tag{3.4}$$

In this case, the output variable $y(k)$ is obtained from an observation equation:

$$y(k) = \begin{bmatrix} 1 & 0 & 0 \end{bmatrix} \begin{bmatrix} x_1(k) \\ x_2(k) \\ x_3(k) \end{bmatrix} \tag{3.5}$$

The relationship between (3.1) and (3.4) is most apparent if we explicitly substitute for the state variables using (3.3), as follows:

$$\mathbf{x}(k) = \begin{bmatrix} y(k) \\ y(k-1) \\ y(k-2) \end{bmatrix} = \begin{bmatrix} -a_1 & -a_2 & -a_3 \\ 1 & 0 & 0 \\ 0 & 1 & 0 \end{bmatrix} \begin{bmatrix} y(k-1) \\ y(k-2) \\ y(k-3) \end{bmatrix} + \begin{bmatrix} 1 \\ 0 \\ 0 \end{bmatrix} u(k-1) \tag{3.6}$$

where the first equation (row) relates directly to equation (3.1) and the remaining two are simply definitions of the delayed output variables (and are necessary to complete the state vector).

Equations (3.6) are illustrated in Figure 3.2, which represents a straightforward decomposition of the TF model in Figure 3.1.

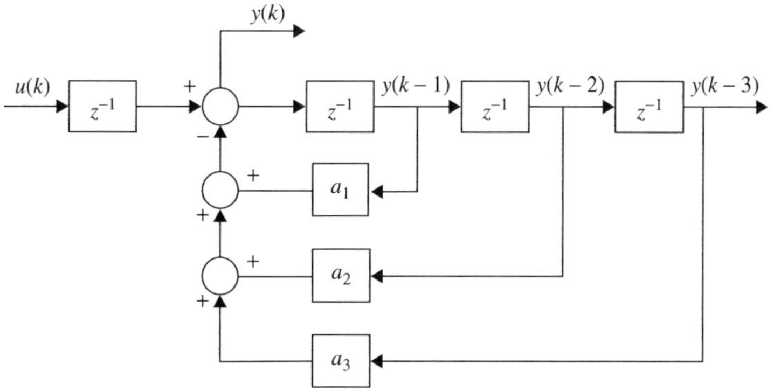

Figure 3.2 State space model described by equations (3.6)

However, if the state vector is now defined as:

$$\boldsymbol{x}(k) = \begin{bmatrix} x_1(k) \\ x_2(k) \\ x_3(k) \end{bmatrix} = \begin{bmatrix} -a_1x_1(k-1) + x_2(k-1) + u(k-1) \\ -a_2x_1(k-1) + x_3(k-1) \\ -a_3x_1(k-1) \end{bmatrix} \tag{3.7}$$

then equation (3.1) or equation (3.2) is described by the state space form:

$$\boldsymbol{x}(k) = \begin{bmatrix} x_1(k) \\ x_2(k) \\ x_3(k) \end{bmatrix} = \begin{bmatrix} -a_1 & 1 & 0 \\ -a_2 & 0 & 1 \\ -a_3 & 0 & 0 \end{bmatrix} \begin{bmatrix} x_1(k-1) \\ x_2(k-1) \\ x_3(k-1) \end{bmatrix} + \begin{bmatrix} 1 \\ 0 \\ 0 \end{bmatrix} u(k-1) \tag{3.8}$$

$$y(k) = \begin{bmatrix} 1 & 0 & 0 \end{bmatrix} \begin{bmatrix} x_1(k) \\ x_2(k) \\ x_3(k) \end{bmatrix} \tag{3.9}$$

The equivalence between {(3.8), (3.9)} and the TF model can be checked by substituting for the state vector (3.7).

Despite their differences, it is clear that the state space representations developed above, namely the equation pairs {(3.4), (3.5)} and {(3.8), (3.9)}, both describe the *same* input–output relationship specified by the TF model. And it is clear that these are not unique representations of the TF model; they are simply two specific examples that illustrate the non-uniqueness of the state space representation.

3.1 Controllable Canonical Form

Let us turn to the general *n*th order deterministic, unit delay control model introduced in Chapter 2. However, for notational simplicity, in this chapter we will assume that the orders of the numerator and denominator polynomials are the same. This assumption does not constrain the model in any sense, since different orders may be considered by simply setting the relevant parameters to zero.

Hence, the system difference equation is given by:

$$y(k) + a_1y(k-1) + \cdots + a_ny(k-n) = b_1u(k-1) + b_2u(k-2) + \cdots + b_nu(k-n) \tag{3.10}$$

where a_i $(i = 1 \cdots n)$ and b_i $(i = 1 \cdots n)$ are constant coefficients. The equivalent TF model is:

$$y(k) = \frac{b_1z^{-1} + b_2z^{-2} + \cdots + b_nz^{-n}}{1 + a_1z^{-1} + a_2z^{-2} + \cdots + a_nz^{-n}}u(k) = \frac{B(z^{-1})}{A(z^{-1})}u(k) \tag{3.11}$$

where $A(z^{-1})$ and $B(z^{-1})$ are appropriately defined polynomials in z^{-1}. To help develop the state space form, we can write equation (3.11) as:

$$y(k) = \frac{\left(b_1 + b_2 z^{-1} + \cdots + b_n z^{-n+1}\right) z^{-1}}{1 + a_1 z^{-1} + a_2 z^{-2} + \cdots + a_n z^{-n}} u(k) \tag{3.12}$$

Furthermore, by introducing an intermediate variable $w(k)$, the TF model can be decomposed into two components, namely:

$$w(k) = \frac{z^{-1}}{1 + a_1 z^{-1} + a_2 z^{-2} + \cdots + a_n z^{-n}} u(k) \tag{3.13}$$

and,

$$y(k) = \left(b_1 + b_2 z^{-1} + \cdots + b_n z^{-n+1}\right) w(k) \tag{3.14}$$

Generalising from equation (3.3), it is clear that defining the state vector:

$$\begin{bmatrix} x_1(k) \\ x_2(k) \\ \vdots \\ x_{n-1}(k) \\ x_n(k) \end{bmatrix} = \begin{bmatrix} w(k) \\ w(k-1) \\ \vdots \\ w(k-n+2) \\ w(k-n+1) \end{bmatrix} \tag{3.15}$$

yields the following state space representation of the TF model (3.11):

$$\begin{bmatrix} x_1(k) \\ x_2(k) \\ \vdots \\ x_{n-1}(k) \\ x_n(k) \end{bmatrix} = \begin{bmatrix} -a_1 & -a_2 & \cdots & -a_{n-1} & -a_n \\ 1 & 0 & \cdots & 0 & 0 \\ \vdots & \ddots & \vdots & \vdots & \vdots \\ 0 & 0 & 1 & 0 & 0 \\ 0 & 0 & 0 & 1 & 0 \end{bmatrix} \begin{bmatrix} x_1(k-1) \\ x_2(k-1) \\ \vdots \\ x_{n-1}(k-1) \\ x_n(k-1) \end{bmatrix} + \begin{bmatrix} 1 \\ 0 \\ \vdots \\ 0 \\ 0 \end{bmatrix} u(k-1) \tag{3.16}$$

$$y(k) = \begin{bmatrix} b_1 & b_2 & \cdots & b_{n-1} & b_n \end{bmatrix} \begin{bmatrix} x_1(k) \\ x_2(k) \\ \vdots \\ x_{n-1}(k) \\ x_n(k) \end{bmatrix} \tag{3.17}$$

Equations {(3.16), (3.17)} are known as the *controllable canonical form*[1] and are illustrated in block diagram form by Figure 3.3.

[1] A standard form: here, the simplest and most significant controllable form possible without loss of generality.

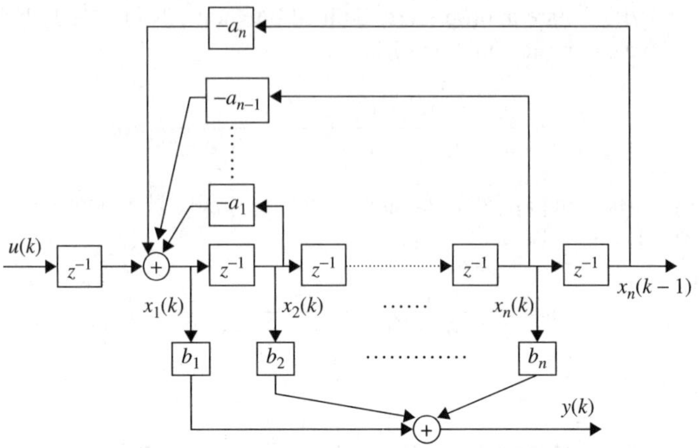

Figure 3.3 Controllable canonical form for the general discrete-time system (3.11)

The controllable canonical form is an example of a *minimal* state space model since there are n states (3.15). The advantage of such a representation is that it is straightforward to construct and yields a convenient structure for the computation of a pole assignment control algorithm (see below). However, one disadvantage of the controllable canonical form is that the state vector is not directly available from the measured output because it consists of the intermediate variable $w(k)$ and its past values.

In this regard, it is clear from equation (3.14) that $w(k)$ is given by:

$$w(k) = \frac{y(k)}{\left(b_1 + b_2 z^{-1} + \cdots + b_n z^{-n+1}\right)} \tag{3.18}$$

Multiplying by z^{-1}/z^{-1} yields:

$$w(k) = \frac{z^{-1}y(k)}{b_1 z^{-1} + b_2 z^{-2} + \cdots + b_n z^{-n}} = \frac{y(k-1)}{B(z^{-1})} \tag{3.19}$$

Therefore, in the controllable canonical form, the minimal state vector is implicitly formed from the delayed and filtered output variable, where the filter is defined by the numerator polynomial of the TF model, i.e. $B(z^{-1})$.

Example 3.2 State Variable Feedback based on the Controllable Canonical Form Consider a second order TF model based on equation (3.11) with $n = 2$, together with illustrative numerical values for the parameters:

$$y(k) = \frac{B(z^{-1})}{A(z^{-1})}u(k) = \frac{b_1 z^{-1} + b_2 z^{-2}}{1 + a_1 z^{-1} + a_2 z^{-2}}u(k) = \frac{0.5z^{-1} - 0.4z^{-2}}{1 - 0.8z^{-1} + 0.15z^{-2}}u(k) \tag{3.20}$$

This model was first introduced in Chapter 2 (Example 2.3). It will be utilised for several examples in this chapter and so is restated here for convenience.

The minimal state vector associated with the TF model (3.20) is:

$$\begin{bmatrix} x_1(k) \\ x_2(k) \end{bmatrix} = \begin{bmatrix} w(k) \\ w(k-1) \end{bmatrix} \tag{3.21}$$

The controllable canonical form is subsequently defined as follows:

$$\begin{bmatrix} x_1(k) \\ x_2(k) \end{bmatrix} = \begin{bmatrix} -a_1 & -a_2 \\ 1 & 0 \end{bmatrix} \begin{bmatrix} x_1(k-1) \\ x_2(k-1) \end{bmatrix} + \begin{bmatrix} 1 \\ 0 \end{bmatrix} u(k-1) \tag{3.22}$$

$$y(k) = \begin{bmatrix} b_1 & b_2 \end{bmatrix} \begin{bmatrix} x_1(k) \\ x_2(k) \end{bmatrix} \tag{3.23}$$

A linear SVF control law for this system is simply a linear combination of the state variables, i.e.

$$u(k) = -l_1 x_1(k) - l_2 x_2(k) + y_d(k) \tag{3.24}$$

where $y_d(k)$ is the command input, while the control gains l_1 and l_2 are chosen by the designer to ensure a satisfactory closed-loop response (see below).

The general form of a SVF control algorithm, such as equation (3.24), will be discussed in Chapter 4. For now, it is sufficient to note that the control input signal $u(k)$ is determined using a negative feedback of each state variable multiplied by a control gain. In the example above, the command input is introduced as a straightforward addition to this feedback law: i.e. using equation (3.19) and equation (3.21), the control law (3.24) is written as:

$$u(k) = -l_1 y^*(k-1) - l_2 y^*(k-2) + y_d(k) \tag{3.25}$$

in which

$$y^*(k) = \frac{1}{\hat{B}(z^{-1})} y(k) \tag{3.26}$$

where $\hat{B}(z^{-1})$ is the estimated numerator polynomial of the control model. The 'hat' notation, i.e. $\hat{B}(z^{-1}) = \hat{b}_1 z^{-1} + \hat{b}_2 z^{-2}$ is introduced here in order to emphasise the inevitable mismatch between the true system represented by $B(z^{-1})/A(z^{-1})$ and the estimated TF model used for control system design, i.e. $\hat{B}(z^{-1})/\hat{A}(z^{-1})$, where $\hat{A}(z^{-1}) = 1 + \hat{a}_1 z^{-1} + \hat{a}_2 z^{-2}$, and $\{\hat{a}_i, \hat{b}_i\}$ $(i = 1, 2)$ are the parameter estimates.

Combining equation (3.25) and equation (3.26) yields:

$$u(k) = -\frac{l_1 z^{-1} + l_2 z^{-2}}{\hat{B}(z^{-1})} y(k) + y_d(k) \tag{3.27}$$

The block diagram of the resultant closed-loop control system, with potential model mismatch, is illustrated by Figure 3.4. Here, the term *plant* is used to represent the notional true system, as distinct from the estimated control model.

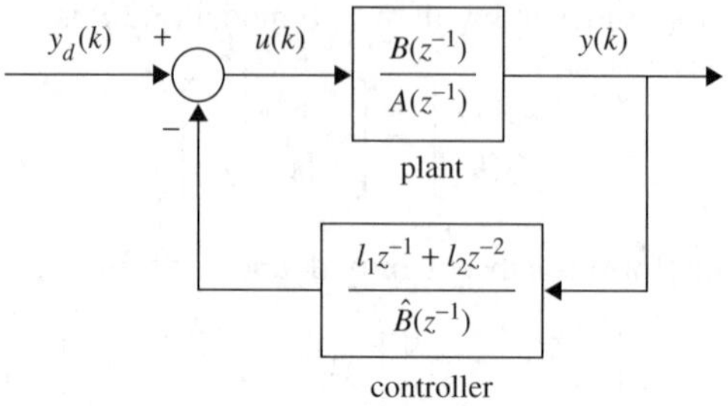

Figure 3.4 The closed-loop control system for Example 3.2 with model mismatch

Example 3.3 State Variable Feedback Pole Assignment based on the Controllable Canonical Form The SVF framework developed in Example 3.2 will now be used to design a pole assignment controller (cf. Example 2.6 and Example 2.7). In this regard, block diagram reduction applied to Figure 3.4 yields the following *Closed-Loop Transfer Function* (CLTF):

$$y(k) = \frac{B(z^{-1})}{A(z^{-1}) + \left(l_1z^{-1} + l_2z^{-2}\right)\dfrac{B(z^{-1})}{\hat{B}(z^{-1})}} y_d(k) \tag{3.28}$$

The denominator of equation (3.28) reveals a cancellation of the numerator polynomial, which is implicit in any SVF controller based on the controllable canonical form. However, using the polynomials defined in equation (3.20) and assuming no plant–model mismatch, i.e. $B(z^{-1}) = \hat{B}(z^{-1}) = b_1z^{-1} + b_2z^{-2}$, the CLTF reduces to:

$$y(k) = \frac{b_1z^{-1} + b_2z^{-2}}{1 + (a_1 + l_1)z^{-1} + (a_2 + l_2)z^{-2}} y_d(k) \tag{3.29}$$

Recall from Chapter 2 that the denominator of a TF defines its stability and transient dynamic behaviour. Therefore, in order to determine the control gains l_1 and l_2, we consider the characteristic equation as follows:

$$1 + (a_1 + l_1)z^{-1} + (a_2 + l_2)z^{-2} = 0 \tag{3.30}$$

Each of the feedback gains are explicitly associated with different model coefficients in equation (3.30) so that, when using the controllable canonical form, it is a trivial matter for the control system designer to obtain a desired closed-loop characteristic equation:

$$D(z^{-1}) = 1 + d_1z^{-1} + d_2z^{-2} = 0 \tag{3.31}$$

where d_1 and d_2 are user chosen coefficients, by setting:

$$l_1 = d_1 - a_1 \text{ and } l_2 = d_2 - a_2 \tag{3.32}$$

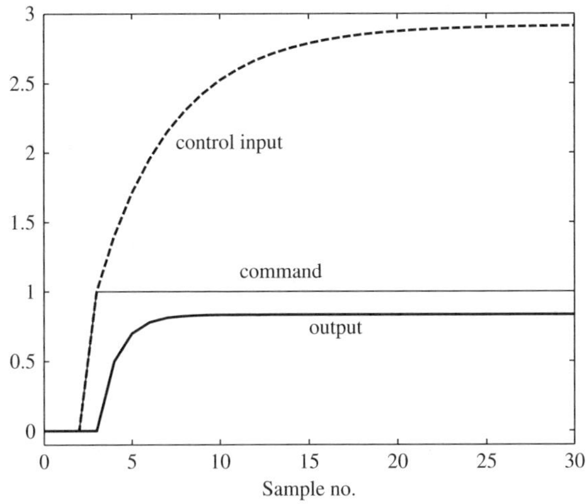

Figure 3.5 Closed-loop unit step response of the SVF control system in Example 3.3

Since the roots of the CLTF are known as the poles of the closed-loop system, this approach is another example of *pole assignment*, as introduced in Chapter 2. In this regard, note that equation (3.31) is equivalent to:

$$D(z) = z^2 + d_1 z + d_2 = 0 \tag{3.33}$$

Therefore, selecting illustrative closed-loop poles of 0.4 and 0.8 on the complex z-plane yields: $(z - 0.4)(z - 0.8) = z^2 - 1.2z + 1.32$, hence $d_1 = -1.2$ and $d_2 = 1.32$. Now, the control gains can be computed from equation (3.32) using the model coefficients $a_1 = -0.8$ and $a_2 = 0.15$, resulting in $l_1 = -0.4$ and $l_2 = 0.17$. The final control system is implemented as shown by Figure 3.4.

The closed-loop response to a unit step in the command input is illustrated by Figure 3.5. For comparison, the open-loop response of the TF model (3.20) to the same input, without any control, is shown by Figure 2.8a. It is clear that, in this simple example, the closed-loop behaviour has been modified considerably. The wider issues of control system performance, as well as the selection of control system gains that will achieve desirable closed-loop performance, will be discussed in later chapters.

However, the basic SVF algorithm considered above is a *Type 0* control system, as discussed in Chapter 2, indicating that it does not have an inherent unity steady-state gain. Hence, the final value of the step response in Figure 3.5 is not equal to the desired level $y_d(k) = 1.0$. This is an obvious disadvantage and its conversion into a *Type 1* control system, with inherent unity steady-state gain, is also discussed in subsequent chapters.

3.1.1 State Variable Feedback for the General TF Model

Following the approach taken in Example 3.3, it is clear that the feedback gains for the general nth order TF model (3.11) are given by:

$$l_i = d_i - a_i \; (i = 1, 2, \cdots n) \tag{3.34}$$

where d_i are the coefficients of the following desired characteristic polynomial:

$$1 + d_1 z^{-1} + d_2 z^{-2} + \cdots + d_n z^{-n} \tag{3.35}$$

This transparency of the pole assignment control design is the main advantage of the controllable canonical form. But, as mentioned previously, while the computation of the control gains is very straightforward, the practical implementation of this control system is more problematic because the state vector is not directly available from the measured output.

In Example 3.3, we have extracted the states using equation (3.26) and then implemented the controller in the form shown by Figure 3.4. Unfortunately, this has disadvantages as a general solution to the control problem. In particular, the control system is very much dependent upon the accuracy of the estimated model, with the associated disadvantage of decreased robustness in the face of uncertainty, as discussed later in Chapter 4. A particularly important example of its limitations arises if the system is non-minimum phase, with zeros (the roots of the numerator polynomial) outside the unit circle in the complex z-plane. In this situation, the feedback loop will have an unstable component, unless there is an exact cancellation of the $B(z^{-1})$ polynomial in equation (3.28), which is highly unlikely in any real, practical example.

3.2 Observable Canonical Form

Consider now a new state vector for the general nth order system (3.11) defined in the following manner [cf. equation (3.7) of Example 3.1]:

$$\begin{bmatrix} x_1(k) \\ x_2(k) \\ \vdots \\ x_{n-1}(k) \\ x_n(k) \end{bmatrix} = \begin{bmatrix} -a_1 x_1(k-1) + x_2(k-1) + b_1 u(k-1) \\ -a_2 x_1(k-1) + x_3(k-1) + b_2 u(k-1) \\ \vdots \\ -a_{n-1} x_1(k-1) + x_n(k-1) + b_{n-1} u(k-1) \\ -a_n x_1(k-1) + b_n u(k-1) \end{bmatrix} \tag{3.36}$$

The associated state space representation then takes the quite simple form:

$$\begin{bmatrix} x_1(k) \\ x_2(k) \\ \vdots \\ x_{n-1}(k) \\ x_n(k) \end{bmatrix} = \begin{bmatrix} -a_1 & 1 & 0 & \cdots & 0 \\ -a_2 & 0 & 1 & \cdots & 0 \\ \vdots & \vdots & \vdots & \ddots & \vdots \\ -a_{n-1} & 0 & 0 & \cdots & 1 \\ -a_n & 0 & 0 & \cdots & 0 \end{bmatrix} \begin{bmatrix} x_1(k-1) \\ x_2(k-1) \\ \vdots \\ x_{n-1}(k-1) \\ x_n(k-1) \end{bmatrix} + \begin{bmatrix} b_1 \\ b_2 \\ \vdots \\ b_{n-1} \\ b_n \end{bmatrix} u(k-1) \tag{3.37}$$

$$y(k) = \begin{bmatrix} 1 & 0 & \cdots & 0 & 0 \end{bmatrix} \begin{bmatrix} x_1(k) \\ x_2(k) \\ \vdots \\ x_{n-1}(k) \\ x_n(k) \end{bmatrix} \tag{3.38}$$

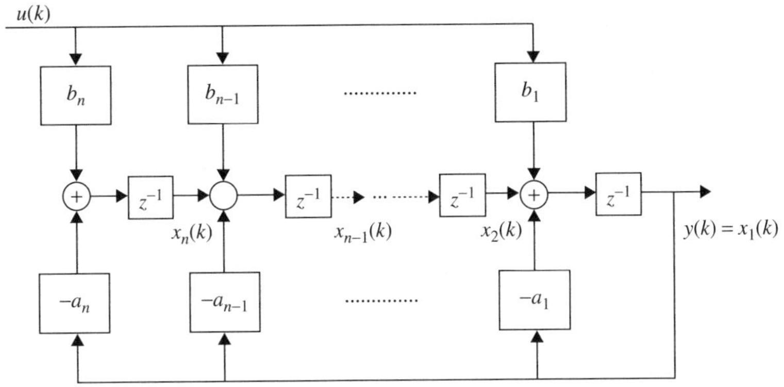

Figure 3.6 Observable canonical form for the general discrete-time system (3.11)

This particular realisation of the system is called the *observable canonical form* and its block diagram structure is shown in Figure 3.6. It is clear from Figure 3.6 that the state vector is formed from a combination of input and output signals, defined by the model parameters.

The term 'observable' canonical form indicates that it is a particularly convenient state space model to employ when considering how state variables can be reconstructed (or estimated in the stochastic situation), based on the measured signals and knowledge of the state space model (see e.g. Luenberger 1971; Kailath 1980; Åström & Wittenmark 1984; Young 2011). Such techniques are potentially important in any implementation of SVF control based on a minimal state space model. However, they are not discussed in detail here, because we obviate their need by exploiting the alternative *Non-Minimal State Space* (NMSS) form that is considered in Chapter 4.

Example 3.4 State Variable Feedback based on the Observable Canonical Form Consider again the second order TF model (3.20). Defining the vector:

$$\begin{bmatrix} x_1(k) \\ x_2(k) \end{bmatrix} = \begin{bmatrix} y(k) \\ -a_2 y(k-1) + b_2 u(k-1) \end{bmatrix} \tag{3.39}$$

the observable canonical form is given by the following second order state space model:

$$\begin{bmatrix} x_1(k) \\ x_2(k) \end{bmatrix} = \begin{bmatrix} -a_1 & 1 \\ -a_2 & 0 \end{bmatrix} \begin{bmatrix} x_1(k-1) \\ x_2(k-1) \end{bmatrix} + \begin{bmatrix} b_1 \\ b_2 \end{bmatrix} u(k-1) \tag{3.40}$$

$$y(k) = \begin{bmatrix} 1 & 0 \end{bmatrix} \begin{bmatrix} x_1(k) \\ x_2(k) \end{bmatrix} \tag{3.41}$$

so that a linear SVF control law for this example is again:

$$u(k) = -l_1 x_1(k) - l_2 x_2(k) + y_d(k) \tag{3.42}$$

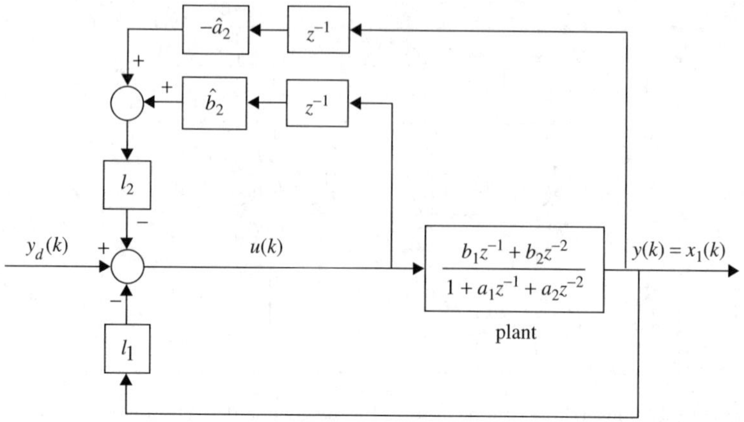

Figure 3.7 The closed-loop control system for Example 3.4 with model mismatch

where $y_d(k)$ is the command input. Note that $x_1(k)$ is equal to the system output $y(k)$ and is, therefore, directly observed (measured). However, $x_2(k)$ is not observed, which once again prevents us from directly implementing the control law. Instead, we employ the model of the system:

$$\frac{\hat{B}(z^{-1})}{\hat{A}(z^{-1})} = \frac{\hat{b}_1 z^{-1} + \hat{b}_2 z^{-2}}{1 + \hat{a}_1 z^{-1} + \hat{a}_2 z^{-2}} \tag{3.43}$$

to provide estimates of the states (3.39): i.e. in this case:

$$u(k) = -l_1 y(k) - l_2 \left(-\hat{a}_2 y(k-1) + \hat{b}_2 u(k-1)\right) + y_d(k) \tag{3.44}$$

and the closed-loop system is represented in the block diagram form by Figure 3.7.

It is interesting to note that, whereas the SVF controller based on a controllable canonical form only involves feedback of the output variable (Figure 3.4), here the control algorithm requires $y(k)$, $y(k-1)$ and $u(k-1)$ (Figure 3.7). This observation will assume greater significance when we introduce NMSS models in Chapter 4.

Assuming no plant–model mismatch, i.e. $B(z^{-1}) = \hat{B}(z^{-1})$ and $A(z^{-1}) = \hat{A}(z^{-1})$, the CLTF associated with Figure 3.7 takes the form:

$$y(k) = \frac{b_1 z^{-1} + b_2 z^{-2}}{1 + (a_1 + l_2 b_2 + b_1 l_1) z^{-1} + (a_2 + l_2 b_2 a_1 + b_2 l_1 - b_1 l_2 a_2) z^{-2}} y_d(k) \tag{3.45}$$

As before, a pole assignment control algorithm can be derived by equating the coefficients of the denominator polynomial with the desired coefficients from equation (3.33). In this case, however, the solution of two simultaneous equations is required to find the unknown control gains l_1 and l_2. Not surprisingly, therefore, the general nth order system would require the solution of n simultaneous equations, making the pole assignment solution more computationally intensive when using this observable canonical form than the previous controllable canonical form.

3.3 General State Space Form

The general state space representation of a linear, discrete-time, SISO system takes the following form:

$$\boldsymbol{x}(k) = \boldsymbol{F}\boldsymbol{x}(k-1) + \boldsymbol{g}u(k-1) \tag{3.46}$$

$$y(k) = \boldsymbol{h}\boldsymbol{x}(k) \tag{3.47}$$

where $\boldsymbol{x}(k)$ is the state vector, $u(k)$ and $y(k)$ are the input and the output variables, respectively; and the state transition matrix $\boldsymbol{F}$, input vector $\boldsymbol{g}$ and output vector $\boldsymbol{h}$ are all of appropriate dimensions. For minimal SISO state space representations, as considered in this chapter, $\boldsymbol{F}$ is a $n \times n$ square matrix, $\boldsymbol{g}$ is a $n \times 1$ column vector and $\boldsymbol{h}$ is a $1 \times n$ row vector. All the state space models discussed earlier (i.e. Example 3.1, Example 3.2 and Example 3.4) take the form {(3.46), (3.47)}, which is illustrated in Figure 3.8.

Note that, as discussed in Chapter 2, discrete-time control system design requires at least one sample time delay in the control model, which is implicit in the state space representation {(3.46), (3.47)} and the TF model (3.11). Notwithstanding this requirement, a state space model with no time delay can be developed by the inclusion of a 'feed-through' $u(k)$ variable in the output equation (3.47), i.e. $y(k) = \boldsymbol{h}\boldsymbol{x}(k) + b_0 u(k)$, where b_0 is a constant coefficient [see equation (2.11)].

3.3.1 *Transfer Function Form of a State Space Model*

Using the state equation (3.46):

$$\boldsymbol{x}(k) = \boldsymbol{F}z^{-1}\boldsymbol{x}(k) + \boldsymbol{g}z^{-1}u(k) \tag{3.48}$$

where, as before, z^{-1} is the backward shift operator. Hence,

$$(\boldsymbol{I} - \boldsymbol{F}z^{-1})\boldsymbol{x}(k) = \boldsymbol{g}z^{-1}u(k) \tag{3.49}$$

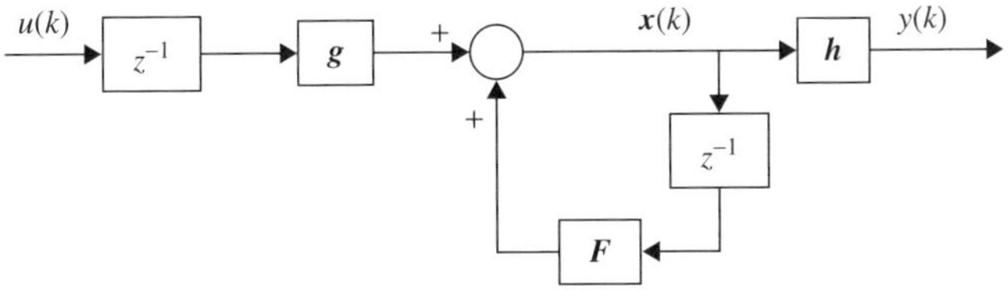

Figure 3.8 Block diagram form for the general state space system description {(3.46), (3.47)}

so that

$$x(k) = (I - Fz^{-1})^{-1} gz^{-1} u(k)$$

and substituting for $x(k)$ in the observation equation (3.47):

$$y(k) = hx(k) = h(I - Fz^{-1})^{-1} gz^{-1} u(k) \tag{3.50}$$

The TF representation $G_1(z^{-1})$ of the state space model {(3.46), (3.47)} is:

$$\frac{y(k)}{u(k)} = G_1(z^{-1}) = h(I - Fz^{-1})^{-1} gz^{-1} \tag{3.51}$$

This TF can be written equivalently in terms of the forward shift operator z, i.e.

$$\frac{y(k)}{u(k)} = G_1(z) = h(zI - F)^{-1} g \tag{3.52}$$

As stated above, the TF $G_1(z)$ completely defines the input–output behaviour of the system and is independent of the state space representation. This becomes apparent if we now define a new state vector $x^*(k) = Tx(k)$, where T is a non-singular transformation matrix.

Replacing $x(k)$ in {(3.46), (3.47)} with $T^{-1}x^*(k)$:

$$x^*(k) = TFT^{-1} x^*(k-1) + Tgu(k-1) \tag{3.53}$$

and

$$y(k) = hT^{-1} x^*(k) \tag{3.54}$$

The input–output TF from {(3.53), (3.54)} is given by:

$$\frac{y(k)}{u(k)} = hT^{-1}(I - TFT^{-1}z^{-1})^{-1} Tgz^{-1} = h(I - Fz^{-1})^{-1} gz^{-1} = G_1(z^{-1}) \tag{3.55}$$

Hence, the TF model $G_1(z^{-1})$ is independent of the state space representation.

Example 3.5 Determining the TF from a State Space Model Substituting the numerical values of the TF model (3.20) into the controllable canonical form {(3.22), (3.23)} yields:

$$x(k) = \begin{bmatrix} 0.8 & -0.15 \\ 1.0 & 0 \end{bmatrix} x(k-1) + \begin{bmatrix} 1 \\ 0 \end{bmatrix} u(k-1) \tag{3.56}$$

$$y(k) = \begin{bmatrix} 0.5 & -0.4 \end{bmatrix} x(k) \tag{3.57}$$

Using (3.51) and the standard rules of matrix inversion (see Appendix A), the TF is derived in terms of the backward shift operator z^{-1}, as follows:

$$G_1(z^{-1}) = \begin{bmatrix} 0.5 & -0.4 \end{bmatrix} \begin{bmatrix} 1-0.8z^{-1} & 0.15z^{-1} \\ -z^{-1} & 1 \end{bmatrix}^{-1} \begin{bmatrix} 1 \\ 0 \end{bmatrix} z^{-1}$$

$$= \frac{z^{-1}}{1-0.8z^{-1}+0.15z^{-2}} \begin{bmatrix} 0.5 & -0.4 \end{bmatrix} \begin{bmatrix} 1 & 0.15z^{-1} \\ -z^{-1} & 1-0.8z^{-1} \end{bmatrix} \begin{bmatrix} 1 \\ 0 \end{bmatrix} \tag{3.58}$$

$$= \frac{0.5z^{-1} - 0.4z^{-2}}{1-0.8z^{-1}+0.15z^{-2}}$$

as expected from equation (3.20). The equivalent forward shift operator form of the TF is:

$$G_1(z) = \frac{0.5z - 0.4}{z^2 - 0.8z + 0.15} \tag{3.59}$$

The following subsection examines the relationship between TF and state space models in more detail.

3.3.2 *The Characteristic Equation, Eigenvalues and Eigenvectors*

Noting the procedure for the inversion of a matrix, the z-operator TF model (3.52) can be rewritten as:

$$G_1(z) = \boldsymbol{h}(z\boldsymbol{I} - \boldsymbol{F})^{-1}\boldsymbol{g} = \frac{\boldsymbol{h}\,(\mathrm{adj}(z\boldsymbol{I} - \boldsymbol{F}))\,\boldsymbol{g}}{|(z\boldsymbol{I} - \boldsymbol{F})|} \tag{3.60}$$

where the determinant $|(z\boldsymbol{I} - \boldsymbol{F})|$ is called the *characteristic polynomial*. The *characteristic equation* of the system is obtained by equating this determinant to zero, i.e.

$$|(z\boldsymbol{I} - \boldsymbol{F})| = 0 \tag{3.61}$$

Evaluating the determinant and equating to zero yields:

$$|(z\boldsymbol{I} - \boldsymbol{F})| = z^n + a_1 z^{n-1} + a_2 z^{n-2} + \cdots + a_n = 0 \tag{3.62}$$

while, in terms of the backward shift operator, the equivalent result is given by:

$$\left|(\boldsymbol{I} - \boldsymbol{F}z^{-1})\right| = 1 + a_1 z^{-1} + a_2 z^{-2} + \cdots + a_n z^{-n} = 0 \tag{3.63}$$

Note that the dimension of the matrix $\boldsymbol{F}$, in this minimal state space case, is always equal to $n \times n$, hence the order of the characteristic equation is also n; and it is clear from equation

(3.62) and equation (3.63) that the characteristic equation of $\boldsymbol{F}$ is equivalent to the characteristic equation of the TF model (3.11).

In mathematics, the characteristic equation of the $n \times n$ matrix $\boldsymbol{F}$ is usually denoted as:

$$|(\lambda \boldsymbol{I} - \boldsymbol{F})| = \lambda^n + a_1\lambda^{n-1} + a_2\lambda^{n-2} + \cdots + a_n = 0 \tag{3.64}$$

The roots of equation (3.64) are called the *eigenvalues* of $\boldsymbol{F}$ (see also Hoffman and Kunze 1971, p. 182; Marcus and Minc 1988, p. 145; O'Neil 2007, pp. 267–276).

As discussed in Chapter 2, the roots of $A(z) = 0$ obtained from the TF model (3.11) are called the *poles* of the system. For a minimal state space model, it follows that the eigenvalues of the matrix $\boldsymbol{F}$ are equal to the poles of the equivalent TF model. By contrast, in Chapter 4, we will show that if the state space representation is *non-minimal*, then the poles of the TF model represent a subset of the eigenvalues of $\boldsymbol{F}$.

Finally, any non-zero vector $\boldsymbol{w}_i = \begin{bmatrix} w_{i1} & w_{i2} & \cdots & w_{in} \end{bmatrix}^T$ which satisfies the following matrix equation:

$$(\lambda_i \boldsymbol{I} - \boldsymbol{F})\boldsymbol{w}_i = \boldsymbol{0} \tag{3.65}$$

where $\lambda_i (i = 1, 2, \ldots, n)$ is the ith eigenvalue of $\boldsymbol{F}$ and $\boldsymbol{0}$ is a null vector (i.e. zeros), is called the *eigenvector* associated with λ_i.

Note that since the eigenvalues of $\boldsymbol{F}$ are the poles of the system TF, it follows that the eigenvalues of $\tilde{\boldsymbol{F}} = \boldsymbol{T}^{-1}\boldsymbol{F}\boldsymbol{T}$ are the same. In other words, the eigenvalues are unaffected by a similarity transformation. To show this, let $\boldsymbol{w}_i$ denote an eigenvector of $\boldsymbol{F}$. Then, by definition, $\boldsymbol{F}\boldsymbol{w}_i = \lambda_i \boldsymbol{w}_i$, where λ_i is the eigenvalue corresponding to $\boldsymbol{w}_i$. Now if we define $\tilde{\boldsymbol{w}}_i = \mathbf{T}^{-1}\boldsymbol{w}_i$ as the transformed eigenvector, then:

$$\tilde{\boldsymbol{F}}\tilde{\boldsymbol{w}}_i = \tilde{\boldsymbol{F}}(\boldsymbol{T}^{-1}\boldsymbol{w}_i) = (\boldsymbol{T}^{-1}\boldsymbol{F}\boldsymbol{T})(\mathbf{T}^{-1}\boldsymbol{w}_i) = \boldsymbol{T}^{-1}\boldsymbol{F}\boldsymbol{w}_i = \boldsymbol{T}^{-1}\lambda_i\boldsymbol{w}_i = \lambda_i\tilde{\boldsymbol{w}}_i$$

Thus, the transformed eigenvector is an eigenvector of the transformed $\boldsymbol{F}$ matrix, and the eigenvalue is unchanged.

Example 3.6 Eigenvalues and Eigenvectors of a State Space Model Returning to the TF model (3.20) and the associated state space form {(3.56), (3.57)}, the state transition matrix $\boldsymbol{F}$ is given by:

$$\boldsymbol{F} = \begin{bmatrix} 0.8 & -0.15 \\ 1.0 & 0 \end{bmatrix} \tag{3.66}$$

so that the characteristic equation (3.64) is simply:

$$|(\lambda \boldsymbol{I} - \boldsymbol{F})| = \begin{vmatrix} \lambda - 0.8 & 0.15 \\ -1.0 & \lambda \end{vmatrix} = 0 \tag{3.67}$$

i.e.

$$\lambda^2 - 0.8\lambda + 0.15 = 0 \quad \text{or} \quad (\lambda - 0.5)(\lambda - 0.3) = 0 \tag{3.68}$$

This shows that the eigenvalues and TF poles are $\lambda_1 = 0.3$ and $\lambda_2 = 0.5$. From equation (3.65), therefore, the eigenvectors of $\boldsymbol{F}$ satisfy:

$$\lambda_i \boldsymbol{w}_i - \boldsymbol{F}\boldsymbol{w}_i = \boldsymbol{0} \quad i = 1, 2 \tag{3.69}$$

In order to find the corresponding eigenvector, $\lambda = 0.3$ is substituted into this equation:

$$0.3\begin{bmatrix} w_{11} \\ w_{12} \end{bmatrix} - \begin{bmatrix} 0.8 & -0.15 \\ 1.0 & 0 \end{bmatrix}\begin{bmatrix} w_{11} \\ w_{12} \end{bmatrix} = \begin{bmatrix} 0 \\ 0 \end{bmatrix} \tag{3.70}$$

so that, letting $w_{12} = 1$, the eigenvector corresponding to $\lambda_1 = 0.3$ is $\boldsymbol{w}_1 = [\,0.3\;\; 1\,]^T$. Similarly, the eigenvector corresponding to $\lambda_2 = 0.5$ can be found from:

$$0.5\begin{bmatrix} w_{21} \\ w_{22} \end{bmatrix} - \begin{bmatrix} 0.8 & -0.15 \\ 1.0 & 0 \end{bmatrix}\begin{bmatrix} w_{21} \\ w_{22} \end{bmatrix} = \begin{bmatrix} 0 \\ 0 \end{bmatrix} \tag{3.71}$$

and this is satisfied when $w_{21} = 1$ and $w_{22} = 2$. Hence, the eigenvector corresponding to $\lambda_1 = 0.5$ is $\boldsymbol{w}_2 = [\,1\;2\,]^T$. The eigenvectors are useful in various ways but, within the present context, they facilitate the transformation of the state space model into a diagonal 'decoupled' form, as shown below.

3.3.3 *The Diagonal Form of a State Space Model*

In general, if the state transition matrix $\boldsymbol{F}$ has distinct (i.e. non-repeated), real eigenvalues, the eigenvectors are linearly independent. Furthermore, if the columns of the matrix $\boldsymbol{T}^{-1}$ in equation (3.53) are chosen as the eigenvectors of $\boldsymbol{F}$, then $\boldsymbol{TFT}^{-1}$will be a diagonal matrix, i.e.

$$\boldsymbol{TFT}^{-1} = \boldsymbol{\Lambda} = \begin{bmatrix} \lambda_1 & 0 & 0 & \cdots & 0 \\ 0 & \lambda_2 & 0 & \cdots & 0 \\ 0 & 0 & \lambda_3 & \cdots & 0 \\ \vdots & \vdots & \vdots & \ddots & \vdots \\ 0 & & 0 & 0 & \lambda_n \end{bmatrix} \tag{3.72}$$

where $\lambda_i (i = 1, 2, \ldots, n)$ are the eigenvalues of $\boldsymbol{F}$, as before. In this case, the state equations in (3.53) are conveniently decoupled (i.e. each state variable appears in only one equation).

The state equations in this decoupled form are written as:

$$\begin{aligned} \boldsymbol{x}^*(k) &= \boldsymbol{\Lambda}\, \boldsymbol{x}^*(k-1) + \boldsymbol{T}\boldsymbol{g}u(k-1) \\ y(k) &= \boldsymbol{h}\boldsymbol{T}^{-1}\boldsymbol{x}^*(k) \end{aligned} \tag{3.73}$$

where $\boldsymbol{x}^*(k) = \boldsymbol{T}\boldsymbol{x}(k)$ is the transformed, decoupled state vector.

Example 3.7 Determining the Diagonal Form of a State Space Model Consider again the system described by the TF model (3.20) and associated state space form {(3.56), (3.57)}. Noting the eigenvectors obtained in the previous section, $\boldsymbol{T}^{-1}$ is defined:

$$\boldsymbol{T}^{-1} = \begin{bmatrix} \boldsymbol{w}_1 & \boldsymbol{w}_2 \end{bmatrix} = \begin{bmatrix} 0.3 & 1 \\ 1 & 2 \end{bmatrix} \tag{3.74}$$

Therefore, the transformation matrix $\boldsymbol{T}$ is:

$$\boldsymbol{T} = \begin{bmatrix} -5.0 & 2.5 \\ 2.5 & -0.75 \end{bmatrix} \tag{3.75}$$

and

$$\boldsymbol{TFT}^{-1} = \begin{bmatrix} -5.0 & 2.5 \\ 2.5 & -0.75 \end{bmatrix} \begin{bmatrix} 0.8 & -0.15 \\ 1.0 & 0 \end{bmatrix} \begin{bmatrix} 0.3 & 1 \\ 1 & 2 \end{bmatrix} = \begin{bmatrix} 0.3 & 0 \\ 0 & 0.5 \end{bmatrix} = \boldsymbol{\Lambda} \tag{3.76}$$

Finally, note that if the eigenvalues of $\boldsymbol{F}$ are not all distinct, then it may not be possible to obtain independent eigenvectors and hence determine the diagonal form. In this case, the $\boldsymbol{F}$ matrix can be transformed into a *Jordan form*, in which the eigenvalues of $\boldsymbol{F}$ lie on the main diagonal, while some of the elements immediately above the main diagonal are unity. However, this form has no importance for the present book and the reader is advised to consult Ayres (1962, p. 206) or Kuo (1980, pp. 195–198) for further information.

3.4 Controllability and Observability

The twin concepts of controllability and observability, first introduced by Kalman (1960), are very important in SVF control system design, since they are concerned with the interconnection between the input, state and the output of a state space model. In this book, the controllability is particularly significant, since it determines whether or not we can control a system's state using the control input variable.

3.4.1 Definition of Controllability (or Reachability[2])

The digital system described by equations {(3.46), (3.47)} is controllable (or reachable) if, for any initial state $\boldsymbol{x}(0)$, it is possible to find an unconstrained control sequence which will transfer it to any final state $\boldsymbol{x}(n)$ (not necessarily the origin) in a finite time (Kuo 1980, p. 423; Middleton and Goodwin 1990, p. 180).

[2] Note that the term reachability is often preferred for discrete-time systems.

3.4.2 Rank Test for Controllability

Using {(3.46), (3.47)}, the desired state $\boldsymbol{x}(n)$, in terms of the initial state $\boldsymbol{x}(0)$ and the input sequence $u(0), u(1), \ldots, u(n-1)$, is given by:

$$\boldsymbol{x}(n) = \boldsymbol{F}^n\boldsymbol{x}(0) + \boldsymbol{g}u(n-1) + \boldsymbol{F}\boldsymbol{g}u(n-2) + \cdots + \boldsymbol{F}^{n-2}\boldsymbol{g}u(1) + \boldsymbol{F}^{n-1}\boldsymbol{g}u(0) \tag{3.77}$$

or

$$\boldsymbol{x}(n) - \boldsymbol{F}^n\boldsymbol{x}(0) = \begin{bmatrix} \boldsymbol{g} & \boldsymbol{F}\boldsymbol{g} & \cdots & \boldsymbol{F}^{n-2}\boldsymbol{g} & \boldsymbol{F}^{n-1}\boldsymbol{g} \end{bmatrix} \begin{bmatrix} u(n-1) \\ u(n-2) \\ \vdots \\ u(1) \\ u(0) \end{bmatrix} \tag{3.78}$$

The left-hand side of equation (3.78) is known (as part of the definition above), while the right-hand side is a function of the input signals. Since equation (3.78) also represents n simultaneous algebraic equations, these equations must be linearly independent for the solution to exist. Hence, a necessary and sufficient condition for controllability, is that the following *controllability matrix*:

$$\boldsymbol{S}_1 = \begin{bmatrix} \boldsymbol{g} & \boldsymbol{F}\boldsymbol{g} & \cdots & \boldsymbol{F}^{n-2}\boldsymbol{g} & \boldsymbol{F}^{n-1}\boldsymbol{g} \end{bmatrix} \tag{3.79}$$

is of full rank n, i.e. that it is non-singular and can be inverted.

3.4.3 Definition of Observability

The digital system described by equations {(3.46), (3.47)} is observable if the state vector $\boldsymbol{x}(k)$, at any sample time k, can be determined from the system model and complete knowledge of the system's input and output variables (Kuo 1980, p. 430; Middleton and Goodwin 1990, p. 196).

3.4.4 Rank Test for Observability

Consider the system {(3.46), (3.47)} with the initial state vector $\boldsymbol{x}(0)$. Since the effect of the known input can always be subtracted from the solution, we will simplify the analysis by assuming $u(k-1) = 0$ in {(3.46), (3.47)}. This yields the following linear algebraic equations for the output:

$$\begin{aligned} y(0) &= \boldsymbol{h}\boldsymbol{x}(0) \\ y(1) &= \boldsymbol{h}\boldsymbol{x}(1) = \boldsymbol{h}\boldsymbol{F}\boldsymbol{x}(0) \\ &\vdots \\ y(n-1) &= \boldsymbol{h}\boldsymbol{F}^{n-1}\boldsymbol{x}(0) \end{aligned} \tag{3.80}$$

or, in vector-matrix form:

$$\begin{bmatrix} \boldsymbol{h} \\ \boldsymbol{hF} \\ \vdots \\ \boldsymbol{hF}^{n-2} \\ \boldsymbol{hF}^{n-1} \end{bmatrix} \boldsymbol{x}(0) = \begin{bmatrix} y(0) \\ y(1) \\ \vdots \\ y(n-2) \\ y(n-1) \end{bmatrix} \tag{3.81}$$

and it is clear that the initial state $\boldsymbol{x}(0)$, can be found if and only if the following *observability matrix*:

$$\boldsymbol{S}_0 = \begin{bmatrix} \boldsymbol{h} \\ \boldsymbol{hF} \\ \vdots \\ \boldsymbol{hF}^{n-2} \\ \boldsymbol{hF}^{n-1} \end{bmatrix} \tag{3.82}$$

is of full rank n. Once the initial state $\boldsymbol{x}(0)$ is known, then the state equation (3.46) is employed to find $\boldsymbol{x}(1), \boldsymbol{x}(2), \ldots, \boldsymbol{x}(n-1)$ and the definition of *observability* is satisfied. Finally, note that the observable canonical form defined in section 3.2 is always observable, but not necessarily controllable, while the opposite is true of the controllable canonical form in section 3.1.

Example 3.8 Rank Tests for a State Space Model Consider the system (3.56). The controllability matrix is:

$$\boldsymbol{S}_1 = \begin{bmatrix} \boldsymbol{g} & \boldsymbol{Fg} \end{bmatrix} = \begin{bmatrix} 1 & 0.8 \\ 0 & 1 \end{bmatrix} \tag{3.83}$$

which is non-singular, i.e. it has rank 2. Hence, the state space representation is completely controllable, as expected for this controllable canonical form. The observability matrix of the system is:

$$\boldsymbol{S}_0 = \begin{bmatrix} \boldsymbol{h} \\ \boldsymbol{hF} \end{bmatrix} = \begin{bmatrix} 0.5 & -0.4 \\ 0 & -0.25 \end{bmatrix} \tag{3.84}$$

which is again non-singular, so the system is also observable.

Now consider an arbitrarily chosen third order system based on {(3.46), (3.47)} with:

$$\boldsymbol{F} = \begin{bmatrix} 0.8 & -0.15 & -0.4 \\ 1 & 0 & 0 \\ 0 & 0 & 0 \end{bmatrix}; \quad \boldsymbol{g} = \begin{bmatrix} 0.5 \\ 0 \\ 1 \end{bmatrix}; \quad \boldsymbol{h} = \begin{bmatrix} 1 & 0 & 0 \end{bmatrix} \tag{3.85}$$

The controllability matrix is:

$$S_1 = \begin{bmatrix} \boldsymbol{g} & \boldsymbol{F}\boldsymbol{g} & \boldsymbol{F}^2\boldsymbol{g} \end{bmatrix} = \begin{bmatrix} 0.5 & 0 & -0.075 \\ 0 & 0.5 & 0 \\ 1 & 0 & 0 \end{bmatrix} \tag{3.86}$$

which is non-singular, i.e. it has rank 3. Hence the state space representation is again controllable. However, the observability matrix of the system is:

$$S_0 = \begin{bmatrix} \boldsymbol{h} \\ \boldsymbol{h}\boldsymbol{F} \\ \boldsymbol{h}\boldsymbol{F}^2 \end{bmatrix} = \begin{bmatrix} 1 & 0 & 0 \\ 0.8 & -0.15 & -0.4 \\ 0.49 & -0.12 & -0.32 \end{bmatrix} \tag{3.87}$$

which is singular and this system is unobservable.

3.5 Concluding Remarks

This chapter has introduced the concept of SVF, based on the definition of various minimal state space models for SISO systems. Such non-unique, minimal state space representations of a TF model always consist of n states, where n is the order of the system (i.e. the highest power of z in the denominator polynomial of the TF model). The advantage of SVF is that it allows us to completely specify the poles of the closed-loop characteristic equation, thereby ensuring the stability and desired transient response of the control system. This concept was briefly demonstrated in the present chapter (Example 3.3), but is developed in a more general form in Chapter 4 and Chapter 5.

While the computation of the control gains is straightforward, the practical implementation of minimal SVF control is more difficult, since the state vector is not normally available directly from the measured output. Therefore, a SVF algorithm is required that generates a *reconstruction*, in the deterministic case, and an *estimate*, in the stochastic case, of the unknown states. In the present book, however, we will concentrate on an alternative NMSS formulation of the control problem, where the entire state is directly available from the input and output measurements, hence state reconstruction and/or estimation are unnecessary complications.

It will suffice to point out, therefore, that one approach to solving the minimal SVF control problem is to formulate it in stochastic terms by invoking the separation principle (Franklin *et al.* 2006, pp. 511–513; Dorf and Bishop 2008, pp. 773–775), as mentioned in Chapter 1, and introducing an optimal Kalman Filter for state estimation (Kalman 1960): see section 6.4 for an example of this approach. The deterministic equivalent is the Luenberger observer (Luenberger 1971). In either case, however, this adds complexity to the design and the controller becomes more dependent on the estimated model of the system, with the associated disadvantage of decreased robustness. One approach that is aimed at improving the robustness is *Loop Transfer Recovery* (LTR: see e.g. Bitmead *et al.* 1990; Lecchini *et al.* 2006): here robustness is obtained by matching the open-loop return ratio of the designed feedback loop with the open-loop return ratio of the Kalman Filter associated with the selected noise model. However, a much more straightforward alternative is the NMSS approach considered in subsequent chapters.

References

Åström, K.J. and Wittenmark, B. (1984) *Computer-Controlled Systems Theory and Design*, Prentice Hall, Englewood Cliffs, NJ.

Ayres, F. Jr. (1962) *Schaum's Outline of Theory and Problems of Matrices*, Schaum, New York.

Bitmead, R.R., Gevers, M. and Wertz, V. (1990) *Adaptive Optimal Control: The Thinking Man's GPC*, Prentice Hall, Harlow.

Bode, H.W. (1940) Feedback amplifier design, *Bell Systems Technical Journal*, **19**, p. 42.

Dorf, R.C. and Bishop, R.H. (2008) *Modern Control Systems*, Eleventh Edition, Pearson Prentice Hall, Upper Saddle River, NJ.

Evans, W.R. (1948) Analysis of control system, *AIEE Transactions*, **67**, pp. 547–551.

Evans, W.R. (1950) Control system synthesis and Root Locus method, *AIEE Transactions*, **69**, pp. 66–69.

Franklin, G.F., Powell, J.D. and Emami-Naeini, A. (2006) *Feedback Control of Dynamic Systems*, Fifth Edition, Pearson Prentice Hall, Upper Saddle River, NJ.

Hoffman, K. and Kunze, R. (1971) *Linear Algebra*, Second Edition, Prentice Hall, NJ.

Kailath, T. (1980) *Linear Systems*, Prentice Hall, Englewood Cliffs, NJ.

Kalman, R.E. (1960) A new approach to linear filtering and prediction problems, *ASME Journal Basic Engineering*, **82**, pp. 34–45.

Kalman, R.E. (1961) On the general theory of control systems, *Proceedings 1st IFAC Congress*, **1**, pp. 481–491.

Kalman, R.E. (1963) Mathematical description of linear dynamical systems, *SIAM Journal of Control*, **1**, pp. 152–192.

Kuo, B.C. (1980) *Digital Control Systems*, Holt, Rinehart and Winston, New York.

Lecchini, A., Gevers, M. and Maciejowski, J. (2006) An iterative feedback tuning procedure for Loop Transfer Recovery, *14th IFAC Symposium on System Identification*, Newcastle, Australia.

Luenberger, D.G. (1971) An introduction to observers, *IEEE Transactions on Automatic Control*, **16**, pp. 596–603.

Marcus, M. and Minc, H. (1988) *Introduction to Linear Algebra*, Dover, New York.

Middleton, R.H. and Goodwin, G.C. (1990) *Digital Control and Estimation: A Unified Approach*, Prentice Hall, Englewood Cliffs, NJ.

Nyquist, H. (1932) Regeneration theory, *Bell Systems Technical Journal*, **11**, pp. 126–147.

Ogata, K. (1970) *Modern Control Engineering*, Prentice Hall, Englewood Cliffs, NJ.

O'Neil, P. V. (2007) *Advanced Engineering Mathematics*, Seventh Edition, Cengage Learning International Edition, Cengage Learning, Andover.

Rosenbrock, H.H. (1970) *State Space and Multivariable Theory*, Nelson, London.

Rosenbrock, H.H. (1974) *Computer-Aided Control System Design*, Academic Press, New York.

Truxal, J.G. (1955) *Automatic Feedback Control System Synthesis*, McGraw-Hill, New York.

Young, P.C. (2011) *Recursive Estimation and Time-Series Analysis: An Introduction for the Student and Practitioner*. Springer-Verlag, Berlin.

Zadeh, L.A. and Desoer, C.A. (1963) *Linear Systems Theory*, McGraw-Hill, New York.

4

Non-Minimal State Variable Feedback

The state space formulation of control system design is, perhaps, the most natural and convenient approach for use with computers. It has the further advantage over *Transfer Function* (TF) design methods of allowing for a simplified and unified treatment of both *Single-Input, Single-Output* (SISO) and *Multi-Input, Multi-Output* (MIMO), multivariable systems (Kailath 1980; Kuo 1980; Skogestad and Postlethwaite 2005). Most importantly, the approach allows for the implementation of powerful *State Variable Feedback* (SVF) control designs, including pole assignment, *Linear Quadratic Gaussian* (LQG) optimal control (i.e. the optimal control of linear stochastic systems with Gaussian disturbances, based on a quadratic cost function) and other optimal 'risk sensitive' procedures, such as H_∞ (e.g. Mustafa and Glover 1990) and the related *Linear Exponential of Quadratic Gaussian* (LEQG) approach of Whittle (1990), which allow for a degree of optimism or pessimism in the optimal control system design.

Unfortunately, the state space approach has one major difficulty: namely, the state vector required for SVF control is not normally available for direct measurement. The standard approach to this problem is the use of state variable estimation and reconstruction algorithms, most notably the Kalman Filter (Kalman 1960) for stochastic systems and the related Luenberger observer (Luenberger 1964, 1971) for deterministic systems. Here, an estimate or reconstruction of the state vector is generated by the algorithm and can be used in the implementation of the control system, in which it replaces the unmeasurable true state vector in the deterministic SVF control law. This separation of the deterministic control system design and state variable estimation functions, followed by their amalgamation in the final control system implementation, is justified theoretically by the *separation*, or *certainty equivalence*, theorem for LQG optimal control (see Chapter 6 and e.g. Wonham 1968; Åström and Wittenmark 1984, pp. 273–275; Green and Limebeer 1995, pp. 179–208; Franklin *et al.* 2006, pp. 511–513; Dorf and Bishop 2008, pp. 773–775; Gopal 2010, pp. 413–428). Unfortunately, as pointed out in Chapter 3, this elegant LQG approach not only adds complexity to the design but, more importantly, the controller becomes more dependent on the estimated model of the system,

True Digital Control: Statistical Modelling and Non-Minimal State Space Design, First Edition.
C. James Taylor, Peter C. Young and Arun Chotai.

with the associated disadvantage of decreased robustness (Maciejowski 1985; Bitmead *et al.* 1990).

In the present chapter, we introduce another, much simpler solution to this problem, by defining a *Non-Minimal State Space* (NMSS) form. Here, the state vector is composed only of those variables that can be measured directly and then stored in the digital computer for use by the control law (section 4.1). In the discrete-time, backward shift operator case, these are the present and past sampled values of the output variable, and the past sampled values of the input variable. Once the controllability conditions have been established for the NMSS model (section 4.2), the non-minimal SVF controller can be implemented straightforwardly, *without resort to state reconstruction* (section 4.3), thus simplifying the control system design and making it more robust to the inevitable uncertainty associated with the estimated model of the system.

The final sections of the chapter elaborate on the relationship between non-minimal and minimal SVF control (section 4.4), while the theoretical and practical advantages of the non-minimal approach are illustrated by worked examples (section 4.5).

4.1 The NMSS Form

The major attraction of the NMSS representation of a linear, discrete-time system is the simplicity of the state vector, which is composed only of the present and past sampled values of the input and output signals, all of which are directly measurable (for standard discrete-time models, see Hesketh 1982, Young *et al.* 1987, Wang and Young 1988 and Taylor *et al.* 2000; and for delta operator discrete-time systems, see Young *et al.* 1998 and Chotai *et al.* 1998). Indeed, the NMSS model is the natural state space description of a discrete-time TF model, since its dimension is dictated by the complete structure of the model, i.e. the denominator polynomial order n, numerator polynomial order m and time delay τ (see below).

This is in contrast to minimal state space descriptions, which only account for the order of the denominator and whose state variables, therefore, usually represent combinations of input and output signals, as exemplified by the observable canonical form of the state space introduced in section 3.2.

4.1.1 The NMSS (Regulator) Representation

A regulation control system, or 'regulator', is a Type 0 closed-loop system (see Examples 2.4 and 3.3) that is concerned only with ensuring closed-loop stability and the maintenance of closed-loop *transient* dynamics that are satisfactory in some sense. It is not intended for maintaining 'set-points' or for tracking the changes in a command input [e.g. the $y_d(k)$ of Chapters 2 and 3], which requires Type 1 servomechanism control.

As we shall see in Chapter 5, it is straightforward to design a NMSS control system that yields a Type 1 closed-loop system with inherent regulatory *and* tracking behaviour. Moreover, such Type 1 NMSS control can be extended quite straightforwardly to multivariable systems, as shown in Chapter 7. For initial tutorial purposes, however, the present chapter is limited to consideration of NMSS control in the SISO regulator case, i.e. a Type 0 control system and with one input–output pathway.

Let us start with the following discrete-time TF model in the backward shift operator z^{-1}:

$$y(k) = \frac{B(z^{-1})}{A(z^{-1})}u(k) \tag{4.1}$$

where

$$A(z^{-1}) = 1 + a_1 z^{-1} + a_2 z^{-2} + \cdots + a_n z^{-n} \qquad (a_n \neq 0) \tag{4.2}$$

$$B(z^{-1}) = b_1 z^{-1} + b_2 z^{-2} + \cdots + b_m z^{-m} \qquad (b_m \neq 0) \tag{4.3}$$

Once again, it is important to note that discrete-time control systems require at least one sample time delay, which is implicit in the model above. As a result, if there is no delay in the model of the system estimated from data, then it is necessary to introduce such a 'false' delay on the control input $u(k)$.

Any additional time delay of $\tau > 1$ samples can be accounted for simply by setting the $\tau - 1$ leading parameters of the $B(z^{-1})$ polynomial to zero, i.e. $b_1 \ldots\ b_{\tau-1} = 0$ (see Example 4.3). Finally, and in contrast to Chapter 3, it is useful to make the distinction here between the denominator polynomial order n and the numerator polynomial order m.

It is straightforward to show (see Example 4.1, Example 4.3 and Example 4.5) that the model (4.1) can be written in the following NMSS form:

$$\boldsymbol{x}(k) = \boldsymbol{F}\boldsymbol{x}(k-1) + \boldsymbol{g}u(k-1) \tag{4.4}$$

with an associated output equation:

$$y(k) = \boldsymbol{h}\boldsymbol{x}(k) \tag{4.5}$$

where the state transition matrix $\boldsymbol{F}$, input vector $\boldsymbol{g}$, and output vector $\boldsymbol{h}$ are defined as follows:

$$\boldsymbol{F} = \begin{bmatrix} -a_1 & -a_2 & \cdots & -a_{n-1} & -a_n & b_2 & b_3 & \cdots & b_{m-1} & b_m \\ 1 & 0 & \cdots & 0 & 0 & 0 & 0 & \cdots & 0 & 0 \\ 0 & 1 & \cdots & 0 & 0 & 0 & 0 & \cdots & 0 & 0 \\ \vdots & \vdots & \ddots & \vdots & \vdots & \vdots & \vdots & \ddots & \vdots & \vdots \\ 0 & 0 & \cdots & 1 & 0 & 0 & 0 & \cdots & 0 & 0 \\ 0 & 0 & \cdots & 0 & 0 & 0 & 0 & \cdots & 0 & 0 \\ 0 & 0 & \cdots & 0 & 0 & 1 & 0 & \cdots & 0 & 0 \\ 0 & 0 & \cdots & 0 & 0 & 0 & 1 & \cdots & 0 & 0 \\ \vdots & \vdots & \ddots & \vdots & \vdots & \vdots & \vdots & \ddots & \vdots & \vdots \\ 0 & 0 & \cdots & 0 & 0 & 0 & 0 & \cdots & 1 & 0 \end{bmatrix} \tag{4.6}$$

$$\boldsymbol{g} = \begin{bmatrix} b_1 & 0 & \cdots & 0 & 0 & 1 & 0 & \cdots & 0 & 0 \end{bmatrix}^T \tag{4.7}$$

$$\boldsymbol{h} = \begin{bmatrix} 1 & 0 & \cdots & 0 & 0 & 0 & 0 & \cdots & 0 & 0 \end{bmatrix} \tag{4.8}$$

The non-minimal state vector $\boldsymbol{x}(k)$ is defined in terms of the present and past sampled outputs and the past sampled inputs, i.e.

$$\boldsymbol{x}(k) = \begin{bmatrix} y(k) & y(k-1) & \cdots & y(k-n+1) & u(k-1) & u(k-2) & \cdots & u(k-m+1) \end{bmatrix}^T \tag{4.9}$$

The non-minimal state vector (4.9) has an order of $(n + m - 1)$. When $m > 1$ (i.e. when the system has zeros and/or pure time delays), the NMSS form has more states than the minimal state space representations discussed in Chapter 3. Hence, as noted above, we refer to this description of the system as a NMSS representation.

In the control literature, the idea of extending the state space from its minimal dimension has a long history: for example, Young and Willems (1972) showed how the addition of *integral-of-error* states introduced inherent Type 1 servomechanism performance in SVF control systems; and the further extension of this idea to include 'input–output' state variables of the above kind, in addition to the integral-of-error states, was suggested by Young *et al.* (1987). The above, purely regulatory form of the NMSS, was suggested by Hesketh (1982) within a pole assignment control context. More recently, a number of different NMSS forms have been suggested for a range of application areas (Taylor *et al.* 1998, 2000, 2009, 2012; Taylor and Shaban 2006; Wang and Young 2006; Gonzalez *et al.* (2009; Exadaktylos and Taylor 2010). They have also appeared in various other areas of science. For example, in the time series literature, Priestley (1988) used an NMSS model in his work on state-dependent parameter estimation, and Burridge and Wallis (1990) use a NMSS model in research on the seasonal adjustment of periodic time series.

Before proceeding further, it is important to note the difference between the output signal $y(k)$ and the output observation vector $\boldsymbol{y}(k) = \boldsymbol{x}(k)$. The latter may be straightforwardly obtained at every sampling instant, provided that the past sampled values of the input and the output are stored in the computer control system. In particular, note that while the vector $\boldsymbol{h}$ in equation (4.8) is chosen to highlight the essential SISO nature of the model, this tends to conceal the fact that all of the state variables in (4.9) can be observed at the kth sampling instant. Thus, for some analytical purposes, it makes sense to utilise the following observation equation:

$$\boldsymbol{y}(k) = \boldsymbol{H}\,\boldsymbol{x}(k); \quad \boldsymbol{H} = \boldsymbol{I}_{n+m-1} \tag{4.10}$$

In equation (4.10), $\boldsymbol{I}_{n+m-1}$ is the $(n + m - 1)$th order identity matrix, i.e.

$$\begin{bmatrix} y(k) \\ y(k-1) \\ \vdots \\ y(k-n+1 \\ u(k-1) \\ \vdots \\ u(k-m+1 \end{bmatrix} = \begin{bmatrix} 1 & 0 & \cdots & 0 & 0 & \cdots & 0 \\ 0 & 1 & \cdots & 0 & 0 & \cdots & 0 \\ \vdots & \vdots & \ddots & \vdots & \vdots & \vdots & \vdots \\ 0 & 0 & 0 & 1 & 0 & 0 & 0 \\ 0 & 0 & 0 & 0 & 1 & 0 & 0 \\ \vdots & \vdots & \vdots & \vdots & \vdots & \ddots & \vdots \\ 0 & 0 & 0 & 0 & 0 & 0 & 1 \end{bmatrix} \boldsymbol{x}(k) \tag{4.11}$$

In this manner, we explicitly acknowledge that all of the state variables can be observed at the kth sampling instant, so that the NMSS form is trivially observable! Moreover, if the system can also be shown to be controllable, then any SVF design will yield a control system that can be directly implemented in practice, utilising only the measured input and output variables, together with their stored past values. In contrast to the SVF examples considered in Chapter 3, therefore, a model-based estimation or reconstruction of the states is not required.

4.1.2 The Characteristic Polynomial of the NMSS Model

Using equation (3.64), the characteristic polynomial $S(\boldsymbol{\lambda})$ of the NMSS representation is given by:

$$S(\lambda) = |\boldsymbol{I}\lambda - \boldsymbol{F}| = \left|\begin{array}{ccccc|ccccc} \lambda + a_1 & a_2 & \cdots & a_{n-1} & a_n & -b_2 & -b_3 & \cdots & -b_{m-1} & -b_m \\ -1 & \lambda & \cdots & 0 & 0 & 0 & 0 & \cdots & 0 & 0 \\ 0 & -1 & \cdots & 0 & 0 & 0 & 0 & \cdots & 0 & 0 \\ \vdots & \vdots & \ddots & \vdots & \vdots & \vdots & \vdots & \ddots & \vdots & \vdots \\ 0 & 0 & \cdots & -1 & \lambda & 0 & 0 & \cdots & 0 & 0 \\ \hline 0 & 0 & \cdots & 0 & 0 & \lambda & 0 & \cdots & 0 & 0 \\ 0 & 0 & \cdots & 0 & 0 & -1 & \lambda & \cdots & 0 & 0 \\ 0 & 0 & \cdots & 0 & 0 & 0 & -1 & \cdots & 0 & 0 \\ \vdots & \vdots & \ddots & \vdots & \vdots & \vdots & \vdots & \ddots & \vdots & \vdots \\ 0 & 0 & \cdots & 0 & 0 & 0 & 0 & \cdots & -1 & \lambda \end{array}\right| \tag{4.12}$$

Evaluating the determinant yields:

$$S(\lambda) = |\boldsymbol{I}\lambda - \boldsymbol{F}| = \lambda^{m-1}\left\{\lambda^n + a_1\lambda^{n-1} + \cdots + a_n\right\} = \lambda^{m-1}A^*(\lambda) \tag{4.13}$$

where $A^*(\lambda) = \lambda^n + a_1\lambda^{n-1} + \cdots + a_n$. Note that $S(\boldsymbol{\lambda})$ is the product of the characteristic polynomial of the original TF model (or the equivalent minimal state space representation) and a term $\boldsymbol{\lambda}^{m-1}$, where the latter is due to the additional states in the NMSS model.

Example 4.1 Non-Minimal State Space Representation of a Second Order TF Model
Consider the following second order TF model from Chapter 2 and Chapter 3:

$$y(k) = \frac{b_1z^{-1} + b_2z^{-2}}{1 + a_1z^{-1} + a_2z^{-2}}u(k) = \frac{0.5z^{-1} - 0.4z^{-2}}{1 - 0.8z^{-1} + 0.15z^{-2}}u(k) \tag{4.14}$$

In the context of the general TF model (4.1), $n = m = 2$ and the system has unity time delay ($\tau = 1$). The non-minimal state vector has an order of $n + m - 1 = 3$, i.e. there are three state variables. Using equation (4.4), equation (4.5), equation (4.6), equation (4.7), equation (4.8) and equation (4.9), the NMSS model is defined as follows:

$$\begin{bmatrix} y(k) \\ y(k-1) \\ u(k-1) \end{bmatrix} = \begin{bmatrix} -a_1 & -a_2 & b_2 \\ 1 & 0 & 0 \\ 0 & 0 & 0 \end{bmatrix} \begin{bmatrix} y(k-1) \\ y(k-2) \\ u(k-2) \end{bmatrix} + \begin{bmatrix} b_1 \\ 0 \\ 1 \end{bmatrix} u(k-1) \tag{4.15}$$

$$y(k) = \begin{bmatrix} 1 & 0 & 0 \end{bmatrix} \begin{bmatrix} y(k) \\ y(k-1) \\ u(k-1) \end{bmatrix} \tag{4.16}$$

Examination of the difference equation obtained from equation (4.14), i.e.

$$y(k) = -a_1 y(k-1) - a_2 y(k-2) + b_1 u(k-1) + b_2 u(k-2) \tag{4.17}$$

verifies that $\{(4.15),(4.16)\}$ holds as one particular state space representation of the system but one where the coefficients are defined directly from the coefficients in the TF model. Indeed, it can be considered as the most obvious state space representation of the TF model. Substituting for the numerical values (4.14), the state transition matrix $\boldsymbol{F}$, input vector $\boldsymbol{g}$ and output vector $\boldsymbol{h}$ are given by:

$$\boldsymbol{F} = \begin{bmatrix} 0.8 & -0.15 & -0.4 \\ 1 & 0 & 0 \\ 0 & 0 & 0 \end{bmatrix}; \quad \boldsymbol{g} = \begin{bmatrix} 0.5 \\ 0 \\ 1 \end{bmatrix}; \quad \boldsymbol{h} = \begin{bmatrix} 1 & 0 & 0 \end{bmatrix} \tag{4.18}$$

The observation matrix (4.10) is a 3×3 dimension identity matrix, i.e. $\boldsymbol{H} = \boldsymbol{I}_3$.

The characteristic equation for this NMSS model is determined from equation (4.12):

$$|(\lambda \boldsymbol{I} - \boldsymbol{F})| = \begin{vmatrix} \lambda - 0.8 & 0.15 & 0.4 \\ -1 & \lambda & 0 \\ 0 & 0 & \lambda \end{vmatrix} = 0 \tag{4.19}$$

Evaluating the determinant yields $\boldsymbol{\lambda}(\boldsymbol{\lambda}^2 - 0.8\boldsymbol{\lambda} + 0.15) = 0$, hence the eigenvalues are 0.5, 0.3 and 0. The eigenvalues 0.5 and 0.3 are equivalent to those of the minimal state space model (Example 3.6) and are also the poles of the original TF model (Example 2.3). The eigenvalue at the origin owes its existence to the additional state $u(k-1)$.

4.2 Controllability of the NMSS Model

If the NMSS model is to be used as the basis for the design of SVF control systems, it is important to evaluate the controllability conditions of the model. Since the NMSS form differs from the conventional minimal approach, these are fully developed in Appendix C using the *Popov, Belevitch and Hautus* (PBH) test (Kailath 1980). However, we will see that the standard results for minimal systems discussed in Chapter 3 are also applicable.

Theorem 4.1 Controllability of the NMSS Representation Given a SISO, discrete-time system described by the TF model (4.1), the NMSS representation (4.4), as defined by the pair $[\boldsymbol{F}, \boldsymbol{g}]$, is completely controllable (or reachable) if and only if the polynomials $A(z^{-1})$ and $B(z^{-1})$ are coprime (Appendix C).

Comment on the Controllability Conditions of Theorem 4.1 Appendix C shows that the modes of the system specified by $\boldsymbol{\lambda} = 0$ are always controllable in the NMSS representation. In fact, from the formulation of the NMSS model, we realise that these modes have been deliberately introduced into the system by $u(k-1)$, $u(k-2)$, …, $u(k-m+1)$. These are obviously controllable since $u(k)$ represents our control input signal.

The coprime condition is equivalent to the standard requirement for SVF control systems, namely that the TF model (4.1) should have no pole-zero cancellations. Consequently, nothing has been lost in controllability terms by moving to the NMSS description, except to increase the effective order of the system from n to $n+m-1$. At the same time, an enormous advantage has been gained: we now have direct access to SVF control based only on the sampled input and output signals.

Finally, the controllability conditions of Theorem 4.1 are equivalent to the normal requirement that the controllability matrix associated with the NMSS representation:

$$\boldsymbol{S}_1 = \begin{bmatrix} \boldsymbol{g} & \boldsymbol{F}\boldsymbol{g} & \boldsymbol{F}^2\boldsymbol{g} & \cdots & \boldsymbol{F}^{n+m-2}\boldsymbol{g} \end{bmatrix} \tag{4.20}$$

has full rank $(n+m-1)$, i.e. that it is non-singular. Equation (4.20) takes a similar form to equation (3.79), here revised to account for the order of the NMSS model.

Example 4.2 Ranks Test for the NMSS Model Returning to the numerical example described by equations (4.18), since the TF model (4.14) has no pole-zero cancellations, the NMSS model is clearly controllable. However, this result can be verified by examining the controllability matrix:

$$S_1 = \begin{bmatrix} \boldsymbol{g} & \boldsymbol{F}\boldsymbol{g} & \boldsymbol{F}^2\boldsymbol{g} \end{bmatrix} = \begin{bmatrix} 0.5 & 0 & -0.075 \\ 0 & 0.5 & 0 \\ 1 & 0 & 0 \end{bmatrix} \tag{4.21}$$

which is non-singular, i.e. it has rank 3.

4.3 The Unity Gain NMSS Regulator

The SVF control law associated with the above NMSS model can be written in the following vectorised form:

$$u(k) = -\boldsymbol{k}^T \boldsymbol{x}(k) + k_d y_d(k) \tag{4.22}$$

where $y_d(k)$ is the command input, k_d is a control gain for the command input and $\boldsymbol{k}$ is the SVF gain vector, defined as:

$$\boldsymbol{k} = \begin{bmatrix} f_0 & f_1 & \cdots & f_{n-1} & g_1 & \cdots & g_{m-1} \end{bmatrix}^T \tag{4.23}$$

in which the feedback control gains $f_0, f_1, \ldots, f_{n-1}, g_1, \ldots, g_{m-1}$ are selected by the designer to achieve desired closed-loop characteristics, as illustrated by Example 4.4 (and, more generally, by the various case studies and design approaches considered elsewhere in this book). As also shown later, the control gain on the command input, k_d is chosen to produce a designer-specified steady-state gain (normally unity to provide command input tracking). Expanding the terms in (4.22) yields:

$$\begin{aligned} u(k) = &-f_0 y(k) - f_1 y(k-1) - \cdots - f_{n-1} y(k-n+1) \\ &-g_1 u(k-1) - \cdots - g_{m-1} u(k-m+1) + k_d y_d(k) \end{aligned} \tag{4.24}$$

Equation (4.22) and equation (4.24) represent the *NMSS regulator control law.*

Example 4.3 Regulator Control Law for a NMSS Model with Four State Variables
Consider the following second order TF model with three numerator parameters:

$$y(k) = \frac{b_1 z^{-1} + b_2 z^{-2} + b_3 z^{-3}}{1 + a_1 z^{-1} + a_2 z^{-2}} u(k) \tag{4.25}$$

In the context of the general TF model (4.1), it is clear that $n = 2$, $m = 3$ and $\tau = 1$. The non-minimal state vector has an order of $n + m - 1 = 4$, i.e. there are four state variables. Utilising equation (4.4), equation (4.5), equation (4.6), equation (4.7), equation (4.8) and equation (4.9), the NMSS model is defined by:

$$\begin{bmatrix} y(k) \\ y(k-1) \\ u(k-1) \\ u(k-2) \end{bmatrix} = \begin{bmatrix} -a_1 & -a_2 & b_2 & b_3 \\ 1 & 0 & 0 & 0 \\ 0 & 0 & 0 & 0 \\ 0 & 0 & 1 & 0 \end{bmatrix} \begin{bmatrix} y(k-1) \\ y(k-2) \\ u(k-2) \\ u(k-3) \end{bmatrix} + \begin{bmatrix} b_1 \\ 0 \\ 1 \\ 0 \end{bmatrix} u(k-1) \tag{4.26}$$

The observation matrix (4.10) is a 4×4 dimension identity matrix, i.e. $\boldsymbol{H} = \boldsymbol{I}_4$.

An instructive special case of the TF model (4.25) is when $b_1 = 0$, representing a pure time delay of two samples ($\tau = 2$). In this regard, consider the following numerical example based on equation (4.25) with $a_1 = -1.7$, $a_2 = 1.0$, $b_1 = 0$, $b_2 = -1$ and $b_3 = 2.0$, i.e.

$$y(k) = \frac{-1.0z^{-2} + 2.0z^{-3}}{1 - 1.7z^{-1} + 1.0z^{-2}} u(k) = \frac{-1.0 + 2.0z^{-1}}{1 - 1.7z^{-1} + 1.0z^{-2}} u(k-2) \tag{4.27}$$

This second order, non-minimum phase oscillator has been studied by Clarke *et al.* (1985) and Young *et al.* (1987). In fact, a unity time-delay version of the model is considered in Example 2.3, with the marginally stable, oscillatory open-loop step response shown in Figure 2.8c.

The NMSS model takes the same form as equation (4.26). Substituting for the numerical values yields the following state transition matrix $\boldsymbol{F}$ and input vector $\boldsymbol{g}$:

$$\boldsymbol{F} = \begin{bmatrix} 1.7 & -1 & -1 & 2 \\ 1 & 0 & 0 & 0 \\ 0 & 0 & 0 & 0 \\ 0 & 0 & 1 & 0 \end{bmatrix}; \quad \boldsymbol{g} = \begin{bmatrix} 0 \\ 0 \\ 1 \\ 0 \end{bmatrix} \tag{4.28}$$

In general, a pure time delay of τ samples is automatically incorporated into the NMSS description by simply setting the first $\tau - 1$ parameters of the numerator polynomial to zero. When $\tau = 2$, as in the present example, the parameter $b_1 = 0$ in both the TF model (4.25) and the $\boldsymbol{g}$ vector (4.26). It is clear, therefore, that the NMSS formulation has no difficulty handling pure time delays. The relationship between this approach and more conventional methods of dealing with long time delays, such as the Smith Predictor (Smith 1957, 1959), is discussed in Chapter 6.

The NMSS regulator control law (4.24) for the NMSS model (4.26) is given by:

$$u(k) = -f_0 y(k) - f_1 y(k-1) - g_1 u(k-1) - g_2 u(k-2) + k_d y_d(k) \tag{4.29}$$

Utilising the backward shift operator:

$$u(k) = -\left(f_0 + f_1 z^{-1}\right) y(k) - \left(g_1 z^{-1} + g_2 z^{-2}\right) u(k) + k_d y_d(k) \tag{4.30}$$

so that the closed-loop system is represented by Figure 4.1.

However, to simplify the block diagram and subsequent analysis for higher order systems, the inner feedback loop of Figure 4.1 is usually evaluated first in order to obtain an equivalent forward path input filter, as illustrated by Figure 4.2.

This example is instructive: it reveals the simple block diagram structure of the NMSS regulator, from which we can easily deduce the general NMSS regulator structure for an arbitrary, controllable system.

But what degree of control do we have over these coefficients? In the minimal case (Example 3.3 and Example 3.4), it is possible to compute values for the gains, in order to specify the coefficients of the closed-loop characteristic equation and, thereby, assign the poles to desired positions on the complex z-plane. However, Theorem 4.1 states that this should also be possible here, so let us continue with the present example and compute the SVF gains for the NMSS regulator.

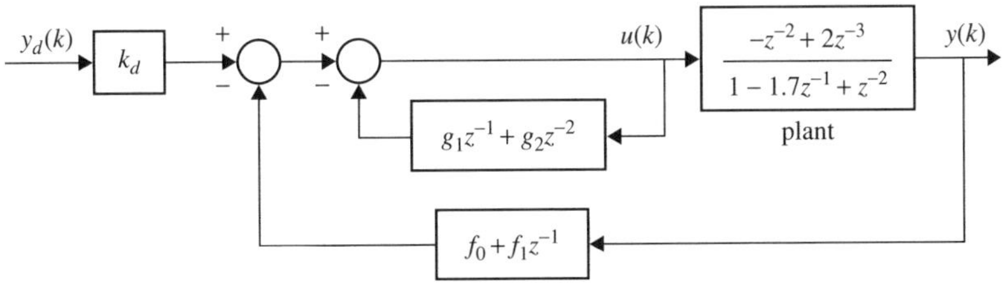

Figure 4.1 NMSS regulator control of Example 4.3 showing an explicit feedback of the input states

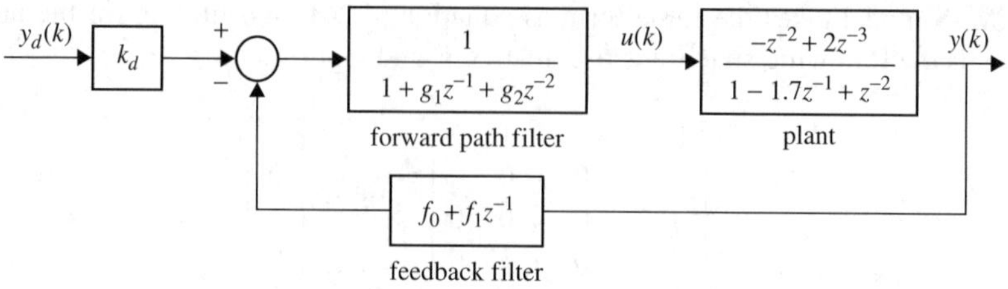

Figure 4.2 Simplified NMSS regulator control of Example 4.3 with a forward path filter

Example 4.4 Pole Assignment for the Fourth Order NMSS Regulator Using block diagram reduction on either Figure 4.1 or Figure 4.2 produces the following closed-loop TF describing the relationship between $y(k)$ and $y_d(k)$:

$$\frac{y(k)}{y_d(k)} = \frac{k_d(-z^{-2}+2z^{-3})}{1+(g_1-1.7)z^{-1}+(1-1.7g_1+g_2-f_0)z^{-2}+(g_1-1.7g_2+2f_0-f_1)z^{-3}+(g_2+2f_1)z^{-4}} \tag{4.31}$$

showing clearly how the control gains affect all the coefficients of the closed-loop TF denominator. Suppose that the design requirement is for the closed-loop system to be critically damped, with no possibility of oscillation. This can be achieved, for example, by assigning four poles to 0.5 on the real axis of the complex z-plane, i.e.

$$D(z) = (z-0.5)^4 = z^4 - 2z^3 + 1.5z^2 - 0.5z + 0.0625 \tag{4.32}$$

or, in terms of z^{-1}:

$$D(z^{-1}) = (1-0.5z^{-1})^4 = 1 - 2z^{-1} + 1.5z^{-2} - 0.5z^{-3} + 0.0625z^{-4} \tag{4.33}$$

Here, $D(z) = 0$ or $D(z^{-1}) = 0$ represent the desired characteristic equation. In this case, the SVF control gains are obtained by simply equating coefficients in the denominator polynomial of equation (4.31) to like coefficients in $D(z^{-1})$, i.e.

$$\begin{aligned} g_1 - 1.7 &= -2 \\ 1 - 1.7g_1 + g_2 - f_0 &= 1.5 \\ g_1 - 1.7g_2 + 2f_0 - f_1 &= -0.5 \\ g_2 + 2f_1 &= 0.0625 \end{aligned} \tag{4.34}$$

These equations are conveniently written in the following vector-matrix form:

$$\begin{bmatrix} 1 & 0 & 0 & 0 \\ -1.7 & 1 & -1 & 0 \\ 1 & -1.7 & 2 & -1 \\ 0 & 1 & 0 & 2 \end{bmatrix} \begin{bmatrix} g_1 \\ g_2 \\ f_0 \\ f_1 \end{bmatrix} = \begin{bmatrix} -0.3 \\ 0.5 \\ -0.5 \\ 0.0625 \end{bmatrix} \tag{4.35}$$

Since the system is controllable, the matrix in equation (4.35) can be inverted, yielding the following unique solution to these simultaneous equations:

$$g_1 = -0.3; \quad g_2 = -0.2359; \quad f_0 = -0.2259; \quad f_1 = 0.1492 \tag{4.36}$$

Substituting these values into equation (4.31) and setting $z^{-1} = 1$, the steady-state gain of the closed-loop system is $k_d/0.0625$. Therefore, selecting a command input gain $k_d = 0.0625$ will result in unity gain closed-loop dynamics.

This example illustrates how, in general, the basic minimal and NMSS regulators, without the input gain k_d, have a non-unity steady-state gain. The controller with the gain k_d chosen to correct this deficiency will be referred to as a *unity gain regulator*. Note, however, that this does *not* provide Type 1 servomechanism performance because the unity gain is not inherent in the closed-loop system: this and other deficiencies of the NMSS regulator design are considered in Example 4.6. In the unlikely event that the model and system are identical, however, the output converges to the command input in a time that is defined by the designed closed-loop dynamics, as illustrated in Figure 4.3.

Contrast this stable response with the highly oscillatory behaviour of the open-loop system (without control) shown by Figure 2.8c. However, some aspects of the open-loop system have been retained: in addition to the time delay being unaffected, the non-minimum phase behaviour, characterised by the output moving initially in the wrong direction, remains. This is because the closed-loop numerator polynomial is only affected by the input gain k_d, so that the dynamic behaviour associated with the numerator polynomial is exactly the same in the open- and closed-loop cases: compare equation (4.27) and equation (4.31).

Example 4.5 Unity Gain NMSS Regulator for the Wind Turbine Simulation The following third order TF model is introduced in Example 2.2:

$$y(k) = \frac{b_1 z^{-1} + b_2 z^{-2} + b_3 z^{-3}}{1 + a_1 z^{-1} + a_2 z^{-2} + a_3 z^{-3}} u(k) \tag{4.37}$$

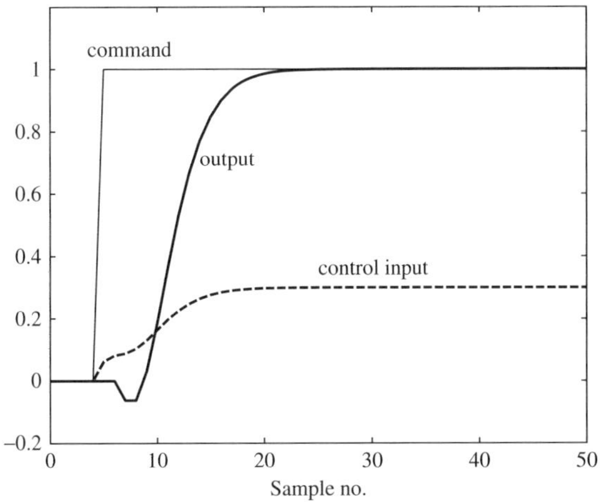

Figure 4.3 Closed-loop unit step response using the unity gain NMSS regulator of Example 4.4

The model is used to represent a wind turbine system, in which $y(k)$ is the generator reaction torque and $u(k)$ the pitch angle of the wind turbine blade. Since $n = m = 3$, there are five state variables in the NMSS model, which is defined as follows:

$$\begin{bmatrix} y(k) \\ y(k-1) \\ y(k-2) \\ u(k-1) \\ u(k-2) \end{bmatrix} = \begin{bmatrix} -a_1 & -a_2 & -a_3 & b_2 & b_3 \\ 1 & 0 & 0 & 0 & 0 \\ 0 & 1 & 0 & 0 & 0 \\ 0 & 0 & 0 & 0 & 0 \\ 0 & 0 & 0 & 1 & 0 \end{bmatrix} \begin{bmatrix} y(k-1) \\ y(k-2) \\ y(k-3) \\ u(k-2) \\ u(k-3) \end{bmatrix} + \begin{bmatrix} b_1 \\ 0 \\ 0 \\ 1 \\ 0 \end{bmatrix} u(k-1) \quad (4.38)$$

The NMSS unity gain regulator control law (4.24) for the model above is:

$$u(k) = -f_0 y(k) - f_1 y(k-1) - f_2 y(k-2) - g_1 u(k-1) - g_2 u(k-2) + k_d y_d(k) \quad (4.39)$$

The control system is implemented in a similar way to Figure 4.2 but here with a higher order feedback filter of the form: $f_0 + f_1 z^{-1} + f_2 z^{-2}$. Although increasingly time consuming to analyse by hand, the pole assignment problem is solved in a similar way to Example 4.4, but with five closed-loop poles and an associated set of five simultaneous equations. Note that Chapter 5 will develop a general computer algorithm to carry out these rather tedious manual calculations for such high order systems.

4.3.1 The General Unity Gain NMSS Regulator

Following from Example 4.3, Example 4.4 and Example 4.5 and, in particular Figure 4.2, it is clear that the block diagram form for the general unity gain NMSS regulator takes the structure shown in Figure 4.4. Here, the feedback and forward path filters are defined as follows:

$$G(z^{-1}) = 1 + g_1 z^{-1} + \cdots + g_{m-1} z^{-m+1} \quad (4.40)$$

$$F(z^{-1}) = f_0 + f_1 z^{-1} + \cdots + f_{n-1} z^{-n+1} \quad (4.41)$$

where n and m are the order of the denominator and numerator polynomials of the TF model (4.1), respectively.

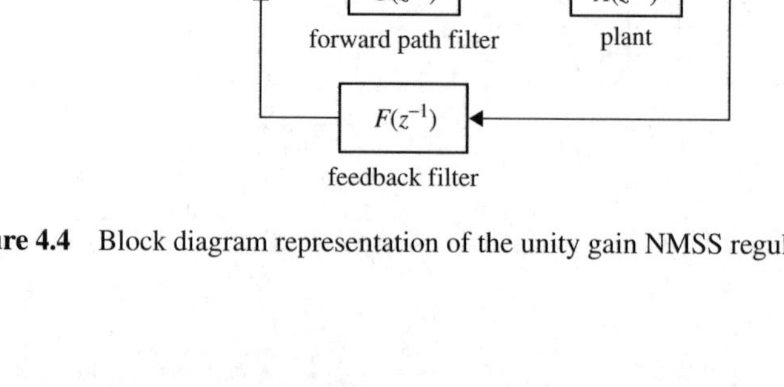

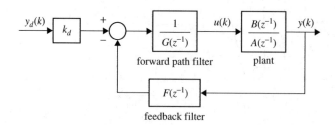

Figure 4.4 Block diagram representation of the unity gain NMSS regulator

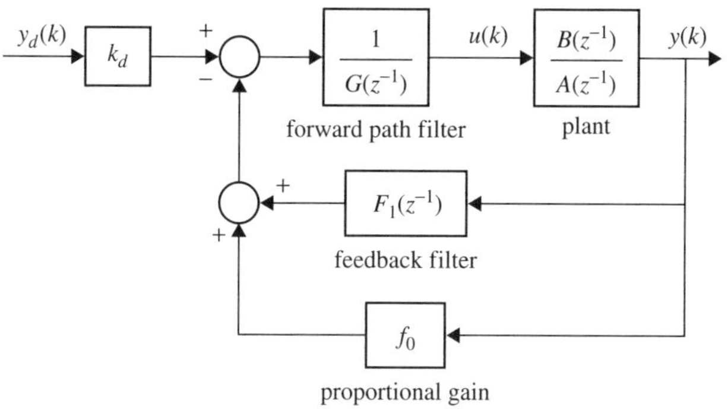

Figure 4.5 Unity gain NMSS regulator with separate proportional gain

By contrast, Figure 4.5 shows how the feedback filter can be divided into two components, with a proportional feedback gain f_0 and an additional filter $F_1(z^{-1})$ operating on the present value of the output $y(k)$, where

$$F_1(z^{-1}) = f_1 z^{-1} + \cdots + f_{n-1} z^{-n+1} \tag{4.42}$$

Figure 4.5 shows how the structure of the NMSS controller compares with conventional digital control systems that employ simple proportional feedback. One useful interpretation of the differences is that the additional feedback and forward path compensators $F_1(z^{-1})$ and $G(z^{-1})$ are introduced to ensure the controllability of the general TF model and allow for SVF control.

The limitations of the unity gain NMSS regulator, indeed of unity gain regulators in general, are discussed in Example 4.6.

Example 4.6 Mismatch and Disturbances for the Fourth Order NMSS Regulator

Returning to the control system shown in Figure 4.2, suppose there is some uncertainty associated with the model used to estimate the gains. In particular, let us assume that the plant numerator polynomial is actually $-z^{-2} + 1.8z^{-3}$, i.e. there is a 10% error in the second parameter. In this case, the closed-loop step response is shown in Figure 4.6, which is obtained using the SVF control gains (4.36) and $k_d = 0.0625$ as previously calculated.

Although the system remains stable and well controlled, despite the error in the model parameter, the steady-state gain of the closed-loop system has changed, so that $y(k)$ no longer converges to the final command level of $y_d(k) = 1$, as required. In fact, the new steady-state gain is 0.642. In other words, a 10% error associated with a single parameter results in a 36% reduction in the steady-state gain!

The regulator has another important limitation, as illustrated in Figure 4.7. This shows how the control system, *now designed with exact knowledge of the system parameters*, reacts to the imposition of an unmeasured (i.e. unexpected) step disturbance applied to the input of the system at the 25th sample: the initial response is identical to that in Figure 4.3 but, following the application of the disturbance, the output moves away from the unity set point defined by $y_d(k)$ and settles at a new level of 1.37, an offset of 37%.

These examples reveal clearly a major shortcoming of the regulator formulation: namely, that it yields Type 0 control systems that do not have an inherent unity (or user specified)

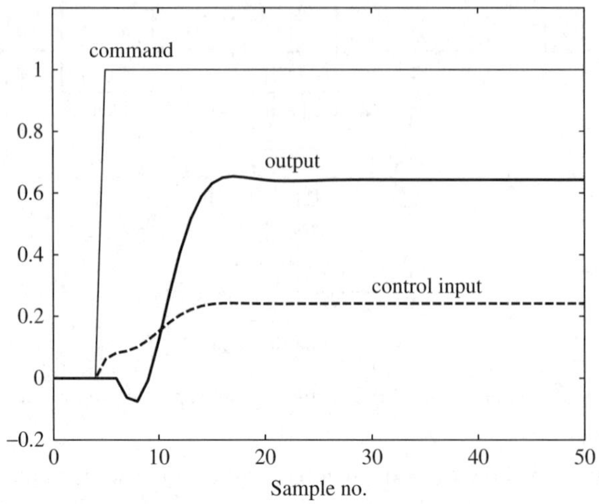

Figure 4.6 Closed-loop unit step response using the unity gain NMSS regulator of Example 4.4 when the model has a 10% error in one parameter

steady-state gain in the closed-loop. In other words, the output does not automatically 'follow' the command input $y_d(k)$ (or achieve a specified set point) and, in the presence of either model uncertainty or unmeasured input disturbances, the controlled system will exhibit steady-state error, i.e. an offset to step commands or constant set-point inputs. Similar problems occur with load level disturbances at the system output, although these are not considered in this example. These results apply to both the minimal and non-minimal regulator solutions.

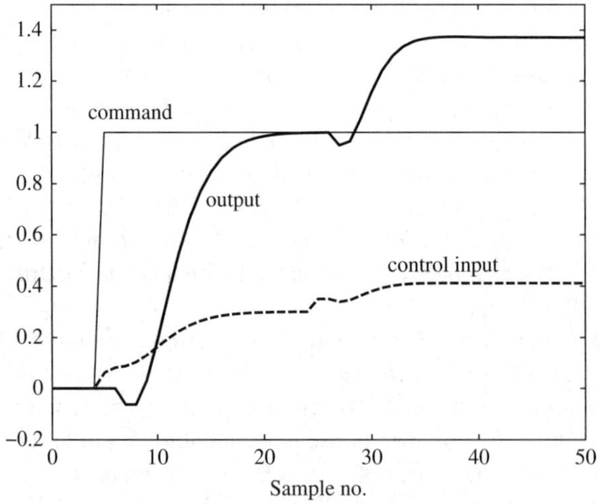

Figure 4.7 Closed-loop unit step response using the unity gain NMSS regulator of Example 4.4, with an input step disturbance of 0.05 at the 25th sample

It is possible to alleviate this problem to some extent by introducing a self-tuning or self-adaptive system, where the control system continuously estimates the model parameters and the input gain k_d is adjusted online in an attempt to counteract offsets. Even so, this is still not very satisfactory: any estimate is always characterised by a degree of uncertainty so that, in general, a steady-state error will normally be present (albeit usually much smaller than without adaption).

In most realistic practical circumstances, a control systems designer would demand inherent Type 1 servomechanism performance, obtained using integral control action, as illustrated by Example 2.5 and Example 2.6. Fortunately, it is straightforward to introduce such characteristics into the NMSS control system: Chapter 5 will show how the NMSS regulator can be augmented by an *integral-of-error* state variable, in order to guarantee the required Type 1 performance. Before moving onto this subject, however, it is useful to consider the simpler regulator form further, in order to examine some of the theoretical and practical consequences of the NMSS design.

4.4 Constrained NMSS Control and Transformations

Equation (4.13) shows that the NMSS model includes the eigenvalues of the minimal model in open-loop. This section demonstrates how the NMSS closed-loop control system similarly encompasses any minimal design as a special constrained case.

In the first instance, note that it is always possible to transform the NMSS model into other (minimal) controllable state space forms describing the same input–output relationship. Theorem 4.2 describes one such transformation from a non-minimal to the observable canonical form, but it is clear that similar transformations are possible for other state space representations.

Theorem 4.2 Transformation from Non-Minimal to Minimal State Vector Here, for simplicity, it is assumed that the order of the numerator polynomial is equal to that of the denominator, i.e. $m = n$, although this condition is not strictly necessary.

Suppose the TF model (4.1) is represented by the following observable canonical form:

$$\boldsymbol{x}_{\min}(k) = \boldsymbol{F}_{\min}\boldsymbol{x}_{\min}(k-1) + \boldsymbol{g}_{\min}u(k-1) \tag{4.43}$$

$$y(k) = \boldsymbol{h}\,\boldsymbol{x}_{\min}(k) \tag{4.44}$$

where

$$\boldsymbol{x}_{\min}(k) = \begin{bmatrix} x_1(k) \\ x_2(k) \\ \vdots \\ x_{n-1}(k) \\ x_n(k) \end{bmatrix}; \quad \boldsymbol{F}_{\min} = \begin{bmatrix} -a_1 & 1 & 0 & \cdots & 0 \\ -a_2 & 0 & 1 & \cdots & 0 \\ \vdots & \vdots & \vdots & \vdots & \vdots \\ -a_{n-1} & 0 & 0 & \cdots & 1 \\ -a_n & 0 & 0 & \cdots & 0 \end{bmatrix} \tag{4.45}$$

$$\boldsymbol{g}_{\min} = \begin{bmatrix} b_1 & b_2 & \cdots & b_{n-1} & b_n \end{bmatrix}^T; \quad \boldsymbol{h}_{\min} = \begin{bmatrix} 1 & 0 & \ldots & 0 & 0 \end{bmatrix} \tag{4.46}$$

The n dimensional minimal state vector $\boldsymbol{x}_{\min}(k)$ is then related to the $2n-1$ dimensional non-minimal state vector $\boldsymbol{x}(k)$ by:

$$\boldsymbol{x}_{\min}(k) = \boldsymbol{T}_1 \cdot \boldsymbol{x}(k) \tag{4.47}$$

Here, the n by $2n-1$ transformation matrix $\boldsymbol{T}_1$ is defined:

$$\boldsymbol{T}_1 = \begin{bmatrix} 1 & 0 & \cdots & 0 & 0 & \cdots & 0 \\ 0 & -a_2 & \cdots & -a_n & b_2 & \cdots & b_n \\ \vdots & \vdots & \iddots & \vdots & \vdots & \iddots & \vdots \\ 0 & -a_n & \cdots & 0 & b_n & \cdots & 0 \end{bmatrix} \tag{4.48}$$

Note that the NMSS regulator model has $2n-1$ states since we are assuming that $m=n$.

Proof of Theorem 4.2 (Taylor *et al.* 2000) As shown by equation (3.36), the states in the minimal case are:

$$\begin{aligned} x_1(k) &= -a_1x_1(k-1) + x_2(k-1) + b_1u(k-1) \\ x_2(k) &= -a_2x_1(k-1) + x_3(k-1) + b_2u(k-1) \\ &\vdots \\ x_n(k) &= -a_nx_1(k-1) + b_nu(k-1) \end{aligned} \tag{4.49}$$

From the set of equations (4.49), it is clear that $x_1(k) = y(k)$ and Theorem 4.2 follows by straightforward algebra.

Corollary 4.1 Transformation from Minimal to Non-Minimal SVF Gain Vector Any minimal state space regulator, given by the following SVF control law:

$$u_{\min}(k) = -\boldsymbol{k}_{\min}^T\boldsymbol{x}_{\min}(k) + k_d y_d(k) \tag{4.50}$$

where the minimal control gain vector:

$$\boldsymbol{k}_{\min} = \begin{bmatrix} l_1 & l_2 & \cdots & l_n \end{bmatrix}^T \tag{4.51}$$

may be converted into an equivalent NMSS form:

$$u(k) = -\boldsymbol{k}^T\boldsymbol{x}(k) + k_d y_d(k) \tag{4.52}$$

where $\boldsymbol{k}$ is the non-minimal control gain vector (4.23), by the following transformation:

$$\boldsymbol{k}^T = \boldsymbol{k}_{\min}^T\boldsymbol{T}_1 \tag{4.53}$$

Note again that the NMSS regulator control gain vector is of dimension $2n-1$.

Proof of Corollary 4.1 Noting from Theorem 4.2 that $\boldsymbol{x}_{\min}(k) = \boldsymbol{T}_1 \cdot \boldsymbol{x}(k)$, equation (4.50) can be written as:

$$u_{\min}(k) = -\boldsymbol{k}_{\min}^T \boldsymbol{T}_1 \boldsymbol{x}(k) + k_d y_d(k)$$

Corollary 4.1 demonstrates that any minimal SVF design may be exactly duplicated by an equivalent non-minimal structure.

Crucially, however, it should be noted that the reverse is not true: there is no minimal equivalent of a *general, unconstrained, non-minimal SVF controller*, since the latter is of a higher dimension. Specifically, the transformation matrix $\boldsymbol{T}_1$ cannot be inverted in equation (4.53), even if padded with zeros to ensure that it is a square matrix.

4.4.1 Non-Minimal State Space Design Constrained to yield a Minimal SVF Controller

Corollary 4.1 provides a direct transformation from the minimal gain vector to the NMSS gain vector. However, we can similarly constrain the NMSS solution to yield the same control algorithm as the minimal case *at the design stage*, by using either pole assignment or linear quadratic optimal design. The latter approach is discussed in Chapter 6 (in the context of *Generalised Predictive Control* or GPC) while, in the former case, n poles are simply assigned to their minimal positions and the additional $(m - 1)$ NMSS poles are set to zero, i.e. to lie at the origin of the complex z-plane.

Obviously the pole assignment approach will ensure equivalent characteristic equations and, since the closed-loop numerator polynomial is simply $B(z^{-1})$ in both cases, then the minimal and non-minimal solutions will yield exactly the same closed-loop TF when there is no model mismatch. Furthermore, if the minimal controller has the same structure as the NMSS case, then the two control algorithms will be *exactly* the same and will yield an identical response, even in the presence of model mismatch or disturbance inputs to the system. In this regard, it is clear that SVF based on the controllable canonical form takes a very different structure to NMSS design, since it only utilises feedback of the output variable. However, as hinted at in Chapter 3 and illustrated in Example 4.7, the NMSS and observable canonical forms do yield an equivalent structure and, therefore, an identical control algorithm when the additional $(m - 1)$ poles in the NMSS case are assigned to zero.

It should be stressed that assigning the extra poles to zero is not necessarily the optimal solution: one advantage of the NMSS approach is that the $(m - 1)$ extra poles can be assigned to desirable locations anywhere on the complex z-plane. This is a particularly useful feature in the linear quadratic optimal case, as discussed in later chapters.

Also, it should be pointed out that minimal state space forms cannot automatically handle the case when $m > n$, i.e. long time delays or high order numerator polynomials, whereas Example 4.3 shows that the NMSS form has no problems in this regard. This is why, in Chapter 3, the examples were selected deliberately so that $m = n$. In fact, examination of equations {(3.16), (3.17)} and {(3.37), (3.38)} shows that the minimal formulation of the problem can only deal with cases where $m > n$, by changing the state vector to dimension m and setting the trailing denominator parameters $a_{n+1}, \ldots, a_m$ to zero. Of course, such an approach is itself non-minimal and so it makes much more sense to explicitly acknowledge this fact by utilising the NMSS model introduced above.

Example 4.7 Transformations between Minimal and Non-Minimal Returning to the system described in Example 4.1, where the TF model and NMSS form are given by equation (4.14) and equation (4.15), respectively, the non-minimal state vector is:

$$\begin{bmatrix} x_1(k) \\ x_2(k) \\ x_3(k) \end{bmatrix} = \begin{bmatrix} y(k) \\ y(k-1) \\ u(k-1) \end{bmatrix} \tag{4.54}$$

and the NMSS regulator control law (4.24) is given by:

$$u(k) = -f_0 y(k) - f_1 y(k-1) - g_1 u(k-1) + k_d y_d(k) \tag{4.55}$$

where f_0, f_1 and g_1 are the SVF control gains and k_d is the command input gain.

By contrast, the state vector for a (minimal) observable canonical form is:

$$\begin{bmatrix} x_1(k) \\ x_2(k) \end{bmatrix}_{\min} = \begin{bmatrix} y(k) \\ -1.5y(k-1) + 0.4u(k-1) \end{bmatrix} \tag{4.56}$$

yielding the following SVF control law:

$$u_{\min}(k) = -\boldsymbol{k}_{\min}^T \boldsymbol{x}_{\min}(k) + k_d y_d(k) \tag{4.57}$$

Hence,

$$u_{\min}(k) = -\begin{bmatrix} l_1 & l_2 \end{bmatrix} \begin{bmatrix} x_1(k) \\ x_2(k) \end{bmatrix}_{\min} + k_d y_d(k) \tag{4.58}$$

where l_1 and l_2 are the SVF control gains, while k_d is the command input gain. Expanding the terms and substituting for the state vector (4.56) yields:

$$u_{\min}(k) = -l_1 y(k) + 1.5 l_2 y(k-1) - 0.4 l_2 u(k-1) + k_d y_d(k) \tag{4.59}$$

Equation (4.55) and equation (4.59) take the same structural form, although in the NMSS case there are three independent control gains, compared with just two in the minimal controller.

The 2×3 transformation matrix $\boldsymbol{T}_1$ defined by equation (4.48) is:

$$\boldsymbol{T}_1 = \begin{bmatrix} 1 & 0 & 0 \\ 0 & -1.5 & 0.4 \end{bmatrix} \tag{4.60}$$

and it is easy to show, from equation (4.54) and equation (4.56), that

$$\begin{bmatrix} x_1(k) \\ x_2(k) \end{bmatrix}_{\min} = \begin{bmatrix} 1 & 0 & 0 \\ 0 & -1.5 & 0.4 \end{bmatrix} \begin{bmatrix} x_1(k) \\ x_2(k) \\ x_3(k) \end{bmatrix} \tag{4.61}$$

If we wish to constrain the NMSS solution into exactly the same control law as the minimal design, then Corollary 4.1 can be utilised to transform the gains, i.e.

$$\boldsymbol{k}^T = \boldsymbol{k}_{\min}^T \boldsymbol{T}_1 = \begin{bmatrix} f_0 & f_1 & g_1 \end{bmatrix} = \begin{bmatrix} l_1 & l_2 \end{bmatrix} \begin{bmatrix} 1 & 0 & 0 \\ 0 & -1.5 & 0.4 \end{bmatrix} \tag{4.62}$$

so that

$$f_0 = l_1, \, f_1 = -1.5 l_2 \quad \text{and} \quad g_1 = 0.4 l_2.$$

4.5 Worked Example with Model Mismatch

The significance of the NMSS formulation is not always immediately apparent from a cursory examination of the state space matrices. For this reason, it is worth re-examining one of our examples to help clarify the discussion.

Example 4.8 The Order of the Closed-loop Characteristic Polynomial Consider again the second order TF model (4.14). Since $n = 2$, minimal SVF controllers are based on the feedback of two state variables $x_1(k)$ and $x_2(k)$, as shown by Example 4.7 for the case of an observable canonical form. For minimal SVF design based on the controllable canonical form, refer to Example 3.2. Although an input gain k_d is not included in the latter example, it is straightforward to do so now.

In this regard, Figure 4.8 represents the unity gain minimal SVF regulator based on the controllable canonical form. Here, the distinction needs to be made between the plant (i.e. the notional true system) represented by $B(z^{-1})/A(z^{-1})$ and the estimated control model represented by $\hat{B}(z^{-1})/\hat{A}(z^{-1})$. In particular, $\hat{B}(z^{-1}) = \hat{b}_1 z^{-1} + \hat{b}_2 z^{-2}$ where $\hat{b}_1$ and $\hat{b}_2$ are the estimated numerator parameters of the control model.

Considering now the NMSS case, the model is based on the three state variables in (4.54). Converting equation (4.55) into the z^{-1} operator form:

$$u(k) = -\left(f_0 + f_1 z^{-1}\right) y(k) - g_1 z^{-1} u(k) + k_d y_d(k) \tag{4.63}$$

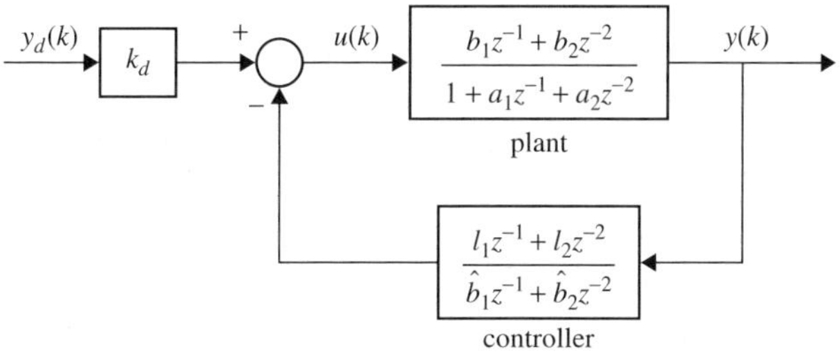

Figure 4.8 Unity gain (minimal) SVF regulator for Example 4.8 based on the controllable canonical form (cf. Figure 3.4)

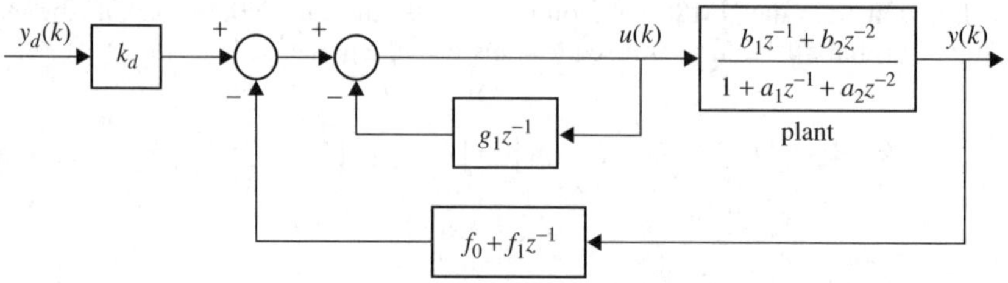

Figure 4.9 Unity gain NMSS regulator for Example 4.8

Hence, the unity gain NMSS regulator is illustrated in block diagram form by Figure 4.9.

Block diagram or algebraic reduction of Figure 4.9 produces the following closed-loop TF:

$$\frac{y(k)}{y_d(k)}=\frac{k_d\left(b_1z^{-1}+b_2z^{-2}\right)}{1+(a_1+g_1+b_1f_0)z^{-1}+(a_2+a_1g_1+b_2f_0+b_1f_1)z^{-2}+(a_2g_1+b_2f_1)z^{-3}} \tag{4.64}$$

In this NMSS case, the closed-loop characteristic polynomial is third order ($n+m-1=3$), one more than for minimal design. Initially, this may appear to be a weakness of the NMSS solution, since higher order systems are potentially more difficult to analyse (at least before the advent of relatively cheap digital computers) and we have to design for third order closed-loop dynamic behaviour. However, this argument is fundamentally flawed, because the problem of model mismatch has not yet been considered. In fact, evaluation of the closed-loop TF relationship for the minimal design in Figure 4.8 shows that:

$$\frac{y(k)}{y_d(k)}=\frac{k_d\left(b_1z^{-1}+\left(\frac{b_1\hat{b}_2}{\hat{b}_1}+b_2\right)z^{-2}+\left(\frac{b_2\hat{b}_2}{\hat{b}_1}\right)z^{-3}\right)}{1+\left(\frac{\hat{b}_2}{\hat{b}_1}+a_1+\frac{l_1b_1}{\hat{b}_1}\right)z^{-1}+\left(\frac{a_1\hat{b}_2}{\hat{b}_1}+a_2+\frac{l_1b_2}{\hat{b}_1}+\frac{l_2b_1}{\hat{b}_1}\right)z^{-2}+\left(\frac{a_2\hat{b}_2}{\hat{b}_1}+\frac{l_2b_2}{\hat{b}_1}\right)z^{-3}} \tag{4.65}$$

and the closed-loop is third order!

Furthermore, the closed-loop TF in the NMSS case, with model mismatch, is the much simpler third order TF in (4.64); and the parameter estimates do not appear at all in equation (4.64), since the model is not utilised in the NMSS controller of Figure 4.9.

This result is clearly significant in practical terms because there will always be mismatch between the estimated model and the real world, and the 'minimal' solution will, in practice, yield a closed-loop TF with a higher order than n. This is one of the key advantages of the NMSS approach: the order of the true closed-loop TF is a function of the order of the open loop numerator polynomial m and the time delay τ, as well as the denominator. Therefore, in the NMSS formulation, the impact of m and τ is acknowledged *at the design stage*.

Example 4.9 Numerical Comparison between NMSS and Minimal SVF Controllers

Continuing from Example 4.8, let us now assign the numerical values shown in equation (4.14). For pole assignment design, it is assumed that there is no model mismatch and so the closed-loop characteristic equation for the minimal SVF controller, based on a controllable canonical form, is simply:

$$1 + (l_1 - 0.8)z^{-1} + (l_2 + 0.15)z^{-2} = 0 \tag{4.66}$$

where l_1 and l_2 are the control gains. Equation (4.66) is obtained by substituting the parameter values $a_1 = -0.8$ and $a_2 = 0.15$ into equation (3.30). If the two poles are assigned to values of, say, 0.7 and 0.8 on the real axis of the complex z-plane, then the desired characteristic polynomial is given by:

$$D(z^{-1}) = (1 - 0.7z^{-1})(1 - 0.8z^{-1}) = 1 - 1.5z^{-1} + 0.56z^{-2} = 0 \tag{4.67}$$

In this case, equating like coefficients from (4.66) and (4.67), produces the solution:

$$l_1 = -0.7; \quad l_2 = 0.41 \tag{4.68}$$

Using these gains and assuming no model mismatch, i.e. $\hat{b}_1 = b_1 = 0.5$ and $\hat{b}_2 = b_2 = -0.4$, equation (4.65) reduces to:

$$\frac{y(k)}{y_d(k)} = \frac{k_d\left(0.5z^{-1} - 0.8z^{-2} + 0.32z^{-3}\right)}{1 - 2.3z^{-1} + 1.76z^{-2} - 0.448z^{-3}} = \frac{k_d\left(0.5z^{-1} - 0.4z^{-2}\right)}{1 - 1.5z^{-1} + 0.56z^{-2}} \tag{4.69}$$

The poles of the left-hand side TF in equation (4.69) are 0.8, 0.8 and 0.7, while there are also two zeros at 0.8 on the real axis of the complex z-plane. In this zero mismatch case, one of the zeros exactly cancels with a pole, yielding the second order system shown by the right-hand side TF in equation (4.69). The steady-state gain of this TF is $k_d/0.6$ so that selecting $k_d = 0.6$ ensures that the closed-loop system has unity gain *when there is no model mismatch.*

For the purposes of this example, the NMSS solution will be constrained to yield exactly the same closed-loop TF as equation (4.69). This is achieved straightforwardly by assigning two of its poles to 0.7 and 0.8 (as in the minimal case) and the additional pole to zero, i.e. the origin of the complex z-plane. Equating the denominator coefficients of equation (4.64) with equation (4.67) in the usual manner yields:

$$g_1 = -0.8; \quad f_0 = 0.2; \quad f_1 = -0.3 \tag{4.70}$$

In this ideal, zero mismatch case, the NMSS and the minimal SVF controllers produce identical closed-loop responses. However, consider the effect of the following mismatched values for the system parameters:

$$a_1 = -0.84, \; a_2 = 0.1425, \; b_1 = 0.525, \text{ and } b_2 = -0.38 \tag{4.71}$$

The new closed-loop poles and zeros shown in Table 4.1 are obtained by substituting the control gains (4.68) and (4.70), together with the above mismatched system parameters, into

Table 4.1 Comparison of SVF based on NMSS and the (minimal) controllable canonical form

Design poles		Actual poles		Actual zeros	
Minimal	NMSS	Minimal	NMSS	Minimal	NMSS
0.8	0.8	1.1573	0.8573	0.8000	0.7238
0.7	0.7	0.6632	0.6777	0.7238	—
N/A	0	0.5545	0.0000	—	—

the respective closed-loop TF models given by equation (4.64) and equation (4.65). In the latter case, the control model coefficients $\hat{b}_1 = 0.5$ and $\hat{b}_2 = -0.4$ are unchanged.

In this realistic case with model mismatch, the pole-zero cancellation of the minimal solution does not occur and the closed-loop is third order. Furthermore, one of the poles is outside the unit circle, so that the closed-loop system is unstable! By contrast, the NMSS poles are all within the unit circle and the response remains stable.

This example shows that the NMSS formulation accounts for the higher order of the closed-loop system *at the design stage* and that we can assign the additional poles to desirable locations, in the above case at the origin. As pointed out previously, however, assigning the extra pole to the origin is not necessarily the optimal solution: the advantage of the NMSS approach is that the $(m - 1)$ extra poles can be assigned anywhere on the complex z-plane (Chapter 6 presents examples that illustrate the benefits of this). Finally, although Table 4.1 applies for a single simulation example, the results presage the likely increased robustness of the NMSS controller in more general terms, as discussed below.

Example 4.10 Model Mismatch and its effect on Robustness The control algorithms discussed above may be further evaluated by taking advantage of *Monte Carlo Simulation* (MCS), as described in Chapter 8. MCS is used here to infer the effects of model parameter uncertainty, based on the estimate of the parametric error covariance matrix $\boldsymbol{P}^*(k)$, obtained when the statistical identification algorithms discussed in Chapter 8 are used to identify the control model from the analysis of noisy input–output data. With the current wide availability of powerful desktop computers, MCS provides one of the simplest and most attractive approaches for assessing the sensitivity of a controller to parametric uncertainty. Here, the model parameters for each realisation in the MCS, are selected randomly from the joint probability distribution defined by $\boldsymbol{P}^*(k)$ and the sensitivity of the controlled system to this parametric uncertainty is evaluated from the ensemble of resulting closed-loop response characteristics.

Figure 4.10 and Figure 4.11 compare 200 MCS realisations using the minimal and NMSS designs, respectively, from Example 4.9. Here, we have constrained the NMSS solution to yield exactly the same control algorithm as the minimal designs when there is no model mismatch. However, as already suggested by Table 4.1, when these control systems are implemented with mismatch, the minimal state space version, based on the controllable canonical form (Figure 4.10), is much less robust than the NMSS equivalent (Figure 4.11), with numerous unstable realisations.

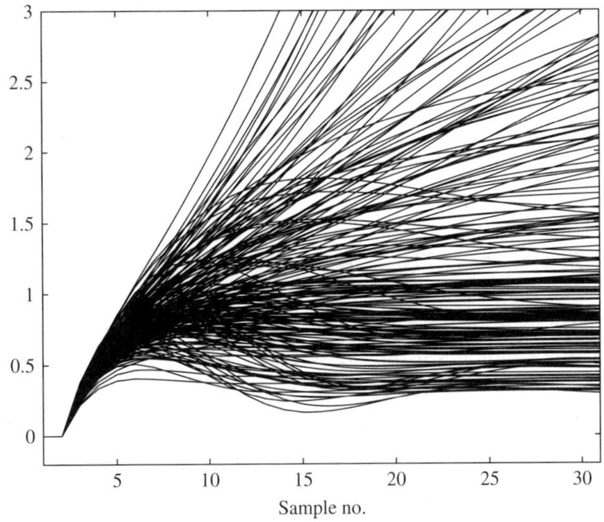

Figure 4.10 MCS using the unity gain controllable canonical form regulator with gains (4.68)

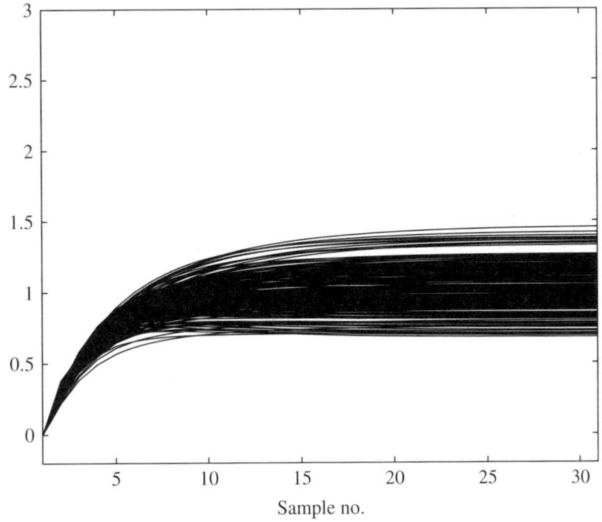

Figure 4.11 MCS using the unity gain NMSS regulator with gains (4.70)

4.6 Concluding Remarks

This chapter has introduced some of the most fundamental results in the theory of NMSS control system design. Several examples have been used to demonstrate how the approach differs from the more conventional minimal SVF approach. In particular, since the state vector is composed only of the present and past sampled values of the output variable, and the past sampled values of the input variable, the NMSS controller is very easy to implement in practice. It does not require any form of state reconstruction, since it always involves full state feedback based solely on measured input and output signals. In fact, the NMSS form is the most obvious and straightforward way to represent the TF model (4.1) in state space terms. Minimal state space models, based on selected canonical forms, represent the same system in a less intuitive manner, requiring each state to be formed from various, often rather abstract, combinations of the input and output signals that are dictated by the canonical form.

This chapter has shown the relationship between the NMSS and minimal state space model forms and how the NMSS model can be transformed into a minimal form, with the associated SVF control law constrained to yield exactly the same closed-loop system as a minimal design *when there is no model mismatch*. Even in this constrained form, however, the non-minimal design can be more robust than SVF based on the minimal controllable canonical form. By contrast, the observable canonical form yields a SVF control system with a similar structure to the NMSS based controller. In this case, when the additional poles in the NMSS case are assigned to the origin of the complex z-plane, the control algorithms are exactly the same. As will become apparent from case studies in later chapters, however, it is not a good idea to constrain NMSS design in this way. Rather, it is advantageous to utilise the additional poles in the NMSS case to provide extra flexibility and design freedom.

In Example 4.7, for instance, when the minimal design shown by equation (4.59) is implemented, there are three control parameters, as in the NMSS controller defined by equation (4.55). But the minimal control gains are really just combinations of two coefficients. This can be contrasted with NMSS design, where there is complete freedom to either manually assign the extra $m - 1$ poles as required, or to employ optimal control with a wider range of possible solutions. And the NMSS approach has a further, very important, advantage: it can inherently handle high order numerator polynomials and long time delays, i.e. situations when the model identification determines that $m > n$.

In conclusion, NMSS design provides a flexible and logical approach to SVF control system design based on TF models, because the NMSS model is formulated in the most obvious way and, at the same time, provides useful degrees of freedom for the control system designer. Unfortunately, the *regulator* SVF control systems considered in the present chapter do not necessarily track the command input $y_d(k)$ or converge on a required set-point with zero steady-state error. In Chapter 5, therefore, we consider how to remove this limitation by augmenting the NMSS model with an *integral-of-error* state variable, so guaranteeing inherent Type 1 servomechanism performance with zero steady-state error.

References

Åström, K.J. and Wittenmark, B. (1984) *Computer-Controlled Systems Theory and Design*, Prentice Hall, Englewood Cliffs, NJ.

Bitmead, R.R., Gevers, M. and Wertz, V. (1990) *Adaptive Optimal Control: The Thinking Man's GPC*, Prentice Hall, Harlow.

Burridge, P. and Wallis, K.F. (1990) Seasonal adjustments and Kalman filtering: extension to periodic variances, *Journal of Forecasting*, **9**, pp. 109–118.

Chotai, A., Young, P.C., McKenna, P.G. and Tych, W. (1998) Proportional-Integral-Plus (PIP) design for delta operator systems: Part 2, MIMO systems, *International Journal of Control*, **70**, pp. 149–168.

Clarke, D.W., Kanjilal, P.P. and Mohtadi, C. (1985) A generalized LQG approach to self-tuning control. Part I Aspects of design, *International Journal of Control*, **41**,pp. 1509–1523.

Dorf, R.C. and Bishop, R.H. (2008) *Modern Control Systems*, Eleventh Edition, Pearson Prentice Hall, Upper Saddle River, NJ.

Exadaktylos, V. and Taylor, C.J. (2010) Multi-objective performance optimisation for model predictive control by goal attainment, *International Journal of* Control, **83**, pp. 1374–1386.

Franklin, G.F., Powell, J.D. and Emami-Naeini, A. (2006) *Feedback Control of Dynamic Systems*, Fifth Edition, Pearson Prentice Hall, Upper Saddle River, NJ.

Gonzalez, A.H., Perez, J.M. and Odloak, D. (2009) Infinite horizon MPC with non-minimal state space feedback, *Journal of Process Control*, **19**, pp. 473–481.

Gopal, M. (2010) *Digital Control and State Variable Methods*, McGraw-Hill International Edition, McGraw-Hill, Singapore.

Green, M. and Limebeer, D.J.N. (1995) *Linear Robust Control*, Prentice Hall, Englewood Cliffs, NJ.

Hesketh, T. (1982) State-space pole-placing self-tuning regulator using input–output values, *IEE Proceedings: Control Theory and Applications*, **129**, pp. 123–128.

Kalman, R.E. (1960) A new approach to linear filtering and prediction problems. *ASME Journal of Basic Engineering*, **82**, pp. 34–45.

Kailath, T. (1980) *Linear Systems*, Prentice Hall, Englewood Cliffs, NJ.

Kuo, B.C. (1980) *Digital Control Systems*, Holt, Rinehart and Winston, New York.

Luenberger, D.G. (1964) Observing the state of a linear system, *IEEE Transactions on Military Electronics*, **8**, pp. 74–80

Luenberger, D.G. (1971) An introduction to observers, *IEEE Transactions on Automatic Control*, **16**, pp. 596–603.

Maciejowski, J.M. (1985) Asymptotic recovery for discrete–time systems, *IEEE Transactions on Automatic Control*, **30**, pp. 602–605.

Mustafa, D. and Glover, K. (1990) *Minimum Entropy* H_∞*Control*, Springer-Verlag, Berlin.

Priestley, M.B. (1988) *Non-Linear and Non–Stationary Time Series Analysis*, Academic Press, London.

Skogestad, S. and Postlethwaite, I. (2005) *Multivariable Feedback Control: Analysis and Design*, Second Edition, John Wiley & Sons, Ltd, London.

Smith, O.J. (1959) A controller to overcome dead time, *ISA Journal*, **6**, pp. 28–33.

Smith, O.J.M. (1957) Closer control of loops with dead time, *Chemical Engineering Progress*, **53**, pp. 217–219.

Taylor, C.J., Chotai, A. and Cross, P. (2012) Non-minimal state variable feedback decoupling control for multivariable continuous-time systems, *International Journal of Control*, **85**, pp. 722–734.

Taylor, C.J., Chotai, A. and Young, P.C. (2000) State space control system design based on non-minimal state-variable feedback: Further generalisation and unification results, *International Journal of Control*, **73**, pp. 1329–1345.

Taylor, C.J., Chotai, A. and Young, P.C. (2009) Nonlinear control by input–output state variable feedback pole assignment, *International Journal of Control*, **82**, pp.1029–1044.

Taylor, C.J. and Shaban, E.M. (2006) Multivariable Proportional-Integral-Plus (PIP) control of the ALSTOM nonlinear gasifier simulation, *IEE Proceedings: Control Theory and Applications*, **153**, pp. 277–285.

Taylor, C.J., Young, P.C., Chotai, A. and Whittaker, J. (1998) Non-minimal state space approach to multivariable ramp metering control of motorway bottlenecks, *IEE Proceedings: Control Theory and Applications*, **145**, pp. 568–574.

Wang, C.L. and Young, P.C. (1988) Direct digital control by input–output, state variable feedback: Theoretical background, *International Journal of Control*, **47**, pp. 97–109.

Wang, L. and Young, P. (2006) An improved structure for model predictive control using non-minimal state space realisation, *Journal of Process Control*, **16**, pp. 355–371.

Whittle, P. (1990) *Risk-Sensitive Optimal Control*, John Wiley & Sons, Ltd, London.

Wonham, W. (1968) On the separation theorem of stochastic control. *SIAM Journal on Control and Optimization*, **6**, pp. 312–326.

Young, P.C., Behzadi, M.A., Wang, C.L. and Chotai, A. (1987) Direct Digital and Adaptive Control by input-output, state variable feedback pole assignment, *International Journal of Control*, **46**, pp. 1867–1881.

Young, P.C., Chotai, A., McKenna, P.G. and Tych, W. (1998) Proportional-Integral-Plus (PIP) design for delta operator systems: Part 1, SISO systems, *International Journal of Control*, **70**, pp. 123–147.

Young, P.C. and Willems, J.C. (1972) An approach to the linear multivariable servomechanism problem, *International Journal of Control*, **15**, pp. 961–979.

5

True Digital Control for Univariate Systems

Chapter 4 has demonstrated the theoretical and practical advantages of thc *Non-Minimal State Space* (NMSS) model where, in the discrete-time backward shift operator case, the state variables consist only of the present and past sampled values of the input and output signals. In the present chapter, the full power of non-minimal *state variable feedback* (SVF) is realised, with the introduction of an *integral-of-error* state variable to ensure Type 1 *servomechanism* performance (section 5.1), i.e. if the closed-loop system is stable, the output will converge asymptotically to a constant command input specified by the user. In this manner, the steady-state errors that were encountered when using the unity gain regulator solutions discussed in the previous two chapters, are eliminated.

Since the SVF control law in the NMSS case involves only the measured input and output variables and their past values, it avoids the need for an explicit state reconstruction filter (observer). It can be compared directly with a 'dynamic output feedback' system that utilises additional dynamic compensation elements (control filters) in both the feedback and forward pathways (see e.g. Kucera 1979; Rosenthal and Wang 1996). Moreover, we will see that it can be interpreted as a logical extension of conventional *Proportional-Integral* (PI) and *Proportional-Integral-Derivative* (PID) controllers, here with additional feedback and input compensators introduced when the process has either second or higher order dynamics; or pure time delays greater than one sampling interval. Hence, the SVF control law obtained in this manner is usually called a *Proportional-Integral-Plus* (PIP) controller (section 5.2).

The structure of the PIP controller is quite general in form, and resembles that of various other digital control systems developed previously. For example, it is closely related to the *Generalised Predictive Control* (GPC) of Clarke *et al.* (1987) and can be interpreted directly in such model-based predictive control terms. The advantage of the PIP controller over GPC and other earlier, more ad hoc, dynamic output feedback designs, lies in its inherent state space formulation. This guarantees closed-loop stability in the fully deterministic case and allows not only for SVF pole assignment (section 5.3) or optimal *Linear Quadratic* (LQ) control

True Digital Control: Statistical Modelling and Non-Minimal State Space Design, First Edition.
C. James Taylor, Peter C. Young and Arun Chotai.

(section 5.4), but also for straightforward extension to the control of stochastic (Chapter 6) and multivariable (Chapter 7) dynamic systems.

First of all, in order to introduce the idea of an integral-of-error state variable and to highlight the structural similarity between PIP and PI control, let us consider a simple example.

Example 5.1 Proportional-Integral-Plus Control of a First Order TF Model The following first order TF model was introduced in Chapter 2:

$$y(k) = \frac{b_1 z^{-1}}{1 + a_1 z^{-1}} u(k) \tag{5.1}$$

where $y(k)$ is the output variable and $u(k)$ is the control input. The model has one pole, hence the associated minimal state space model consists of one state variable. Furthermore, since the model also has one numerator parameter and unity time delay, the regulator NMSS form (4.9) also consists of one state variable, i.e. $y(k)$.

By contrast, consider the following non-minimal 'servomechanism' state vector:

$$\boldsymbol{x}(k) = [y(k) \quad z(k)]^T \tag{5.2}$$

where $z(k)$ represents the discrete-time integral (summation) of the error between the output $y(k)$ and the command input $y_d(k)$: i.e. an integral-of-error state variable. In recursive terms, this can be written as:

$$z(k) = z(k-1) + (y_d(k) - y(k)) \tag{5.3}$$

or, in TF form:

$$z(k) = \frac{1}{1 - z^{-1}} (y_d(k) - y(k)) \tag{5.4}$$

where the TF:

$$\frac{1}{1 - z^{-1}}$$

will be recognised as a discrete-time integrator. Comparison of equation (5.4) with equation (2.33) indicates how the $z(k)$ state will introduce integral action into the final control system.

Noting from the TF model (5.1) that $y(k) = -a_1 y(k-1) + b_1 u(k-1)$, equation (5.3) can be replaced with:

$$z(k) = z(k-1) + y_d(k) + a_1 y(k-1) - b_1 u(k-1) \tag{5.5}$$

Hence, the NMSS model is defined by the following pair of state and observation equations:

$$\boldsymbol{x}(k) = \begin{bmatrix} y(k) \\ z(k) \end{bmatrix} = \begin{bmatrix} -a_1 & 0 \\ a_1 & 1 \end{bmatrix} \begin{bmatrix} y(k-1) \\ z(k-1) \end{bmatrix} + \begin{bmatrix} b_1 \\ -b_1 \end{bmatrix} u(k-1) + \begin{bmatrix} 0 \\ 1 \end{bmatrix} y_d(k) \tag{5.6}$$

$$y(k) = \begin{bmatrix} 1 & 0 \end{bmatrix} \boldsymbol{x}(k) \tag{5.7}$$

As in Chapter 4, the SVF control law associated with the above NMSS model is based on a negative feedback of the state vector:

$$u(k) = -\boldsymbol{k}^T \boldsymbol{x}(k) \tag{5.8}$$

where the control gain vector:

$$\boldsymbol{k}^T = [f_0 \quad -k_I] \tag{5.9}$$

In contrast to the regulator NMSS controller (4.22), the command input $y_d(k)$ enters the control algorithm by means of the integral-of-error state variable $z(k)$ and so a separate term for $y_d(k)$ is not required in equation (5.8).

Evaluating the SVF control algorithm (5.8) yields:

$$u(k) = -\begin{bmatrix} f_0 & -k_I \end{bmatrix} \begin{bmatrix} y(k) \\ z(k) \end{bmatrix} = -f_0 y(k) + k_I z(k) \tag{5.10}$$

Using equation (5.4) and equation (5.10), the control law for this system is given by:

$$u(k) = -f_0 y(k) + \frac{k_I}{1 - z^{-1}} (y_d(k) - y(k)) \tag{5.11}$$

In order to distinguish equation (5.8) or its TF equivalent (5.11) from the unity gain NMSS regulator introduced previously, this control law will be termed the PIP controller. In this simple first order case, however, it is obvious that equation (5.11) is identical to the digital PI control algorithm (2.38) introduced in Chapter 2. Indeed, the block diagram form of this control system has already been illustrated by Figure 2.15, in which f_0 represents the proportional control gain and k_I the integral control gain.

In other words, for the simplest TF model (5.1), the PIP formulation is equivalent to a digital PI control system, albeit one implemented in the output feedback arrangement shown by Figure 2.15, rather than a more conventional parallel arrangement (see e.g. Franklin *et al.* 2006, p. 187; Dorf and Bishop 2008, p. 445). In contrast to classical PI design, however, the vagaries of manual tuning are replaced by the systematic application of SVF control theory to yield pole assignment control; or, as discussed later, optimal control. Example 2.7 has already demonstrated how to use pole assignment to determine numerical values for the control gains, f_0 and k_I, with Figure 2.17 and Figure 2.18 showing illustrative closed-loop simulation results.

Note that the negative sign associated with k_I in equation (5.9) is utilised for consistency with the basic PI controller previously discussed. It ensures that the integral action appears in the forward path of the negative feedback control system. Finally, as shown by equation (2.41), the steady-state gain of the closed-loop system is unity, as required for Type 1 servomechanism performance.

Example 5.2 Implementation Results for Laboratory Excavator Bucket Position Returning to the laboratory robot excavator mentioned in Example 2.1, the TF model (5.1) with $a_1 = -1$ and $b_1 = 0.0498$ has been identified from experimental data (Shaban 2006; Taylor *et al.* 2007). In this case, $y(k)$ represents the bucket joint angle while $u(k)$ is a scaled control

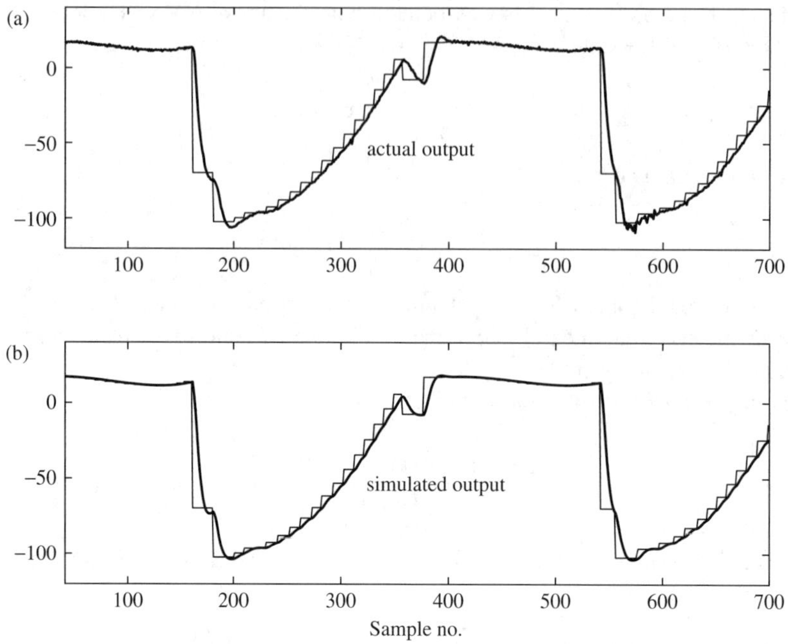

Figure 5.1 Closed-loop response using the PIP controller of Example 5.2, showing the output bucket joint angle in degrees (thick traces) and the time varying command input (thin traces). (a) Experimental data collected from the laboratory excavator and (b) the equivalent simulated response based on the TF model (5.1). The sampling rate is 0.11s

input voltage. Solving the pole assignment problem for the following complex conjugate pair of poles $p_{1,2} = 0.814 \pm 0.154j$, yields $f_0 = 6.30$ and $k_I = 1.17$ (these particular pole positions are utilised here because they are the closed-loop pole positions obtained using optimal PIP control, as considered later in section 5.5). The response of the closed-loop system to a time varying command input is illustrated by Figure 5.1 and Figure 5.2, which show the bucket position and voltage input, respectively. The command input has been obtained as part of the high-level control objectives for this device, i.e. a practical experiment for digging a trench in a sandpit.

Using the PIP control system (implemented here in PI form, so the PIP and PI are exactly equivalent), the measured bucket joint angle closely follows this command input. Experimental and simulated data are very similar in both cases, confirming that the TF model (5.1) adequately represents the laboratory excavator bucket angle and that the PIP control system is robust to modelling errors in this example.

But what have we achieved in this first order example? At first sight, the PIP controller used to generate these results is also the ubiquitous PI controller. So at the superficial level, we do not seem to have made much progress since Chapter 2 of this book! Nevertheless, one advantage of the PIP approach is already clear: although the control algorithm (5.11) retains the straightforward implementation structure of classical PI control, it has been derived within the powerful framework of SVF design. It is this underlying framework for the PIP design that

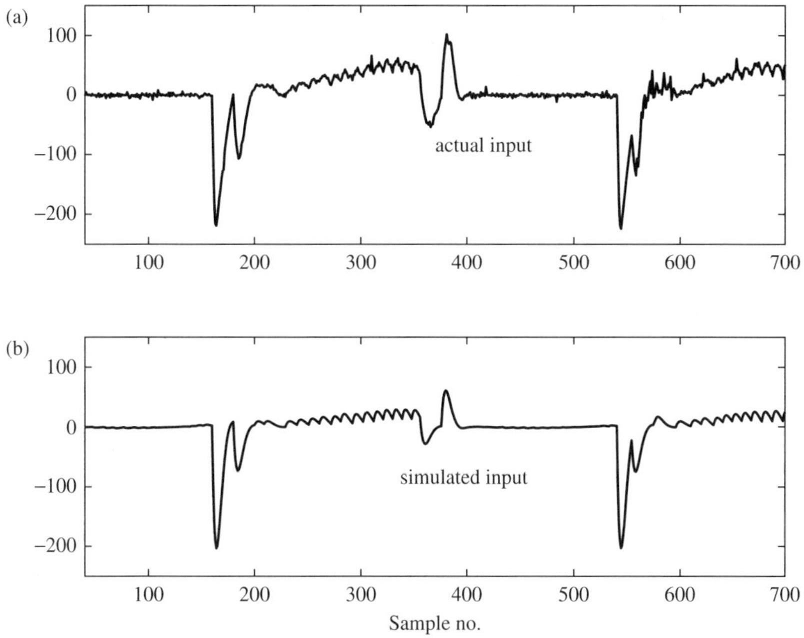

Figure 5.2 Control input signals associated with Figure 5.1, showing the scaled voltage. (a) Experimental data collected from the laboratory excavator and (b) the equivalent simulated response based on the TF model (5.1)

differentiates the controllers and foreshadows the advantages of PIP control system design when applied to higher order processes with time delays greater than unity. In other words, by following a similar approach to Chapter 4, we should now be able to develop an equivalent NMSS servomechanism form for TF models of any dimension and hence solve the general pole assignment problem.

5.1 The NMSS Servomechanism Representation

Consider again the general discrete-time TF model with unity time delay:

$$y(k) = \frac{B(z^{-1})}{A(z^{-1})}u(k) = \frac{b_1 z^{-1} + \cdots + b_m z^{-m}}{1 + a_1 z^{-1} + \cdots + a_n z^{-n}}u(k) \tag{5.12}$$

It is straightforward to show (see Example 5.1, Example 5.3 and Example 5.4) that the model (5.12) can be represented by the following NMSS equations (Young *et al.* 1987; Wang and Young 1988):

$$\left.\begin{aligned} \boldsymbol{x}(k) &= \boldsymbol{F}\boldsymbol{x}(k-1) + \boldsymbol{g}u(k-1) + \boldsymbol{d}y_d(k) \\ y(k) &= \boldsymbol{h}\boldsymbol{x}(k) \end{aligned}\right\} \tag{5.13}$$

The $n + m$ dimensional state vector $\boldsymbol{x}(k)$ consists of the present and past sampled values of the output variable $y(k)$ and the past sampled values of the input variable $u(k)$, now enhanced by the integral-of-error state variable introduced to ensure Type 1 servomechanism performance, i.e.

$$\boldsymbol{x}(k) = \begin{bmatrix} y(k) & y(k-1) & \cdots & y(k-n+1) & u(k-1) & u(k-2) & \cdots & u(k-m+1) & z(k) \end{bmatrix}^{\mathrm{T}} \tag{5.14}$$

where $z(k)$ is defined by equation (5.3). The associated state transition matrix $\boldsymbol{F}$, input vector $\boldsymbol{g}$, command input vector $\boldsymbol{d}$ and output vector $\boldsymbol{h}$ are:

$$\boldsymbol{F} = \begin{bmatrix} -a_1 & -a_2 & \cdots & -a_{n-1} & -a_n & b_2 & b_3 & \cdots & b_{m-1} & b_m & 0 \\ 1 & 0 & \cdots & 0 & 0 & 0 & 0 & \cdots & 0 & 0 & 0 \\ 0 & 1 & \cdots & 0 & 0 & 0 & 0 & \cdots & 0 & 0 & 0 \\ \vdots & \vdots & \ddots & \vdots & \vdots & \vdots & \vdots & \ddots & \vdots & \vdots & \vdots \\ 0 & 0 & \cdots & 1 & 0 & 0 & 0 & \cdots & 0 & 0 & 0 \\ 0 & 0 & \cdots & 0 & 0 & 0 & 0 & \cdots & 0 & 0 & 0 \\ 0 & 0 & \cdots & 0 & 0 & 1 & 0 & \cdots & 0 & 0 & 0 \\ 0 & 0 & \cdots & 0 & 0 & 0 & 1 & \cdots & 0 & 0 & 0 \\ \vdots & \vdots & \ddots & \vdots & \vdots & \vdots & \vdots & \ddots & \vdots & \vdots & \vdots \\ 0 & 0 & \cdots & 0 & 0 & 0 & 0 & \cdots & 1 & 0 & 0 \\ a_1 & a_2 & \cdots & a_{n-1} & a_n & -b_2 & -b_3 & \cdots & -b_{m-1} & -b_m & 1 \end{bmatrix} \tag{5.15}$$

$$\left.\begin{aligned} \boldsymbol{g} &= [b_1 \;\; 0 \;\; 0 \;\; \cdots \;\; 0 \;\; 1 \;\; 0 \;\; 0 \;\; \cdots \;\; 0 \;\; -b_1]^{\mathrm{T}} \\ \boldsymbol{d} &= [0 \;\; 0 \;\; 0 \;\; \cdots \;\; 0 \;\; 0 \;\; 0 \;\; 0 \;\; \cdots \;\; 0 \;\; 1]^{\mathrm{T}} \\ \boldsymbol{h} &= [1 \;\; 0 \;\; \cdots \;\; 0 \;\; 0 \;\; 0 \;\; 0 \;\; \cdots \;\; 0 \;\; 0 \;\; 0] \end{aligned}\right\} \tag{5.16}$$

In this NMSS model, the dimension is increased from the minimum n up to $(n + m)$, one more than the regulator NMSS form discussed in Chapter 4. It must be emphasised, however, that the NMSS form is not limited to this particular structure and other additional states can be added in order to introduce other features than can enhance the control system design, as shown by further examples considered in later chapters.

The output vector $\boldsymbol{h}$ in (5.16) simply extracts the output variable $y(k)$ from the state vector and is chosen to highlight the essential *Single-Input, Single-Output* (SISO) nature of the model (5.12). But, as in the regulator NMSS case, the observation matrix $\boldsymbol{H}$ can be defined also as an identity matrix of dimension $n + m$ by $n + m$, in a similar manner to that shown by equation (4.11), because all of the state variables at any sampling instant k can be stored in the digital computer (or special purpose micro-controller) that is being used to implement the PIP control system. In other words, equations (5.13), equation (5.14), equation (5.15) and equations (5.16) provide a 'snapshot' physical description of the SISO system at the kth sampling instant; while the replacement of $\boldsymbol{h}$ by $\boldsymbol{H}$ in the observation equation provides the 'control' description, revealing that the whole of the NMSS state vector is accessible and available for exploitation because of the stored past values $y(k-1)\cdots y(k-n+1)$ and $u(k-1)\cdots u(k-m+1)$ of $y(k)$ and $u(k)$, respectively, that are being retained in the control computer.

5.1.1 Characteristic Polynomial of the NMSS Servomechanism Model

Using equation (3.64), the open-loop characteristic polynomial $S(\boldsymbol{\lambda})$ of the NMSS representation (5.13) is given by:

$$S(\lambda) = |\lambda \boldsymbol{I} - \boldsymbol{F}| = \begin{bmatrix} \lambda + a_1 & a_2 & \cdots & a_{n-1} & a_n & -b_2 & -b_3 & \cdots & -b_{m-1} & -b_m & 0 \\ -1 & \lambda & \cdots & 0 & 0 & 0 & 0 & \cdots & 0 & 0 & 0 \\ 0 & -1 & \cdots & 0 & 0 & 0 & 0 & \cdots & 0 & 0 & 0 \\ \vdots & \vdots & \ddots & \vdots & \vdots & \vdots & \vdots & \ddots & \vdots & \vdots & \vdots \\ 0 & 0 & \cdots & -1 & \lambda & 0 & 0 & \cdots & 0 & 0 & 0 \\ 0 & 0 & \cdots & 0 & 0 & \lambda & 0 & \cdots & 0 & 0 & 0 \\ 0 & 0 & \cdots & 0 & 0 & -1 & \lambda & \cdots & 0 & 0 & 0 \\ 0 & 0 & \cdots & 0 & 0 & 0 & -1 & \cdots & 0 & 0 & 0 \\ \vdots & \vdots & \ddots & \vdots & \vdots & \vdots & \vdots & \ddots & \vdots & \vdots & \vdots \\ 0 & 0 & \cdots & 0 & 0 & 0 & 0 & \cdots & -1 & \lambda & 0 \\ -a_1 & -a_2 & \cdots & -a_{n-1} & -a_n & b_2 & b_3 & \cdots & b_{m-1} & b_m & \lambda - 1 \end{bmatrix} \tag{5.17}$$

Equation (5.17) can be written as:

$$|\lambda \boldsymbol{I} - \boldsymbol{F}| = (\lambda - 1) \left[\begin{array}{ccccc|ccccc} \lambda + a_1 & a_2 & \cdots & a_{n-1} & a_n & -b_2 & -b_3 & \cdots & -b_{m-1} & -b_m \\ -1 & \lambda & \cdots & 0 & 0 & 0 & 0 & \cdots & 0 & 0 \\ 0 & -1 & \cdots & 0 & 0 & 0 & 0 & \cdots & 0 & 0 \\ \vdots & \vdots & \ddots & \vdots & \vdots & \vdots & \vdots & \ddots & \vdots & \vdots \\ 0 & 0 & \cdots & -1 & \lambda & 0 & 0 & \cdots & 0 & 0 \\ \hline 0 & 0 & \cdots & 0 & 0 & \lambda & 0 & \cdots & 0 & 0 \\ 0 & 0 & \cdots & 0 & 0 & -1 & \lambda & \cdots & 0 & 0 \\ 0 & 0 & \cdots & 0 & 0 & 0 & -1 & \cdots & 0 & 0 \\ \vdots & \vdots & \ddots & \vdots & \vdots & \vdots & \vdots & \ddots & \vdots & \vdots \\ 0 & 0 & \cdots & 0 & 0 & 0 & 0 & \cdots & -1 & \lambda \end{array}\right] \tag{5.18}$$

Evaluating the determinant yields the following the open-loop characteristic polynomial:

$$S(\lambda) = |\lambda \boldsymbol{I} - \boldsymbol{F}| = (\lambda - 1)\lambda^{m-1}(\lambda^n + a_1\lambda^{n-1} + \cdots + a_n) \tag{5.19}$$

Note that $S(\lambda)$ is the product of the characteristic polynomial of the original system (or the equivalent minimal state space representation), a term λ^{m-1} due to the additional input states in the NMSS case and a third term $(\lambda - 1)$ associated with the integral-of-error state.

As discussed in Chapter 3 and Chapter 4, SVF design requires that the state space model should be controllable, the conditions for which are stated by Theorem 5.1.

Theorem 5.1 Controllability of the NMSS Servomechanism Model Given a SISO, discrete-time system described by a TF model (5.12), the NMSS representation (5.13) defined by the pair $[\boldsymbol{F}, \boldsymbol{g}]$, is completely controllable if and only if the following two conditions are satisfied (Wang and Young 1988):

(i) the polynomials $A(z^{-1})$ and $B(z^{-1})$ are coprime;
(ii) $b_1 + b_2 + \cdots + b_m \neq 0$.

The first condition restates the normal requirement that the TF representation should have no pole-zero cancellations. The second condition states that there should be no zeros at unity, which would otherwise cancel with the unity pole introduced to give integral action. The proof of this theorem is very similar to that discussed in Appendix C for the regulator NMSS form, so it will not be repeated here.

Finally, the above controllability conditions are equivalent to the normal requirement that the controllability matrix:

$$\boldsymbol{S}_1 = [\boldsymbol{g} \quad \boldsymbol{F}\boldsymbol{g} \quad \boldsymbol{F}^2\boldsymbol{g} \quad \cdots \quad \boldsymbol{F}^{n+m-1}\boldsymbol{g}] \tag{5.20}$$

has full rank, i.e. for controllability in this NMSS servomechanism case, $\boldsymbol{S}_1$ should have rank $n + m$.

Example 5.3 Non-Minimal State Space Servomechanism Representation of a Second Order TF Model Let us return to the following second order TF model with three numerator parameters, which was introduced in Chapter 4 (Example 4.3):

$$y(k) = \frac{b_1 z^{-1} + b_2 z^{-2} + b_3 z^{-3}}{1 + a_1 z^{-1} + a_2 z^{-2}} u(k) \tag{5.21}$$

or, in difference equation form:

$$y(k) = -a_1 y(k) - a_2 y(k-1) + b_1 u(k-1) + b_2 u(k-2) + b_3 u(k-3) \tag{5.22}$$

The integral-of-error state variable (5.3) can be replaced with:

$$z(k) = z(k-1) + y_d(k) + a_1 y(k) + a_2 y(k-1) - b_1 u(k-1) - b_2 u(k-2) - b_3 u(k-3) \tag{5.23}$$

In the context of the general TF model (5.12), it is clear that $n = 2$ and $m = 3$, hence the non-minimal servomechanism state vector $\boldsymbol{x}(k)$ consists of $n + m = 5$ state variables:

$$\boldsymbol{x}(k) = [y(k) \quad y(k-1) \quad u(k-1) \quad u(k-2) \quad z(k)]^T \tag{5.24}$$

It is straightforward to confirm that the following NMSS servomechanism model represents equation (5.22) and equation (5.23):

$$\begin{bmatrix} y(k) \\ y(k-1) \\ u(k-1) \\ u(k-2) \\ z(k) \end{bmatrix} = \begin{bmatrix} -a_1 & -a_2 & b_2 & b_3 & 0 \\ 1 & 0 & 0 & 0 & 0 \\ 0 & 0 & 0 & 0 & 0 \\ 0 & 0 & 1 & 0 & 0 \\ a_1 & a_2 & -b_2 & -b_3 & 1 \end{bmatrix} \begin{bmatrix} y(k-1) \\ y(k-2) \\ u(k-2) \\ u(k-3) \\ z(k-1) \end{bmatrix} + \begin{bmatrix} b_1 \\ 0 \\ 1 \\ 0 \\ -b_1 \end{bmatrix} u(k-1) + \begin{bmatrix} 0 \\ 0 \\ 0 \\ 0 \\ 1 \end{bmatrix} y_d(k) \tag{5.25}$$

As pointed out above, although the observation equation is usually defined as follows:

$$y(k) = [\,1 \quad 0 \quad 0 \quad 0 \quad 0\,]\boldsymbol{x}(k) \tag{5.26}$$

all the state variables are readily stored in a digital computer, so that the system is trivially observable and the observation matrix can be defined also as $\boldsymbol{y}(k) = \boldsymbol{H}\boldsymbol{x}(k)$, where, in this example, $\boldsymbol{H} = \boldsymbol{I}_5$.

Example 5.4 Rank Test for the NMSS Model Consider again the following non-minimum phase oscillator with a pure time delay of two sampling intervals:

$$y(k) = \frac{-z^{-2} + 2.0z^{-3}}{1 - 1.7z^{-1} + 1.0z^{-2}} u(k) \tag{5.27}$$

The TF model (5.27) is based on equation (5.21) with $a_1 = -1.7$, $a_2 = 1.0$, $b_1 = 0$, $b_2 = -1$ and $b_3 = 2.0$. In a similar manner to Example 4.3, the pure time delay of two samples is addressed by simply setting b_1 to zero in both the TF model and associated NMSS description. Hence, the NMSS servomechanism representation is algebraically equivalent to (5.25) or, equivalently, by substituting the numerical parameter values directly into equation (5.15) and equation (5.16):

$$\boldsymbol{F} = \begin{bmatrix} 1.7 & -1 & -1 & 2 & 0 \\ 1 & 0 & 0 & 0 & 0 \\ 0 & 0 & 0 & 0 & 0 \\ 0 & 0 & 1 & 0 & 0 \\ -1.7 & 1 & 1 & -2 & 1 \end{bmatrix}; \; \boldsymbol{g} = \begin{bmatrix} 0 \\ 0 \\ 1 \\ 0 \\ 0 \end{bmatrix}; \quad \text{and} \quad \boldsymbol{h} = [\,1 \quad 0 \quad 0 \quad 0 \quad 0\,] \tag{5.28}$$

Since the TF model (5.27) has no pole-zero cancellations and $b_1 + b_2 + b_3 = 1 \neq 0$, the NMSS model is clearly controllable. This result can be verified formally by examining the 5×5 dimension controllability matrix:

$$\boldsymbol{S}_1 = \begin{bmatrix} \boldsymbol{g} & \boldsymbol{F}\boldsymbol{g} & \boldsymbol{F}^2\boldsymbol{g} & \boldsymbol{F}^3\boldsymbol{g} & \boldsymbol{F}^4\boldsymbol{g} \end{bmatrix} = \begin{bmatrix} 0 & -1 & 0.3 & 1.51 & 2.267 \\ 0 & 0 & -1 & 0.3 & 1.51 \\ 1 & 0 & 0 & 0 & 0 \\ 0 & 1 & 0 & 0 & 0 \\ 0 & 1 & 0.7 & -0.81 & -3.077 \end{bmatrix} \tag{5.29}$$

which has full rank 5 and so is non-singular.

5.2 Proportional-Integral-Plus Control

The SVF control law associated with the NMSS model (5.13) takes the usual form:

$$u(k) = -\boldsymbol{k}^T \boldsymbol{x}(k) \tag{5.30}$$

where $\boldsymbol{k}$ is the $n+m$ dimensional SVF control gain vector:

$$\boldsymbol{k}^T = [f_0 \quad f_1 \quad \cdots \quad f_{n-1} \quad g_1 \quad \cdots \quad g_{m-1} \quad -k_I] \tag{5.31}$$

in which the feedback control gains are selected by the designer to achieve desired closed-loop characteristics, as discussed in section 5.3 and section 5.4. Expanding the terms in (5.30):

$$\begin{aligned} u(k) = &-f_0 y(k) - f_1 y(k-1) - \cdots - f_{n-1} y(k-n+1) \\ &- g_1 u(k-1) - \cdots - g_{m-1} u(k-m+1) + k_I z(k) \end{aligned} \tag{5.32}$$

Equation (5.30) and equation (5.32) will be referred to as the NMSS servomechanism controller or, more commonly, as the *Proportional-Integral-Plus* or *PIP control law*.

Following a similar approach to Example 5.1, the negative sign associated with the integral gain k_I in equation (5.31) is introduced to allow the integral feedback term to take on the form used in conventional PI and PID control laws. The block diagram representation of the closed-loop system in Figure 5.3 shows why the PIP controller can be considered as a logical extension of the PI controller. Here, the proportional action f_0 and integral action $k_I(1-z^{-1})^{-1}$, are enhanced by the higher order input $G_1(z^{-1})$ and feedback $F_1(z^{-1})$ compensators (the 'Plus' elements in the PIP controller), defined as follows:

$$\left.\begin{aligned} F_1(z^{-1}) &= f_1 z^{-1} + \cdots + f_{n-1} z^{-n+1} \\ G_1(z^{-1}) &= g_1 z^{-1} + \cdots + g_{m-1} z^{-m+1} \end{aligned}\right\} \tag{5.33}$$

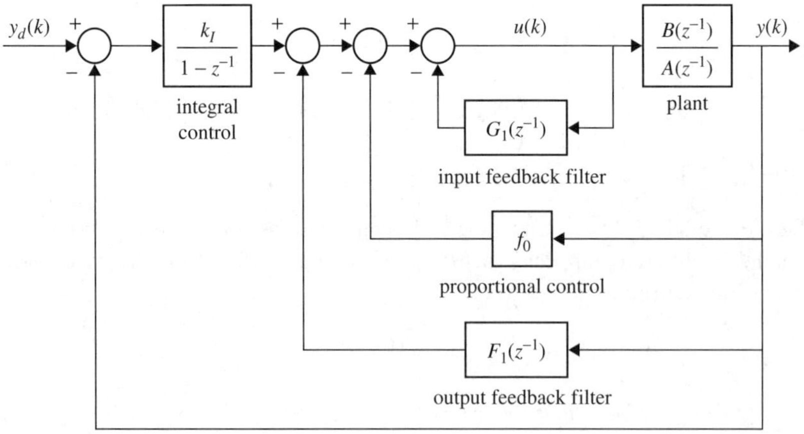

Figure 5.3 Block diagram of the univariate PIP control system explicitly showing the proportional and integral control action

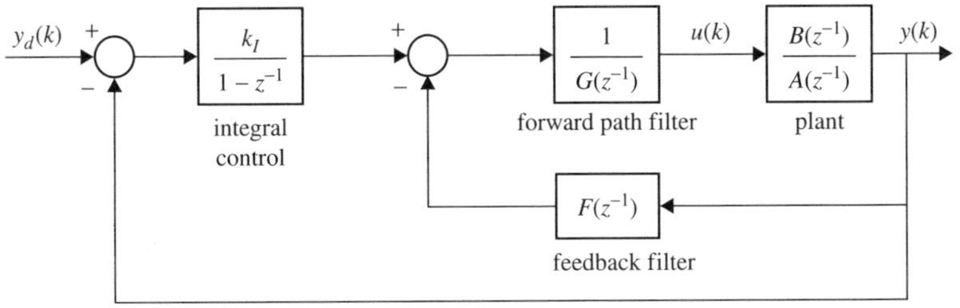

Figure 5.4 Block diagram of the univariate PIP control system in reduced form

These additional feedback and input compensators are introduced when the process has either second order or higher dynamics, or pure time delays greater than one sampling interval, i.e. when either n or m are greater than unity in equation (5.12). Straightforward algebra or block diagram analysis shows that Figure 5.3 may be represented equivalently by the simplified form illustrated in Figure 5.4, where:

$$\left.\begin{aligned} F(z^{-1}) &= f_0 + F_1(z^{-1}) = f_0 + f_1 z^{-1} + \cdots + f_{n-1} z^{-n+1} \\ G(z^{-1}) &= 1 + G_1(z^{-1}) = 1 + g_1 z^{-1} + \cdots + g_{m-1} z^{-m+1} \end{aligned}\right\} \tag{5.34}$$

In order to derive $G(z^{-1})$, the inner feedback loop involving $G_1(z^{-1})$, as shown in Figure 5.3, is reduced to a forward path filter, using equation (2.26).

Finally, note that the 'Plus' term in the PIP name replaces the 'Derivative' term associated with the conventional 'three-term' PID controller. Appendix D shows that the PIP controller can be transformed into a form that explicitly includes derivative action. This analysis reveals that the NMSS control strategy has resulted in a control system that can be related structurally to more conventional PI and PID controllers. However, as discussed below, it is inherently much more flexible and sophisticated than such conventional designs. In particular, it is able to exploit the power of SVF for closed-loop control system design and so introduce *inherent* higher order derivative action.

5.2.1 The Closed-Loop Transfer Function

The closed-loop control system can be obtained in TF form directly from Figure 5.4 by block diagram reduction:

$$y(k) = \frac{k_I B(z^{-1})}{\left(1 - z^{-1}\right)\left(G(z^{-1})A(z^{-1}) + F(z^{-1})B(z^{-1})\right) + k_I B(z^{-1})} y_d(k) \tag{5.35}$$

The closed-loop characteristic equation of the PIP control system is, therefore:

$$\left(1 - z^{-1}\right)\left(G(z^{-1})A(z^{-1}) + F(z^{-1})B(z^{-1})\right) + k_I B(z^{-1}) = 0 \tag{5.36}$$

Assuming that the closed-loop system is stable and that the command input takes a constant value $\bar{y}_d$, then as $k \to \infty$ the output will reach an equilibrium level where $y(k) = y(k-1)$. In other words, the output converges asymptotically to a steady-state value of $y(\infty)$, where $y(\infty)$

is found by setting $z^{-1} = 1$ in equation (5.35). In this case, the difference operator $1 - z^{-1}$ in the denominator of (5.35) becomes zero and the closed-loop TF function reduces to:

$$y(\infty) = \frac{k_I B(z^{-1})}{k_I B(z^{-1})} \bar{y}_d = \bar{y}_d \tag{5.37}$$

where $\bar{y}_d$ is the time-invariant command input. As required, therefore, the steady-state gain of the closed-loop TF is unity and Type 1 servomechanism performance is achieved.

The device of introducing integral action into the control law by adjoining an additional integral-of-error state variable was suggested by a number of authors in the late 1960s and early 1970s, including Young and Willems (1972). It probably represents the first intentional exploitation of a non-minimal concept and, therefore, seems particularly apposite for utilisation within the present much more general context.

Example 5.5 Proportional-Integral-Plus Control System Design for NMSS Model with Five State Variables Let us return to the non-minimum phase oscillator (5.27) considered in the previous example. Using equation (5.4), equation (5.24), equation (5.30) and equation (5.31), the PIP control law for this system is given by:

$$u(k) = -f_0 y(k) - f_1 y(k-1) - g_1 u(k-1) - g_2 u(k-2) + \frac{k_I}{1 - z^{-1}} (y_d(k) - y(k)) \tag{5.38}$$

The control system is represented in block diagram form by Figure 5.5, in which the control filters (5.34) are defined as follows:

$$F(z^{-1}) = f_0 + f_1 z^{-1} \tag{5.39}$$

$$G(z^{-1}) = 1 + g_1 z^{-1} + g_2 z^{-2} \tag{5.40}$$

Alternatively, using equation (2.1) and rearranging (5.38), yields the difference equation form of the PIP algorithm:

$$\begin{aligned} u(k) = u(k-1) + k_I(y_d(k) - y(k)) - f_0(y(k) - y(k-1)) \\ - f_1(y(k-1) - y(k-2)) - g_1(u(k-1) - u(k-2)) \\ - g_2(u(k-2) - u(k-3)) \end{aligned} \tag{5.41}$$

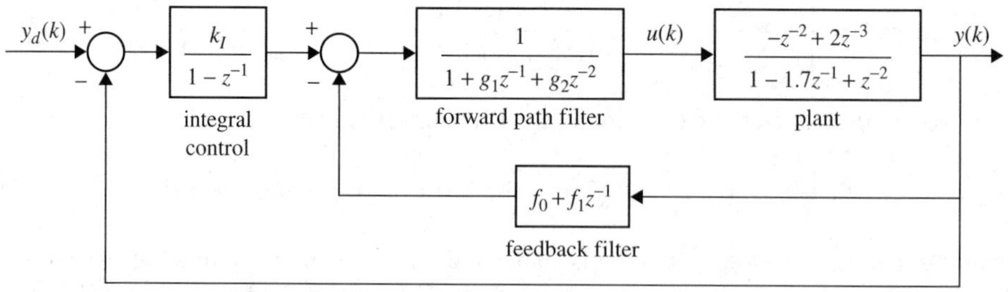

Figure 5.5 PIP control of the non-minimum phase oscillator in Example 5.5

which can be coded straightforwardly for practical applications using a digital computer.

Evaluation of the characteristic equation (5.36) for this example reveals that the closed-loop TF is fifth order. The presence of integral control action has increased the order of the system by one degree, when compared with the equivalent NMSS regulator solution (4.31). However, there are five control gains so that, once more, the poles of the characteristic equation can be assigned to desired locations in the complex z-plane, here by solving a set of five simultaneous equations. As the order of the system increases, so it becomes more time consuming to determine the algebraic solution by hand. Fortunately, as discussed below, the pole assignment solution for the general TF model (5.12) may be written conveniently in vector-matrix form, and it is then straightforward to compute using standard software packages.

5.3 Pole Assignment for PIP Control

In the present section, the automatic control objective is to design a SVF control law such that the closed-loop poles lie at pre-assigned positions in the complex z-plane.

Many algorithms have been proposed for calculating the SVF pole assignment control gains (e.g. Luenberger 1967; Gopinath 1971; Munro 1979; Kailath 1980). However, the special form of the NMSS representation allows us to compute the gains in a particularly straightforward manner. Indeed, it is possible to develop a general pole assignment algorithm for PIP control in one of two main ways that naturally yield the same result:

(i) by polynomial algebra, based on the block diagram Figure 5.4, as discussed in Appendix E; or
(ii) by state space analysis using an approach similar to that proposed previously for continuous-time systems (Young and Willems 1972).

The latter approach provides greater insight into the nature of the solution and is the one considered below. Of course, it is not always necessary to utilise these general algorithms: in low order cases the computations are so simple that special solutions may be computationally more efficient and so more appropriate to applications such as adaptive control.

5.3.1 *State Space Derivation*

The closed-loop control system is derived by substituting the PIP control law (5.30) into the NMSS model (5.13). This yields the following closed-loop state space system:

$$\left.\begin{aligned} \boldsymbol{x}(k) &= (\boldsymbol{F}-\boldsymbol{gk})\boldsymbol{x}(k-1)+\boldsymbol{d}y_d(k) \\ y(k) &= \boldsymbol{hx}(k) \end{aligned}\right\} \tag{5.42}$$

Following a similar approach to the analysis of open-loop systems in Chapter 3, the z-operator TF representation of (5.42) can be written as:

$$G(z)=\boldsymbol{h}(z\boldsymbol{I}-\boldsymbol{F}+\boldsymbol{gk}^T)^{-1}\boldsymbol{d}=\frac{\boldsymbol{h}\left(\mathrm{adj}(z\boldsymbol{I}-\boldsymbol{F}+\boldsymbol{gk}^T)\right)\boldsymbol{d}}{\left|(z\boldsymbol{I}-\boldsymbol{F}+\boldsymbol{gk}^T)\right|} \tag{5.43}$$

where $\boldsymbol{I}$ represents a $n+m$ dimension identity matrix, $\text{adj}(z\boldsymbol{I}-\boldsymbol{F}+\boldsymbol{g}\boldsymbol{k}^T)$ denotes the adjoint of the matrix $(z\boldsymbol{I}-\boldsymbol{F}+\boldsymbol{g}\boldsymbol{k}^T)$ and $|(z\boldsymbol{I}-\boldsymbol{F}+\boldsymbol{g}\boldsymbol{k}^T)|$ denotes its determinant. The characteristic equation of the closed-loop PIP control system is obtained by equating this determinant to zero, i.e.

$$|(z\boldsymbol{I}-\boldsymbol{F}+\boldsymbol{g}\boldsymbol{k}^T)| = 0 \tag{5.44}$$

The eigenvalues of $(\boldsymbol{F}-\boldsymbol{g}\boldsymbol{k}^T)$ are the roots of $|(z\boldsymbol{I}-\boldsymbol{F}+\boldsymbol{g}\boldsymbol{k}^T)| = 0$. Provided that the pair $[\boldsymbol{F}, \boldsymbol{g}]$ is completely controllable and we are dealing with a low order system, it is straightforward to compute the elements of $\boldsymbol{k}$ by simply equating the coefficients of equation (5.44) with those of a specified characteristic polynomial having the desired roots, as in Example 5.1. This yields a set of $(n+m)$ linear simultaneous algebraic equations, which can be solved in the normal manner.

In the higher order situation, this approach can be unwieldy and a more systematic procedure seems necessary for general practical applications and for use in *Computer Aided Control System Design* (CACSD) programs. This requires a method for determining the simultaneous algebraic equations *directly* from the NMSS model and the desired characteristic polynomial coefficients.

In order to develop a more general algorithm, we follow the approach of Young (1972) and let $p(z)$ and $r(z)$ denote the open- and closed-loop characteristic polynomials, respectively:

$$p(z) = |z\boldsymbol{I}-\boldsymbol{F}^T| \tag{5.45}$$

$$r(z) = |z\boldsymbol{I}-\boldsymbol{F}+\boldsymbol{g}\boldsymbol{k}^T| \tag{5.46}$$

Note that

$$z\boldsymbol{I}-\boldsymbol{F}+\boldsymbol{g}\boldsymbol{k}^T = (z\boldsymbol{I}-\boldsymbol{F})\left(\boldsymbol{I}+(z\boldsymbol{I}-\boldsymbol{F})^{-1}\boldsymbol{g}\boldsymbol{k}^T\right) \tag{5.47}$$

Therefore, taking the determinant on both sides of (5.47):

$$|z\boldsymbol{I}-\boldsymbol{F}+\boldsymbol{g}\boldsymbol{k}^T| = |z\boldsymbol{I}-\boldsymbol{F}|\,|\boldsymbol{I}+(z\boldsymbol{I}-\boldsymbol{F})^{-1}\boldsymbol{g}\boldsymbol{k}^T| \tag{5.48}$$

where the identity matrices are of appropriate dimensions, and rearranging the terms inside the determinant:

$$|\boldsymbol{I}+(z\boldsymbol{I}-\boldsymbol{F})^{-1}\boldsymbol{g}\boldsymbol{k}^T| = |\boldsymbol{I}+\boldsymbol{g}\boldsymbol{k}^T(z\boldsymbol{I}-\boldsymbol{F})^{-1}| = |\boldsymbol{I}+\boldsymbol{k}^T(z\boldsymbol{I}-\boldsymbol{F})^{-1}\boldsymbol{g}| \tag{5.49}$$

Using equation (5.48) and equation (5.49), we see that

$$|z\boldsymbol{I}-\boldsymbol{F}+\boldsymbol{g}\boldsymbol{k}^T| = |z\boldsymbol{I}-\boldsymbol{F}|\,|\boldsymbol{I}+\boldsymbol{k}^T(z\boldsymbol{I}-\boldsymbol{F})^{-1}\boldsymbol{g}| \tag{5.50}$$

Therefore,

$$r(z) = p(z)\left|\boldsymbol{I} + \boldsymbol{k}^T(z\boldsymbol{I} - \boldsymbol{F})^{-1}\boldsymbol{g}\right| \tag{5.51}$$

Recall that

$$(z\boldsymbol{I} - \boldsymbol{F})^{-1} = \frac{\operatorname{adj}(z\boldsymbol{I} - \boldsymbol{F})}{|z\boldsymbol{I} - \boldsymbol{F}|} \tag{5.52}$$

Let the vector $\boldsymbol{q}(z)$ be defined as:

$$\boldsymbol{q}(z) = \operatorname{adj}(z\boldsymbol{I} - \boldsymbol{F})\,\boldsymbol{g} \tag{5.53}$$

so that the closed-loop characteristic equation $r(z)$ can be written in the form:

$$r(z) = p(z) + \boldsymbol{k}^T\boldsymbol{q}(z) \tag{5.54}$$

Hence,

$$\boldsymbol{k}^T\boldsymbol{q}(z) = r(z) - p(z) \tag{5.55}$$

Thus knowing $\boldsymbol{q}(z)$, $r(z)$ and $p(z)$, the solution of equation (5.55) yields the feedback gain vector $\boldsymbol{k}$, as required.

The open-loop characteristic equation of the NMSS model (5.13) is given by:

$$p(z) = |z\boldsymbol{I} - \boldsymbol{F}| = (z-1)z^{m-1}\left(z^n + a_1 z^{n-1} + \cdots + a_n\right) = z^{n+m} + \sum_{i=1}^{n+m}(a_i - a_{i-1})z^{n+m-i} \tag{5.56}$$

where $a_0 = 1$ and $a_i = 0$ for $i > n$.

Now define a desired closed-loop characteristic polynomial $D(z)$:

$$D(z) = z^{n+m} + d_1 z^{n+m-1} + \cdots + d_{n+m} = z^{n+m} + \sum_{i=1}^{n+m} d_i z^{n+m-i} \tag{5.57}$$

From equation (5.55) and equation (5.56), and replacing $r(z)$ by the desired closed-loop characteristic polynomial $D(z)$, the following equation is obtained:

$$\boldsymbol{k}^T\boldsymbol{q}(z) = d(z) - p(z) = \sum_{i=1}^{n+m}\{d_i - (a_i - a_{i-1})\}z^{n+m-i} \tag{5.58}$$

again with $a_0 = 1$ and $a_i = 0$ for $i > n$. It can be shown that:

$$\mathrm{adj}\,(z\boldsymbol{I} - \boldsymbol{F}) = \sum_{j=1}^{n+m} z^{j-1} \sum_{i=j}^{n+m} (a_{n+m-i} - a_{n+m-i-1})\boldsymbol{F}^{i-j} \tag{5.59}$$

so that equation (5.58) can be written as:

$$\boldsymbol{k}^T \sum_{j=1}^{n+m} z^{j-1} \sum_{i=j}^{n+m} (a_{n+m-i} - a_{n+m-i-1})\boldsymbol{F}^{i-j}\boldsymbol{g} = \sum_{i=1}^{n+m} (d_i - (a_i - a_{i-1}))z^{n+m-i} \tag{5.60}$$

Equating the coefficients of the like powers of z on both sides of (5.60), the following equations can be generated:

$$\begin{aligned}
j = 1,\quad & \boldsymbol{k}^T \sum_{i=1}^{n+m} (a_{n+m-i} - a_{n+m-i-1})\ \boldsymbol{F}^{i-1}\boldsymbol{g} = d_{n+m} - (a_{n+m} - a_{n+m-1}) \\
j = 2,\quad & \boldsymbol{k}^T \sum_{i=2}^{n+m} (a_{n+m-i} - a_{n+m-i-1})\ \boldsymbol{F}^{i-2}\boldsymbol{g} = d_{n+m-1} - (a_{n+m-1} - a_{n+m-2}) \\
\vdots\quad & \qquad\vdots \qquad\qquad\qquad\qquad\qquad \vdots \\
j = n+m-1,\ & \boldsymbol{k}^T \sum_{i=n+m-1}^{n+m} (a_{n+m-i} - a_{n+m-i-1})\ \boldsymbol{F}^{i-(n+m-1)}\boldsymbol{g} = d_2 - (a_2 - a_1) \\
j = n+m,\quad & \boldsymbol{k}^T a_0 \boldsymbol{g} = d_1 - (a_1 - a_0)
\end{aligned} \tag{5.61}$$

Now define the matrix $\boldsymbol{M}$ as follows:

$$\boldsymbol{M} = \begin{bmatrix}
1 & 0 & 0 & \cdots & 0 & 0 & \cdots & 0 & 0 \\
a_1 - 1 & 1 & 0 & \cdots & 0 & 0 & \cdots & 0 & 0 \\
a_2 - a_1 & a_1 - 1 & 1 & \cdots & 0 & 0 & \cdots & 0 & 0 \\
\vdots & \vdots & \vdots & \cdots & \vdots & \vdots & \vdots & \vdots & \vdots \\
-a_n & a_n - a_{n-1} & a_{n-1} - a_{n-2} & \cdots & 1 & 0 & \cdots & 0 & 0 \\
0 & -a_n & a_n - a_{n-1} & \cdots & a_1 - 1 & 1 & \cdots & 0 & 0 \\
\vdots & \vdots & \vdots & \vdots & \vdots & \vdots & \vdots & \vdots & \vdots \\
0 & 0 & 0 & \cdots & a_n - a_{n-1} & a_{n-1} - a_{n-2} & \cdots & 1 & 0 \\
0 & 0 & 0 & \cdots & -a_n & a_n - a_{n-1} & \cdots & a_1 - 1 & 1
\end{bmatrix} \tag{5.62}$$

Since $a_0 = 1$ and $a_i = 0$ for $i > n$, the set of equations (5.61) can be written conveniently in the following vector-matrix equation form:

$$M \begin{bmatrix} g^T \\ g^T F^T \\ g^T \left(F^T\right)^2 \\ \vdots \\ g^T \left(F^T\right)^n \\ g^T \left(F^T\right)^{n+1} \\ \vdots \\ g^T \left(F^T\right)^{n+m-2} \\ g^T \left(F^T\right)^{n+m-1} \end{bmatrix} k = d - p \tag{5.63}$$

where d and p are the vectors of coefficients of the desired closed-loop characteristic polynomial and the open-loop characteristic polynomial of the NMSS model, respectively:

$$d^T = [\, d_1 \quad d_2 \quad d_3 \quad \cdots \quad d_n \quad d_{n+1} \quad \cdots \quad d_{n+m-1} \quad d_{n+m} \,] \tag{5.64}$$

$$p^T = [\, a_1 - 1 \quad a_2 - a_1 \quad a_3 - a_2 \quad \cdots \quad a_n - a_{n-1} \quad -a_n \quad \cdots \quad 0 \quad 0 \,] \tag{5.65}$$

and k is the $(n + m)$ dimensional control gain vector (5.31). Equation (5.63) represents the required set of $(n + m)$ simultaneous equations in the $(n + m)$ unknown SVF gains.

It will be noted that M on the left-hand side of (5.63) is lower triangular with unity elements on the main diagonal, hence it is non-singular. The second matrix is the transpose of the NMSS controllability matrix (5.20). As a result, it is clear that a solution to the simultaneous equations (5.63) exists *if and only if* the NMSS model is controllable, i.e. the controllability matrix S_1 must be of full rank $(n + m)$. We have already seen above that this is the case for the NMSS model, as long as the conditions of Theorem 5.1 hold.

In this controllable case, equation (5.63) can be written:

$$(M S_1^T) k = d - p \tag{5.66}$$

where $M S^T$ is invertible, so that the control gain vector can be obtained from:

$$k = (M S_1^T)^{-1} (d - p) \tag{5.67}$$

This is the required algorithm for computing the PIP gains. For later reference, define a matrix $\Sigma = M S_1^T$.

Theorem 5.2 Pole Assignability of the PIP Controller Given a SISO discrete-time system described by the NMSS representation (5.13), all of the poles of the closed-loop system can be assigned to arbitrary positions in the closed-loop z-plane by the PIP control law (5.30), if and only if the controllability conditions of Theorem 5.1 are fully satisfied.

Proof of Theorem 5.2 If the matrix $\boldsymbol{\Sigma} = \boldsymbol{M}\boldsymbol{S}_1^T$ in equation (5.67) is non-singular, the eigenvalues of the closed-loop system can be arbitrarily assigned by the use of the state feedback algorithm $u(k) = -\boldsymbol{k}\,x(k)$ and the feedback gain vector $\boldsymbol{k}$ for the PIP control system will be uniquely defined. It is not difficult to prove the non-singularity of the Σ matrix, provided that the conditions of Theorem 5.1 are satisfied. In particular, if we add to each row in turn, all of the rows below that row, then $\boldsymbol{\Sigma}$ becomes:

$$\left[\begin{array}{cccccccccc|c}
0 & 0 & \cdots & 0 & 0 & 0 & 0 & \cdots & 0 & 0 & b_1 + \cdots + b_m \\
\hline
-b_1 & 0 & \cdots & 0 & 0 & -1 & 0 & \cdots & 0 & 0 & 0 \\
-b_2 & -b_1 & \cdots & 0 & 0 & -a_1 & -1 & \cdots & 0 & 0 & 0 \\
\vdots & \vdots & \cdots & \vdots & \vdots & \vdots & \vdots & \cdots & \vdots & \vdots & \vdots \\
-b_{m-1} & -b_{m-2} & \cdots & -b_2 & -b_1 & \vdots & \vdots & \cdots & \vdots & -1 & 0 \\
-b_m & -b_{m-1} & \cdots & -b_3 & -b_2 & -a_{n-1} & \vdots & \cdots & -a_2 & -a_1 & 0 \\
0 & -b_m & \cdots & -b_4 & -b_3 & -a_n & -a_{n-1} & \cdots & -a_3 & -a_2 & 0 \\
\vdots & 0 & \cdots & \vdots & \vdots & 0 & -a_n & \cdots & \vdots & \vdots & \vdots \\
\vdots & \vdots & \cdots & \vdots & \vdots & \vdots & 0 & \cdots & \vdots & \vdots & \vdots \\
0 & 0 & \cdots & -b_m & -b_{m-1} & 0 & 0 & \cdots & -a_n & -a_{n-1} & 0 \\
0 & 0 & \cdots & 0 & -b_m & 0 & 0 & \cdots & 0 & -a_n & 0
\end{array}\right]$$

Here, the lower left block matrix will be recognised as the well known Sylvester resultant matrix (see e.g. Kailath 1980), which is non-singular if the polynomials $A(z^{-1})$ and $B(z^{-1})$ are coprime. If, in addition $b_1 + \cdots + b_m \neq 0$, then the whole Σ matrix will be non-singular as required. These are the same conditions as stated in Theorem 5.1.

Thus, provided the system (5.13) satisfies the controllability conditions of Theorem 5.1, we will always be able to compute the specified PIP pole assignment gains via the solution of the set of linear, simultaneous equations (5.67) or their equivalent (E.5) in Appendix E. These conditions for pole assignability are intuitively obvious and quite non-restrictive. But they are, of course, based on purely theoretical results. In practice, like all other linear control system design procedures, PIP pole assignment will only apply strictly if the controlled system is, and continues to be, described adequately by the linear TF model (5.12) and so maintains a linear mode of behaviour. The robustness consequences of parametric uncertainty will be considered in Example 5.6.

Example 5.6 Pole Assignment Design for the NMSS Model with Five State Variables Let us return to the non-minimum phase oscillator (5.27) and PIP controller (5.38) considered in the previous two examples. In order to illustrate the pole assignment approach, let the desired closed-loop characteristic polynomial (5.57) be specified as $D(z) = (z - 0.5)^4$ or, equivalently:

$$D(z^{-1}) = (1 - 0.5z^{-1})^4 = 1 - 2z^{-1} + 1.5z^{-2} - 0.5z^{-3} + 0.0625z^{-4}$$

The desired coefficients (5.64) are then expressed in vector form as follows:

$$\boldsymbol{d}^T = [-2 \quad 1.5 \quad -0.5 \quad 0.0625 \quad 0] \tag{5.68}$$

while the open-loop coefficients (5.65) are:

$$\boldsymbol{p}^T = [-2.7 \quad 2.7 \quad -1 \quad 0 \quad 0] \tag{5.69}$$

and equation (5.66) takes the form:

$$\begin{bmatrix} 0 & 0 & 1 & 0 & 0 \\ -1 & 0 & -2.7 & 1 & -1 \\ 3 & -1 & 2.7 & -2.7 & 2 \\ -2 & 3 & -1 & 2.7 & 0 \\ 0 & -2 & 0 & -1 & 0 \end{bmatrix} \begin{bmatrix} f_0 \\ f_1 \\ g_1 \\ g_2 \\ k_I \end{bmatrix} = \begin{bmatrix} 0.7 \\ -1.2 \\ 0.5 \\ 0.0625 \\ 0 \end{bmatrix} \tag{5.70}$$

Solving (5.70) yields:

$$\boldsymbol{k} = [0.176 \quad -0.464 \quad 0.700 \quad 0.928 \quad -0.063]^T \tag{5.71}$$

Hence, the PIP control law (5.38) is:

$$u(k) = -0.176y(k) + 0.464y(k-1) - 0.7u(k-1) - 0.928u(k-2) + \frac{0.063}{1-z^{-1}}(y_d(k) - y(k)) \tag{5.72}$$

Note that, although the dimension of the NMSS servomechanism system here is five, the closed-loop system has been specified as fourth order by placing the additional pole at the origin of the complex z-plane, i.e. $d_5 = 0$ in equation (5.68). This allows for a direct comparison with the NMSS regulator solution developed in Example 4.3 and Example 4.4. In fact, in the ideal case, the response of the PIP (NMSS servomechanism) and NMSS regulator controllers are exactly the same as that shown in Figure 4.3.

However, one advantage of the PIP controller is highlighted by Figure 5.6, which shows the closed-loop response of the system when an unmeasured output disturbance input is introduced. Here, we see that the addition of the integral-of-error state variable in the PIP control law, not only ensures basic steady-state tracking of the command input but also 'backs-off' the disturbance, i.e. the control input brings the system back to the command level following the imposition of the step disturbance at the 25th sample. For comparison, the equivalent unity gain NMSS regulator response, which exhibits a steady-state error, is illustrated by Figure 4.7.

Now suppose there is some uncertainty associated with the model used to estimate the gains and that the system numerator polynomial is actually $-z^{-2} + 1.8z^{-3}$, i.e. there is a 10% error in the second parameter. The closed-loop PIP control system with this model mismatch is illustrated by Figure 5.7, with the unit step response shown in Figure 5.8. This shows that the system remains well controlled despite the error in the model parameter and that the steady-state gain of the closed-loop system is still unity. For comparison, the equivalent unity gain NMSS regulator response, again with a steady-state error, is illustrated by Figure 4.6.

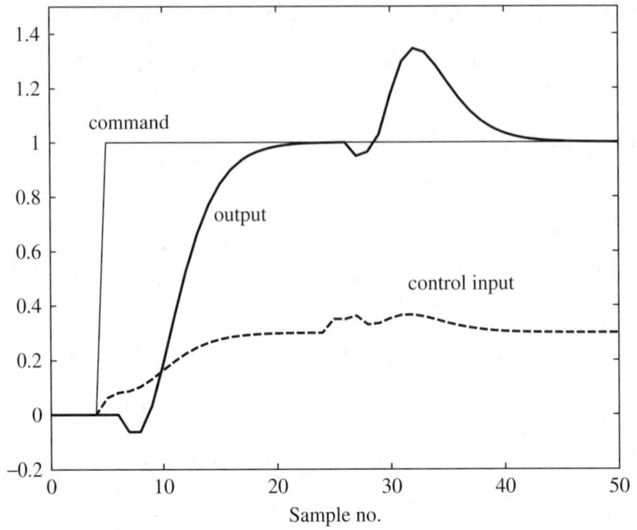

Figure 5.6 Closed-loop unit step response using the PIP controller of Example 5.6, with an input step disturbance of 0.05 at the 25th sample (cf. Figure 4.7 using the unity gain NMSS regulator)

These results are based on the fourth order closed-loop design. In fact, any order $\leq n + m$ could be selected for the closed-loop system, including a dead-beat specification where all the poles are assigned to the origin of the complex z-plane, i.e. $D(z^{-1}) = 1$. As discussed in Chapter 2 (Example 2.7), such a dead-beat design is superficially attractive since, in theory, it yields a very rapid, critically damped response.

In general, however, it represents a rather risky design in practical terms, with numerous potential disadvantages: for example, it can be too 'harsh', often leading to problems such as input saturation and poor disturbance rejection. Furthermore, such dead-beat control systems can also have very low stability margins with considerable sensitivity to parametric uncertainty. Indeed, for the present example, with the model mismatch in Figure 5.7 and Figure 5.8, the dead-beat design yields an unstable closed-loop response.

Example 5.7 Implementation Results for FACE system with Disturbances Let us return to the FACE system mentioned in Example 2.1, as represented by the TF model (5.1) with

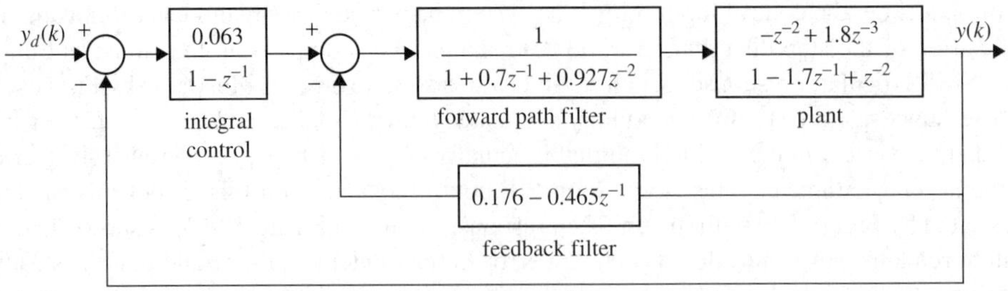

Figure 5.7 PIP control of the non-minimum phase oscillator in Example 5.6 with model mismatch

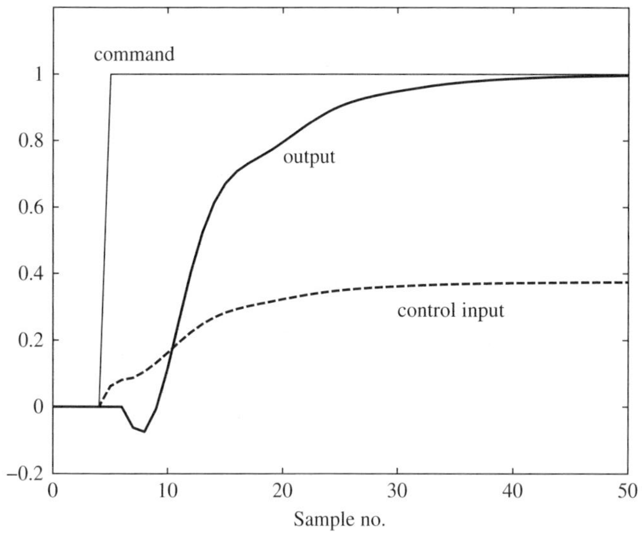

Figure 5.8 Closed-loop unit step response using the PIP controller of Example 5.6, when the model has a 10% error in one of the parameters (cf. Figure 4.6 using the unity gain NMSS regulator)

$a_1 = -0.713$ and $b_1 = 2.25$. Here, $y(k)$ is the CO_2 concentration (parts per million, ppm) and $u(k)$ is the voltage regulating a mass flow valve. Practical FACE systems are dominated by the wind disturbances (Taylor *et al.* 2000), hence the PIP control law offers an essential improvement over the simpler NMSS regulator solution. However, the question of how to choose suitable pole positions remains open. Referring back to Example 2.6 (equivalently Example 5.1), this is the design decision that has to be made when choosing between the dead-beat response illustrated by Figure 2.17 and slower designs such as Figure 2.18.

In fact, practical implementation results for the FACE system, based on dead-beat design, are very poor, with the mass flow value control input oscillating rapidly between its maximum and minimum extremes. By contrast, trial and error experimentation using the practical FACE system suggests that satisfactory closed-loop responses, for a wide range of wind conditions, are obtained by selecting the following second order desired closed-loop characteristic polynomial:

$$D(z^{-1}) = (1 - 0.75z^{-1})^2 = 1 - 1.5z^{-1} + 0.5625z^{-2} \tag{5.73}$$

In this case, solving (5.67) yields:

$$\boldsymbol{k} = [\,0.0669 \quad -0.0278\,]^T \tag{5.74}$$

and the PIP control law takes the PI form shown by Figure 2.15, here with $f_0 = 0.0669$ and $k_I = 0.0278$.

Figure 5.9 illustrates the response for a typical implementation experiment. Considering the relatively gusty wind conditions associated with this experiment (indicated by the rapid

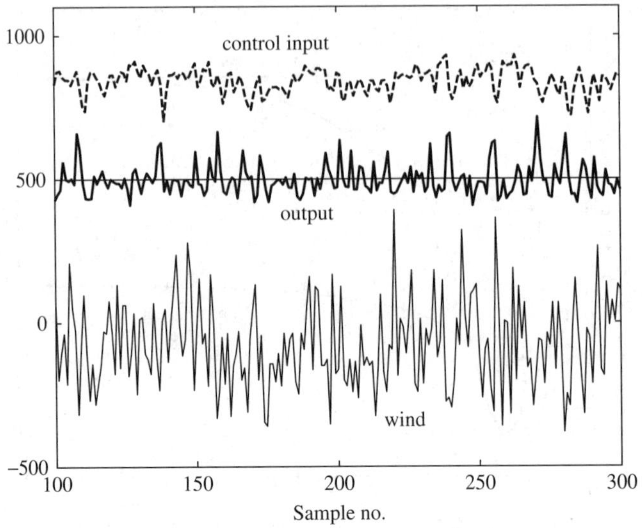

Figure 5.9 Closed-loop response using the PIP controller (5.74), showing the output CO_2 concentration (ppm), the command input at 500 ppm, the control input (a scaled voltage regulating a mass flow valve) and a scaled voltage representing changes in wind velocity (for which zero represents the mean wind speed). The sampling rate is 10 s

changes required in the control input variable), these results show how the basic PI structure can, nonetheless, yield acceptable results. Here, the 1 min mean CO_2 concentrations (based on averaging the 10 s means obtained by the data logger) never deviate from the set point by more than 15%, which is well within the objectives of the study. Note that the wind velocity is represented by an arbitrarily scaled voltage that has not been calibrated in this case – the lower trace in Figure 5.9 is shown only to highlight the time varying nature of this disturbance signal.

5.4 Optimal Design for PIP Control

In this book, the control design specification considered so far has always been pole assignment, an approach in which the designer directly selects the poles of the closed-loop system. In Example 5.7, these poles were chosen by trial and error experimentation, in order to achieve a satisfactory response. In the present section, the PIP control gain vector is chosen to fulfil an alternative closed-loop requirement. Here, we introduce the concept of *optimal control*, i.e. a control system that simultaneously ensures the completion of the system objectives (e.g. stability) and the minimisation of a performance index (see e.g. Zak 2003, p. 225; Dorf and Bishop 2008, p. 781). In the most general terms, the parameters of an optimal control system are adjusted so that a performance index reaches an extremum value; usually, the performance index is defined so that its numerical value decreases as the quality (in some defined sense) of the control increases (Zak 2003).

More specifically, a well known example of SVF optimal control is when the control gain vector is chosen to minimise the *Linear Quadratic* (LQ) type of performance criterion (Kalman

1960a; Wonham 1968). For a SISO control system, the aim is to design a feedback gain vector $\boldsymbol{k}$ that will minimise the following quadratic cost function:

$$J = \sum_{k=0}^{\infty} \boldsymbol{x}(k)^T \boldsymbol{Q}\boldsymbol{x}(k) + r(u(k)^2) \tag{5.75}$$

where $\boldsymbol{Q}$ is a $(n+m)$ by $(n+m)$ symmetric, positive semi-definite matrix and r is a positive scalar. As usual, $\boldsymbol{x}(k)$ and $u(k)$ are the state vector and control input variable, respectively. The quadratic form $\boldsymbol{x}(k)^T\boldsymbol{Q}\boldsymbol{x}(k)$ (see Appendix A) represents a very general or weighted (by the elements of $\boldsymbol{Q}$) 'sum of squares' type operation on the elements of $\boldsymbol{x}(k)$, providing a measure of the 'power' in the state variables that is being balanced by the power in the input signal. As a result, the minimisation of this cost function over the interval $k = 0 \to \infty$ means that a balance is struck between the minimisation of the state variable changes and the amount of input power required to achieve this.

Equation (5.75) is a standard formulation of the infinite time, optimal LQ servomechanism cost function for a SISO system (see e.g. Åström and Wittenmark 1984, pp. 254–281; Zak 2003, pp. 244–314; Franklin *et al.* 2006, pp. 487–488; Dorf and Bishop 2008, pp. 781–791). However, it is worth noting that the special structure of the non-minimal state vector means that the diagonal elements of $\boldsymbol{Q}$ have particularly simple interpretation. In particular, the diagonal elements of $\boldsymbol{Q}$ represent the weights assigned to squared values not only of the current and past values of the output but also the integral-of-error term and the past values of the input, as discussed below.

5.4.1 Linear Quadratic Weighting Matrices

For now, we will follow the usual convention and let $\boldsymbol{Q}$ be a purely diagonal matrix:

$$\boldsymbol{Q} = \text{diag}(\, q_1 \quad q_2 \quad \ldots \quad q_n \quad q_{n+1} \quad \ldots \quad q_{n+m-1} \quad q_{n+m}\,) \tag{5.76}$$

Noting the NMSS state vector (5.14), the user-defined output weighting parameters $q_1, q_2, \ldots, q_n$ and input weighting parameters $q_{n+1}, \ldots, q_{n+m-1}$ are generally set equal to common values of q_y and q_u, respectively, while q_{n+m} is denoted by q_e to indicate that it provides a weighting constraint on the integral-of-error state variable $z(k)$. Hence,

$$\boldsymbol{Q} = \text{diag}(\, q_y \quad \cdots \quad q_y \quad q_u \quad \cdots \quad q_u \quad q_e\,) \tag{5.77}$$

Also, because of the special nature of the NMSS formulation, we typically set $r = q_u$ in equation (5.75). Using this approach, q_y, q_u and q_e represent the partial weightings on the output, input and integral-of-error variables in the NMSS vector $\boldsymbol{x}(k)$, usually defined as follows:

$$q_y = \frac{W_y}{n}; \quad q_u = \frac{W_u}{m}; \quad q_e = W_e \tag{5.78}$$

so that the total weightings are given by scalar weights W_y, W_u and W_e.

These three scalar weighting factors are chosen by the designer to achieve the desired closed-loop performance, as discussed in Example 5.8 and subsequently in section 5.5. In one sense, this formulation is analogous to the PID controller that also has three tuning parameters. Of course, for classical PID control the tuning parameters are simply the proportional, integral and derivative control gains, and changing these directly may yield an unsatisfactory or even an unstable solution. This contrasts with LQ design, in which closed-loop stability is guaranteed in the deterministic case (i.e. no uncertainty in the model coefficients) when the system is represented by the TF model (5.12), as shown in section 5.4.2.

Satisfactory closed-loop performance can often be obtained by straightforward manual tuning of the diagonal LQ weights or simply W_y, W_u and W_e. However, in more difficult situations, for example in the multivariable case, there can be advantages in considering the exploitation of the off-diagonal elements of $\boldsymbol{Q}$. Indeed, we will see that the PIP-LQ control system is ideal for incorporation within a multi-objective optimisation framework. In this more complex situation, satisfactory compromise can be obtained between conflicting objectives, such as multivariable decoupling, robustness, overshoot, rise times and actuator demands. This is achieved by concurrent optimisation of the diagonal and off-diagonal elements of the weighting matrices in the cost function, as discussed in Chapter 7.

For now, however, it is worth noting that the 'default' PIP-LQ controller, which is obtained using total optimal control weights of unity, i.e.

$$W_y = W_u = W_e = 1 \tag{5.79}$$

or, using equations (5.78), equivalently,

$$q_y = \frac{1}{n}; \quad q_u = r = \frac{1}{m}; \quad q_e = 1 \tag{5.80}$$

is often found to work well in practice. If a faster or slower closed-loop response is desirable, then q_e can usually be straightforwardly increased or reduced as required.

5.4.2 The LQ Closed-loop System and Solution of the Riccati Equation

Given the NMSS system description $[\boldsymbol{F}, \boldsymbol{g}]$, the weighting matrix $\boldsymbol{Q}$ and the input weighting r, the SVF gain vector $\boldsymbol{k}$ is given by standard LQ theory (e.g. Willems 1971; Pappas *et al.* 1980; Åström and Wittenmark 1984, p. 260). In particular, the control gains are determined using:

$$\boldsymbol{k}^T = (r + \boldsymbol{g}^T\boldsymbol{P}\boldsymbol{g})^{-1}\boldsymbol{g}^T\boldsymbol{P}\boldsymbol{F} \tag{5.81}$$

where the matrix $\boldsymbol{P}$ is the steady-state solution of the following discrete-time, matrix Riccati equation:

$$\boldsymbol{P} - \boldsymbol{F}^T\boldsymbol{P}\boldsymbol{F} + \boldsymbol{F}^T\boldsymbol{P}\boldsymbol{g}(r + \boldsymbol{g}^T\boldsymbol{P}\boldsymbol{g})^{-1}\boldsymbol{g}^T\boldsymbol{P}\boldsymbol{F} - \boldsymbol{Q} = \boldsymbol{0} \tag{5.82}$$

in which $\boldsymbol{0}$ is a matrix of zeros. Using equation (5.30), the optimal SVF control law takes the form:

$$u(k) = -\left((r + \boldsymbol{g}^T \boldsymbol{P} \boldsymbol{g})^{-1} \boldsymbol{g}^T \boldsymbol{P} \boldsymbol{F}\right) \boldsymbol{x}(k) \tag{5.83}$$

Substituting from (5.83) into (5.42), the closed-loop system becomes:

$$\boldsymbol{x}(k) = \left(\boldsymbol{F} - \boldsymbol{g}(r + \boldsymbol{g}^T \boldsymbol{P} \boldsymbol{g})^{-1} \boldsymbol{g}^T \boldsymbol{P} \boldsymbol{F}\right) \boldsymbol{x}(k-1) + \boldsymbol{d} y_d(k) = 0 \tag{5.84}$$

and the closed-loop poles are obtained from the following characteristic equation:

$$\left|\lambda \boldsymbol{I} - \boldsymbol{F} + \boldsymbol{g}(r + \boldsymbol{g}^T \boldsymbol{P} \boldsymbol{g})^{-1} \boldsymbol{g}^T \boldsymbol{P} \boldsymbol{F}\right| = 0 \tag{5.85}$$

It can be shown (e.g. Kuo 1997) that the closed-loop poles are the $(n + m)$ stable eigenvalues of the following generalised eigenvalue problem:

$$|\boldsymbol{V}_L \lambda - \boldsymbol{V}_R| = 0 \tag{5.86}$$

where

$$\boldsymbol{V}_R = \begin{bmatrix} \boldsymbol{F} & \boldsymbol{0} \\ \boldsymbol{Q} & -\boldsymbol{I} \end{bmatrix}; \quad \boldsymbol{V}_L = \begin{bmatrix} \boldsymbol{I} & \boldsymbol{g} r \boldsymbol{g}^T \\ \boldsymbol{Q} & -\boldsymbol{F}^T \end{bmatrix} \tag{5.87}$$

This latter result is useful in obtaining a computational solution of the matrix Riccati equation, as required to define the PIP control law. Here, we solve the following eigenvalue–eigenvector equation:

$$(\boldsymbol{V}_L \lambda - \boldsymbol{V}_R) \boldsymbol{w} = 0 \tag{5.88}$$

where λ is the eigenvalue, $\boldsymbol{w}$ is the corresponding eigenvector, and the matrices $\boldsymbol{V}_R$ and $\boldsymbol{V}_L$ are given by equations (5.87).

Note that the standard approach used for solving the steady-state matrix Riccati equation, as originally used in the MATLAB®[1] Control System Toolbox (Grace *et al.* 1990), requires that the state transition matrix is non-singular. However, in the NMSS model the $\boldsymbol{F}$ matrix can be singular and so this approach is not applicable. The problem is easily obviated, however, by recourse to one of three methods: the generalised eigenvalue method mentioned above; the technique of Hewer (1971, 1973), which relies on the successive (quasi-linear in matrix terms) approximations of the Riccati equation by Lyapunov equations (the numerical solutions of which are well known); or the direct recursive solution of the matrix Riccati equation discussed below. None of these approaches requires inversion of $\boldsymbol{F}$.

[1] MATLAB®, The MathWorks Inc., Natick, MA, USA.

5.4.3 *Recursive Solution of the Discrete-Time Matrix Riccati Equation*

The direct method of obtaining the feedback gain vector $\boldsymbol{k}$ is given by Bryson and Ho (1969) and Borrie (1986). Here, the equations

$$\boldsymbol{k}^T(i) = (r + \boldsymbol{g}^T \boldsymbol{P}(i+1)\boldsymbol{g})^{-1}\boldsymbol{g}^T \boldsymbol{P}(i+1)\boldsymbol{F} \tag{5.89}$$

and

$$\boldsymbol{P}(i) = \boldsymbol{Q} + \boldsymbol{F}^T \boldsymbol{P}(i+1)\boldsymbol{F} - \boldsymbol{F}^T \boldsymbol{P}(i+1)\boldsymbol{g}\boldsymbol{k}^T(i) \tag{5.90}$$

are solved by a backwards recursion with the boundary conditions ($\boldsymbol{P}(N) = \boldsymbol{0}; \boldsymbol{k}(N) = \boldsymbol{0}$), and with the value of N chosen as large as necessary to reach the steady-state solutions (alternatively, a convergence criterion can be evaluated), as in Example 5.8.

Example 5.8 PIP-LQ Design for the NMSS Model with Five State Variables Consider once again the non-minimum phase oscillator (5.27) and associated PIP controller (5.38). The default PIP-LQ design (5.80) has the following weighting terms: $q_y = 1/2$, $q_u = 1/3$ and $q_e = 1$. Hence,

$$\boldsymbol{Q} = \mathrm{diag}(\,0.5 \quad 0.5 \quad 0.3333 \quad 0.3333 \quad 1\,); \quad r = 0.3333 \tag{5.91}$$

Consider $N = 20$ in (5.89) and (5.90) and set:

$$\boldsymbol{k}^T(20) = [\, f_0 \quad f_1 \quad g_1 \quad g_2 \quad -k_I \,] = [\,0 \quad 0 \quad 0 \quad 0 \quad 0\,]; \quad \boldsymbol{P}(20) = \boldsymbol{0} \tag{5.92}$$

where $\boldsymbol{0}$ is a 5×5 zero matrix. Solving {(5.89), (5.90)} with these initial conditions and the NMSS matrices (5.28), gives $\boldsymbol{k}^T(19) = [\,0 \quad 0 \quad 0 \quad 0 \quad 0\,]$ and $\boldsymbol{P}(19) = \boldsymbol{Q}$. A second iteration yields $\boldsymbol{k}^T(18) = [\,0 \quad 0 \quad 0 \quad 0 \quad 0\,]$ and

$$\boldsymbol{P}(18) = \begin{bmatrix} 5.335 & -2.55 & -2.55 & 5.1 & -1.7 \\ -2.55 & 2 & 1 & -3 & 1 \\ -2.55 & 1.5 & 2.1667 & -3 & 1 \\ 5.1 & -3 & -3 & 6.3333 & -2 \\ -1.7 & 1 & 1 & -2 & 2 \end{bmatrix} \tag{5.93}$$

The next iteration yields:

$$\boldsymbol{k}^T(17) = [\,-1.814 \quad 1.42 \quad 0.22 \quad -2.84 \quad 0.4\,] \tag{5.94}$$

Continuing in the same manner to $N = 0$, yields:

$$\boldsymbol{k}^T(0) = [\,1.462 \quad -1.886 \quad 1.603 \quad 3.772 \quad -0.259\,] \tag{5.95}$$

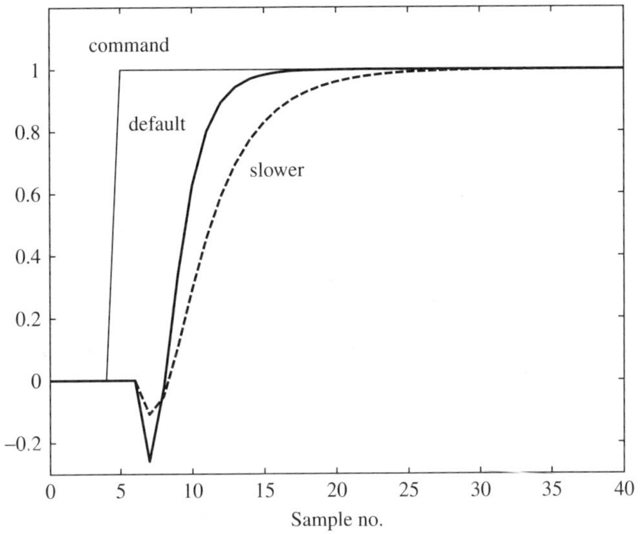

Figure 5.10 Closed-loop unit step response using the default (bold solid trace) and modified slower (dashed trace) PIP-LQ controllers of Example 5.8

Unfortunately, for this TF model, the default PIP controller with this gain vector is rather fast and the non-minimum phase behaviour becomes excessive, as shown by the 7th sample in the closed-loop step response shown in Figure 5.10.

However, using trial and error adjustment of only the integral-of-error weighting to change the speed of response, quickly yields improved closed-loop behaviour. For example, using $q_y = 1/2$, $q_u = 1/3$ and $q_e = 0.1$ yields:

$$\boldsymbol{k}^T(0) = [\,0.597 \quad -1.008 \quad 1.226 \quad 2.017 \quad -0.109\,] \tag{5.96}$$

With these settings, the closed-loop response shown as the dashed trace in Figure 5.10, is similar to that obtained for the PIP pole assignment controller (5.71) and the equivalent NMSS regulator (4.36). The latter two designs are both based on assigning the poles to 0.5 on the real axis of the complex z-plane. It is interesting to note, however, that evaluating equation (5.44) for this example reveals the following different closed-loop pole positions $\lambda_1 = \lambda_2 = 0$, $\lambda_3 = 0.748$ and $\lambda_{4,5} = 0.363 \pm 0.162j$ for the LQ design.

This LQ controller has a higher gain margin (Franklin *et al.* 2006, pp. 353), compared with the pole assignment algorithm (5.71), implying that the latter design is closer to the stability margin and potentially less robust. However, a detailed evaluation of these two particular examples using, for example, Monte Carlo Simulation, is left to the reader. It should be noted, of course, that the poles and weightings selected for this simulation example have been chosen for illustrative purposes, rather than because they met any particular design objectives (such as robustness).

5.5 Case Studies

We are now in a position to solve the pole assignment or LQ optimal control problem for any of the TF models considered so far in this book. For example, returning to the laboratory robot excavator considered in Example 5.2, it is possible to explain how the closed-loop pole positions were chosen. The PIP-LQ approach was utilised, with straightforward trial and error experimentation suggesting that $\boldsymbol{Q} = \text{diag}\,(1\ 20)$ and $r = 10$ provides adequate control of the bucket angle for a wide range of operating conditions (Shaban 2006). These weightings yield $f_0 = 6.30$ and $k_I = 1.17$, while the associated closed-loop poles are $p_{1,2} = 0.814 \pm 0.154j$, as stated previously in Example 5.2.

Using the same sampling rate of 0.1 s, similar first order TF models have been identified for the other degrees-of-freedom (i.e. hydraulically actuated joints) of the laboratory excavator. However, in some cases, the identified TF model includes two samples pure time delay, i.e. $n = 1$ and $m = \tau = 2$ and so the NMSS servomechanism representation is defined by the following third order state vector:

$$x(k) = [\, y(k) \quad u(k-1) \quad z(k)\,]^T \tag{5.97}$$

In this case, the PIP control law takes the form:

$$u(k) = -f_0 y(k) - g_1 u(k-1) + \frac{k_I}{1 - z^{-1}}(y_d(k) - y(k)) \tag{5.98}$$

For more information about the laboratory robot excavator, including PIP control system design for these other joints, see Shaban (2006) and Taylor *et al.* (2007). For PIP control of the full scale, commercial excavator and other examples of PIP control in the construction industry, see e.g. Gu *et al.* (2004) and Shaban *et al.* (2008).

Another previous case study, Example 2.8, considered the control of an axial fan positioned at the outlet of a forced ventilation test chamber. Here, the identified TF model (2.48) for air velocity is also based on $n = 1$ and $m = \tau = 2$. Hence, the PIP control law is obtained similarly using equation (5.97) and equation (5.98). For examples of PIP control of ventilation rate and other micro-climatic variables, such as temperature and CO_2, see Lees *et al.* (1996, 1998), Price *et al.* (1999) and Taylor *et al.* (2000, 2004a, b).

A higher order example is the NMSS servomechanism representation of the third order wind turbine model (2.9), introduced in Example 2.2. This has the following sixth order NMSS state vector:

$$x(k) = [\, y(k) \quad y(k-1) \quad y(k-2) \quad u(k-1) \quad u(k-2) \quad z(k)\,]^T \tag{5.99}$$

so that the PIP control law associated with equation (5.99) takes the form:

$$u(k) = -f_0 y(k) - f_1 y(k-1) - f_2 y(k-2) - g_1 u(k-1) - g_2 u(k-2) + \frac{k_I}{1 - z^{-1}} y_d(k) \tag{5.100}$$

In all these cases, the pole assignment or LQ design procedure follows in exactly the same way as for the previous examples.

Finally, in order to illustrate the practical utility of the PIP approach, Example 5.9 is based on a traditionally rather difficult control problem, namely the control of a real system whose model is an integrator with a long pure time delay.

Example 5.9 PIP-LQ Control of CO_2 in Carbon-12 Tracer Experiments Figure 5.11 and Figure 5.12 illustrate the response of a PIP-LQ controller applied to the carbon dioxide ($^{12}CO_2$) tracer system described by Minchin *et al.* (1994). These experiments, in which the radioactive isotope carbon-12 is employed to study the uptake and transport of carbon in plants, require the tight regulation and control of specified CO_2 levels in the leaf chamber. This system is quite difficult to control because of the rather limited pulse-width-modulated input signal, where the control input represents the duration of time during each 5 s sample that the CO_2 gas is injected into the system.

Furthermore, the system behaves essentially as an integrator with a 4-sample time delay, i.e.

$$y(k) = \frac{b_4 z^{-4}}{1 + a_1 z^{-1}} u(k) = \frac{0.0045 z^{-4}}{1 - z^{-1}} u(k) \tag{5.101}$$

where $y(k)$ is the CO_2 concentration (ppm) and $u(k)$ is the CO_2 injection time (s). The TF model (5.101) has been identified from experimental data using the statistical methods described in Chapter 8 (Taylor *et al.* 1996).

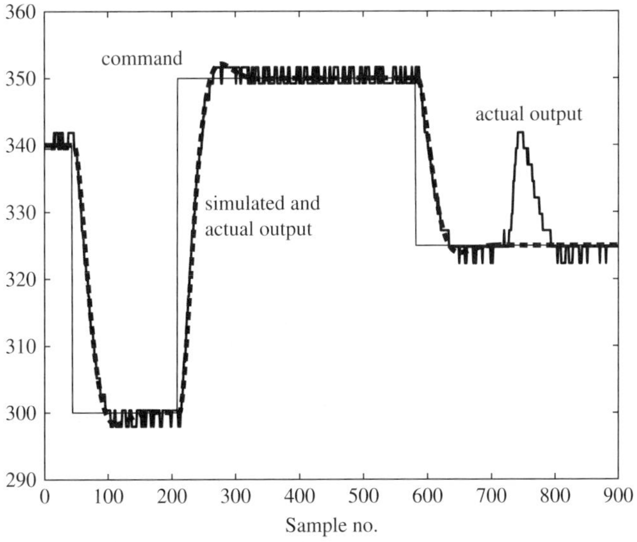

Figure 5.11 Closed-loop response using the PIP controller of Example 5.9, showing the command input (sequence of steps of varying magnitude), together with the simulated (dashed trace) and measured CO_2 concentration (subjected to high frequency chatter). The plant was shaded at the 730th sample

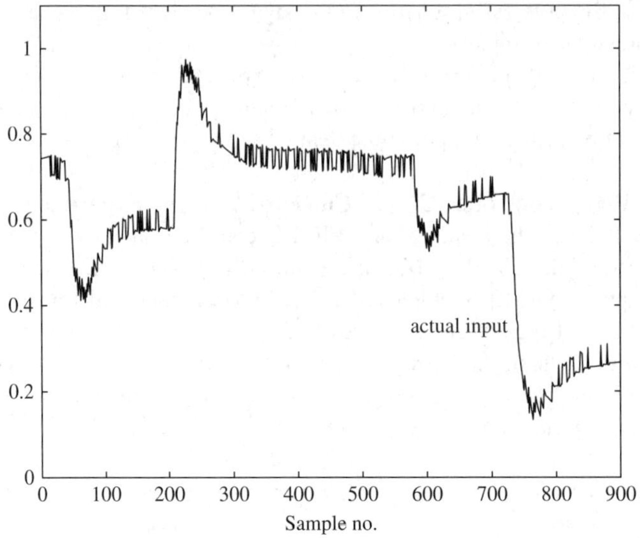

Figure 5.12 Control input signal (seconds of CO_2 injection) associated with Figure 5.11

The NMSS state vector for (5.101) takes the following form:

$$\boldsymbol{x}(k) = [y(k) \quad u(k-1) \quad u(k-2) \quad u(k-3) \quad z(k)]^T \tag{5.102}$$

and the associated NMSS model (5.13) is defined in the usual manner, here with $a_1 = -1$ and $b_4 = 0.0045$. Note that the relatively long time delay of 4 samples (20 s) is automatically incorporated into the NMSS description by setting $b_1 = b_2 = b_3 = 0$ as usual.

Utilising the default PIP-LQ settings (5.80) and solving {(5.89), (5.90)} yields:

$$\boldsymbol{k}^T(0) = [\,22.9987 \quad 0.0949 \quad 0.0992 \quad 0.1035 \quad -0.9536\,] \tag{5.103}$$

The PIP control system is implemented as in equation (5.32).

As can be seen in Figure 5.11 and Figure 5.12, the practical performance of this PIP controller is very good, with the CO_2 concentration kept within 1 or 2 ppm of the set point at steady state. Figure 5.11 also demonstrates the robustness of the controller to changing dynamics in the system: at the 730th sample, the CO_2 level starts to rise following the covering of the leaf chamber with a neutral coloured light filter (this has no preferential absorption of particular wavelengths). This filter has the effect of approximately halving the rate of CO_2 uptake. Despite this major output disturbance, the PIP controller soon returns the concentration to the set point, where the regulatory performance remains good. This demonstrates the practical importance of the integral-of-error state feedback in the PIP controller.

Note that the output signal is quantised due to limitations in the measurement device, hence the input signal in Figure 5.12 also appears noisy. However, it is clear that this does not disrupt the performance of the PIP control system. Indeed, even though the open-loop system is marginally stable (with a pole at unity) and the output measurement is quantised, the practical implementation of this PIP-LQ design is very robust.

5.6 Concluding Remarks

This chapter has introduced the PIP controller, in which NMSS models are formulated so that full SVF control can be implemented directly from the measured input and output signals of the controlled process. We have seen that the PIP controller can be interpreted as a logical extension of conventional PI and PID controllers, with additional dynamic feedback and input compensators introduced automatically when the process has second order or higher dynamics; or pure time delays greater than one sampling interval. In contrast to conventional PI/PID controllers, PIP design has numerous advantages: in particular, its structure exploits the power of SVF methods, where the vagaries of manual tuning are replaced by either pole assignment or LQ design.

Such SVF techniques become particularly important as we move to multivariable systems in Chapter 7. However, even in the simplest SISO case discussed in the present chapter, some of the main advantages of the NMSS approach have become apparent:

- As discussed in Chapter 4, the NMSS form is a particularly obvious and rather natural way of representing the general TF model (5.12) in state space terms.
- In the full servomechanism formulation discussed in the present chapter, the introduction of an integral-of-error state variable provides a logical approach to ensure Type 1 servomechanism performance, i.e. steady state tracking of the set point. Type 2 servomechanism could also be obtained, if required, by introducing a double integration of error state.
- Since all the states are available, either by direct measurement or access to the past input and output values stored by a digital computer or micro-controller, no observer is required to implement the controller. The approach often yields a control law of similar complexity to that of a conventional PID controller. It is clear that full state feedback is always achieved, so neither Kalman filtering (Kalman 1960b) nor Loop Transfer Recovery (Bitmead *et al.* 1990) is required in deterministic PIP design. In this regard, the incorporation of an observer or Kalman Filter is not only more complicated, it can also reduce the robustness of any design.
- Not only is the observer a redundant complexity but, equally importantly, the cost function weighting terms in the PIP case apply only to the present and past values of the output variable and past values of the input variables, rather than the far less intuitive combinations of variables required in the minimal state case. Indeed, the fact that PIP control is straightforward to implement and has intuitive tuning parameters (i.e. W_y, W_u and W_e) is one of its key practical strengths and attractions.

Although one of the original motivations for introducing the PIP controller was simply to avoid the need for a state observer with its inherent problems, it was quickly realised that the non-minimal state vector can form the basis for a wide range of extensions to the basic algorithm. These include an expanded form of GPC, feed-forward control for improved disturbance rejection, stochastic optimal control and risk-sensitive (exponential of quadratic and H_∞) optimal control, as discussed in Chapter 6; and multivariable control design, as considered in Chapter 7.

References

Åström, K.J. and Wittenmark, B. (1984) *Computer-Controlled Systems Theory and Design*, Prentice Hall, Englewood Cliffs, NJ.

Bitmead, R.R., Gevers, M. and Wertz, V. (1990) *Adaptive Optimal Control: The Thinking Man's GPC*, Prentice Hall, Harlow.

Borrie, J.A. (1986) *Modern Control Systems: A Manual of Design Methods*, Prentice Hall International, Englewood Cliffs, NJ.

Bryson, A.E. Jr and Ho, Y.C. (1969) *Applied Optimal Control: Optimization, Estimation, and Control*, Blaisdell, Waltham, MA.

Clarke, D.W., Mohtadi, C. and Tuffs, P.S. (1987) Generalised Predictive Control, *Automatica*, **23**, Part I The basic algorithm, pp. 137–148, Part II Extensions and interpretations, pp. 149–160.

Dorf, R.C. and Bishop, R.H. (2008) *Modern Control Systems*, Eleventh Edition, Pearson Prentice Hall, Upper Saddle River, NJ.

Franklin, G.F., Powell, J.D. and Emami-Naeini, A. (2006) *Feedback Control Of Dynamic Systems*, Fifth Edition, Pearson Prentice Hall, Upper Saddle River, NJ.

Gopinath, B. (1971) On the control of linear multiple input-output systems, *Bell System Technical Journal*, **50**, pp. 1063–1081.

Grace, A., Laub, A.J., Little, J.N. and Thompson, C. (1990) *Matlab: Control System Toolbox Users Guide*, The MathWorks Inc., Natick, MA.

Gu, J. , Taylor, C.J., and Seward, D. (2004) Proportional-Integral-Plus control of an intelligent excavator, *Journal of Computer-Aided Civil and Infrastructure Engineering*, **19**, pp. 16–27.

Hewer, G.A. (1971) An iterative technique for the computation of the steady-state gains for the discrete optimal regulator. *IEEE Transactions on Automatic Control*, **16**, pp. 382–384.

Hewer, G.A. (1973) Analysis of a discrete matrix Riccati equation of a linear control and Kalman filtering, *Journal of Mathematical Analysis and Applications*, **42**, pp. 226–236.

Kailath, T. (1980) *Linear Systems*, Prentice Hall, Englewood Cliffs, NJ.

Kalman, R.E. (1960a) Contributions to the theory of optimal control, *Boletin Sociedad Matemática Mexicana*, **5**, pp. 102–119 (http://garfield.library.upenn.edu/classics1979/A1979HE37100001.pdf).

Kalman, R.E. (1960b) A new approach to linear filtering and prediction problems, *ASME Transactions Journal of Basic Engineering*, **82**, pp. 34–45.

Kucera, V. (1979) *Discrete Linear Control: The Polynomial Equation Approach*. John Wiley & Sons, Ltd, Chichester.

Kuo, B.C. (1997) *Digital Control Systems*, Second Edition, The Oxford Series in Electrical and Computer Engineering, Oxford University Press, New York.

Lees, M.J., Taylor, J., Chotai, A., Young, P.C. and Chalabi, Z.S. (1996) Design and implementation of a Proportional-Integral-Plus (PIP) control system for temperature, humidity, and carbon dioxide in a glasshouse, *Acta Horticulturae (ISHS)*, **406**, pp. 115–124.

Lees, M.J., Taylor, C.J., Young, P.C. and Chotai, A. (1998) Modelling and PIP control design for open top chambers, *Control Engineering Practice*, **6**, pp. 1209–1216.

Luenberger, D.G. (1967) Canonical forms for linear multivariable systems, *IEEE Transactions on Automatic Control*, **12**, pp. 290–293.

Minchin, P.E.H., Farrar, J.F. and Thorpe, M.R. (1994) Partitioning of carbon in split root systems of barley: effect of temperature of the root, *Journal of Experimental Botany*, **45**, pp. 1103–1109.

Munro, N. (1979) Pole-assignment, *IEE Proceedings: Control Theory and Applications*, **126**, pp. 549–554.

Pappas, T., Laub, A.J. and Sandell, J.N.R. (1980) On the numerical solution of the discrete-time algebraic Riccati equation, *IEEE Transactions on Automatic Control*, **25**, pp. 631–641.

Price, L., Young, P., Berckmans, D., Janssens, K. and Taylor, J. (1999) Data-Based Mechanistic Modelling (DBM) and control of mass and energy transfer in agricultural buildings, *Annual Reviews in Control*, **23**, pp. 71–82.

Rosenthal, J. and Wang, X.A. (1996) Output feedback pole placement with dynamic compensators, *IEEE Transactions on Automatic Control*, **41**, pp. 830–842.

Shaban, E.M. (2006) *Nonlinear Control for Construction Robots using State Dependent Parameter Models*, PhD thesis, Engineering Department, Lancaster University.

Shaban, E.M., Ako, S., Taylor, C.J. and Seward, D.W. (2008) Development of an automated verticality alignment system for a vibro-lance, *Automation in Construction*, **17**, pp. 645–655.

Taylor, C.J., Lees, M.J., Young, P.C. and Minchin, P.E.H. (1996) True digital control of carbon dioxide in agricultural crop growth experiments, *International Federation of Automatic Control 13th Triennial World Congress* (IFAC-96), 30 June–5 July, San Francisco, USA, Elsevier, Vol. B, pp. 405–410.

Taylor, C.J., Leigh, P.A., Chotai, A., Young, P.C., Vranken, E. and Berckmans, D. (2004a) Cost effective combined axial fan and throttling valve control of ventilation rate, *IEE Proceedings: Control Theory and Applications*, **151**, pp. 577–584.

Taylor, C.J., Leigh, P., Price, L., Young, P.C., Berckmans, D., Janssens, K., Vranken, E. and Gevers, R. (2004b) Proportional-Integral-Plus (PIP) control of ventilation rate in agricultural buildings, *Control Engineering Practice*, **12**, pp. 225–233.

Taylor, C.J., Shaban, E.M., Stables, M.A. and Ako, S. (2007) Proportional-Integral-Plus (PIP) control applications of state dependent parameter models, *IMECHE Proceedings: Systems and Control Engineering*, **221**, pp. 1019–1032.

Taylor, C.J., Young, P.C., Chotai, A., McLeod, A.R. and Glasock, A.R. (2000) Modelling and proportional-integral-plus control design for free air carbon dioxide enrichment systems, *Journal of Agricultural Engineering Research*, **75**, pp. 365–374.

Wang, C.L. and Young, P.C. (1988) Direct digital control by input–output, state variable feedback: Theoretical background, *International Journal of Control*, **47**, pp. 97–109.

Willems, J.C. (1971) Least squares stationary optimal control and the algebraic Riccati equation, *IEEE Transactions on Automatic Control*, **16**, pp. 621–634.

Wonham, W.M. (1968) On a matrix Riccati equation of stochastic control, *SIAM Journal on Control*, **6**, pp. 681–697.

Young, P.C. (1972) *Lectures on Algebraic Methods of Control System Synthesis:Lecture 2*, Department of Engineering, University of Cambridge Report Number CUED/B–CONTROL/TR24.

Young, P.C., Behzadi, M.A., Wang, C.L. and Chotai, A. (1987) Direct Digital and Adaptive Control by input–output, state variable feedback pole assignment, *International Journal of Control*, **46**, pp. 1867–1881.

Young, P.C. and Willems, J.C. (1972) An approach to the linear multivariable servomechanism problem, *International Journal of Control*, **15**, pp. 961–979.

Zak, S.H. (2003) *Systems and Control*, The Oxford Series in Electrical and Computer Engineering, Oxford University Press, New York.

6

Control Structures and Interpretations

In previous chapters, we have examined several *State Variable Feedback* (SVF) control systems, culminating in Chapter 5 with the introduction of the univariate *Proportional-Integral-Plus* (PIP) controller. Apart from the integral-of-error state variable, which is introduced specifically to ensure Type 1 servomechanism performance, we have seen that the *Non-Minimal State Space* (NMSS) form provides one of the most natural ways to define the *Transfer Function* (TF) model in state space terms. However, there is no reason to limit the design to this basic form. Indeed, one advantage of the NMSS model is that the state vector is readily extended to account for the availability of additional measured or estimated information, and these additional states can be utilised to develop a more sophisticated Generalised PIP controller.

Furthermore, there are a number of different methods available for implementing even the basic PIP control law, in addition to the straightforward SVF form of equation (5.30), i.e. $\boldsymbol{u}(k) = -\boldsymbol{k}^T \boldsymbol{x}(k)$. These are, of course, all based on equation (5.30) but the control input can be generated using different control *structures* that can have various advantages in practical applications. The present chapter employs both simulated and real examples to discuss the robustness and disturbance response characteristics of the main PIP control structures that emerge from the analysis. These include the feedback and forward path forms (section 6.1), incremental forms for practical implementation (section 6.2), links to the Smith Predictor for time-delay systems (section 6.3), stochastic optimal and risk sensitive design (section 6.4), feed-forward control (section 6.5) and predictive control (section 6.6).

6.1 Feedback and Forward Path PIP Control Structures

The structure represented by Figure 5.4 is referred to as PIP control in standard *feedback* form. When considering other structures, it should be noted that the various forms fall into two broad categories: in the first place, there are those structures which, although they may illustrate certain features of the controller in a different manner to that normally used, nevertheless remain identical to the nominal form in practice; secondly, and more importantly, there are

True Digital Control: Statistical Modelling and Non-Minimal State Space Design, First Edition.
C. James Taylor, Peter C. Young and Arun Chotai.

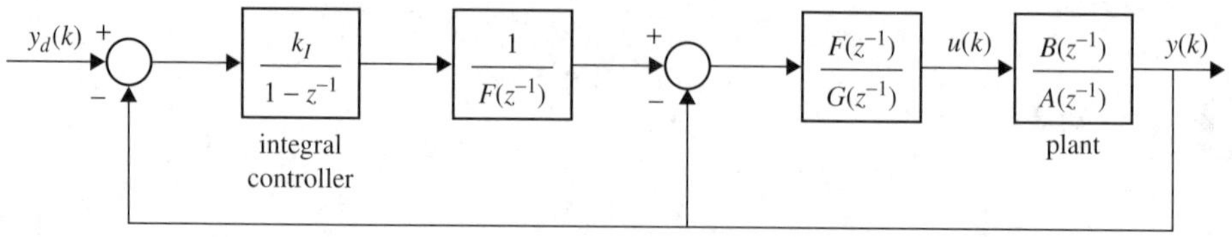

Figure 6.1 Alternative representation of the feedback PIP control system (equivalent to Figure 5.4)

those structures which have *different closed-loop characteristics* to the original form: these differences only become apparent when the system is subjected to disturbances or model mismatch. An example of the former of the two categories is illustrated in Figure 6.1 (Young *et al.* 1987). This alternative structure results when the two control polynomials are combined into a single TF $F(z^{-1})/G(z^{-1})$, which can be positioned in either the feedback or forward paths. Figure 6.1 illustrates the latter case, where the additional forward path filter $1/F(z^{-1})$, is required in order to ensure complete equivalence of the closed-loop system.

In Figure 6.1, the general discrete-time TF model:

$$y(k) = \frac{B(z^{-1})}{A(z^{-1})}u(k) = \frac{b_1 z^{-1} + \cdots + b_m z^{-m}}{1 + a_1 z^{-1} + \cdots + a_n z^{-n}}u(k) \tag{6.1}$$

PIP control polynomials:

$$\left.\begin{aligned} F(z^{-1}) &= f_0 + F_1(z^{-1}) = f_0 + f_1 z^{-1} + \cdots + f_{n-1} z^{-n+1} \\ G(z^{-1}) &= 1 + G_1(z^{-1}) = 1 + g_1 z^{-1} + \cdots + g_{m-1} z^{-m+1} \end{aligned}\right\} \tag{6.2}$$

and integral gain k_I are all defined in Chapter 5 but are repeated here for convenience. Similarly, the closed-loop TF obtained from Figure 5.4 or Figure 6.1 is:

$$y(k) = \frac{k_I B(z^{-1})}{\Delta\left(G(z^{-1})A(z^{-1}) + F(z^{-1})B(z^{-1})\right) + k_I B(z^{-1})} y_d(k) \tag{6.3}$$

where $\Delta = 1 - z^{-1}$ denotes the difference operator. Note that Δ is utilised frequently in this chapter and that $1/\Delta$ represents an integrator, as shown in Chapter 2 (Example 2.5).

Although Figure 5.4 and Figure 6.1 appear to show two quite different structures, in practice both forms are analytically identical, and yield the same response even in the presence of model mismatch or disturbance inputs to the system. However, Figure 6.1 does have important ramifications for the case of PIP controllers implemented using other operators. For example, the δ-operator PIP algorithm (discussed briefly in Chapter 9) involves discrete derivatives of the input and output signals and is not necessarily realisable in practice, hence alternative structures (similar to Figure 6.1) are required in this case (see Chapter 9 and Young *et al.* 1998)[1].

[1] A continuous-time equivalent of δ-operator PIP control is also possible, where full time derivatives have to be avoided in a similar manner (see e.g. Taylor *et al.* 1998b, 2012; Gawthrop *et al.* 2007). Although such a continuous-time form of PIP control is not considered in this book because it does not fall within our definition of True Digital Control, its implementation follows the same approach used in the δ-operator case.

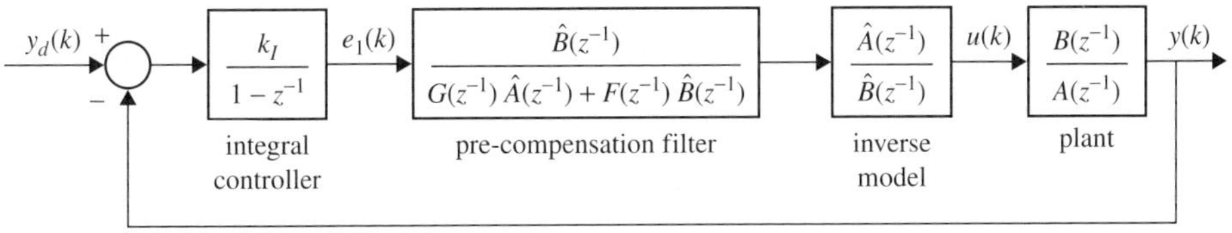

Figure 6.2 Forward path PIP control with inverse model

6.1.1 Proportional-Integral-Plus Control in Forward Path Form

It is straightforward to eliminate the inner loop of Figure 5.4 completely, to form a single forward path TF, in which the control algorithm is based on a unity feedback of the output variable. To derive this new structure, consider the inner loop of the feedback PIP system illustrated in Figure 5.4, consisting only of the plant $B(z^{-1})/A(z^{-1})$, feedback filter $F(z^{-1})$ and input filter $G(z^{-1})$. Using the rules of block diagram analysis, this component of the closed-loop system may be reduced to a single TF representation, labelled the 'pre-compensation filter' in Figure 6.2. In practice, however, the system is always distinct from the controller and its behaviour can be determined only from the estimated TF model. Hence, the inverse of the TF model is also required in order to cancel out the plant dynamics. As in earlier chapters, the circumflex notation is introduced to distinguish the estimated TF model $\hat{B}(z^{-1})/\hat{A}(z^{-1})$ from the nominal system or plant.

Note that a non-minimum phase system does not cause any problems of internal instability here, despite the inverse model, since the $\hat{B}(z^{-1})$ polynomial in the denominator of the inverse model may be analytically cancelled before practical implementation. In fact, noting the explicit cancellation of the estimated $\hat{B}(z^{-1})$ polynomials in Figure 6.2, the block diagram can be simplified to yield a single pre-compensation filter, as shown by Figure 6.3. This new structure is referred to as the *forward path* PIP controller because of the unity feedback, with a single pre-compensation filter operating on the error signal between the command input and the measured system output (Taylor *et al.* 1998a).

An equivalent forward path arrangement is illustrated in Figure 6.4 (Lees 1996), which appears superficially similar to the feedback PIP form of Figure 5.4, with one major difference: the inner feedback loop is based on the model output rather than the measured output $y(k)$. In Figure 6.4, we see how the estimated TF model acts as a source of information on the output state variables, but that the actual measured output from the system is also employed to ensure Type 1 servomechanism behaviour.

One advantage of using an 'internal model' of this type (see later comments), is that measurement noise does not pass through the feedback filter $F(z^{-1})$. Instead, the noisy output variable is only utilised by the integral component of the control algorithm. Such integral action

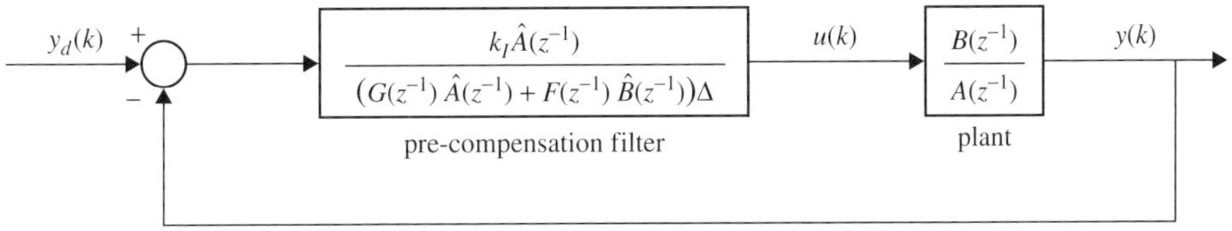

Figure 6.3 Forward path PIP control in reduced form

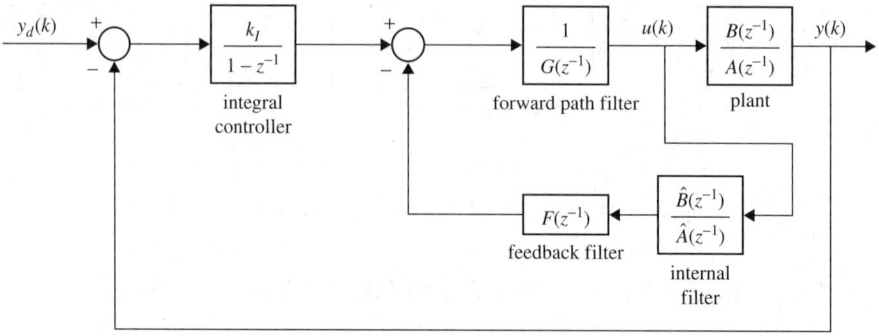

Figure 6.4 Forward path PIP control with feedback of an internal model

is effectively a 'low pass' filter, i.e. it acts to attenuate any high frequency noise entering via the output measurement. Therefore, the actuator signal from the forward path implementation is generally smoother than that produced by an equivalent feedback form (see Example 6.3).

Yet another way of representing the same controller is illustrated in Figure 6.5. In this case, the denominator of the TF labelled 'pre-compensation filter', is the same as the denominator of the desired closed-loop system (6.3), whilst the feedback loop consists of an error signal, i.e. the difference between the measured plant output and the response of the internal model. Figure 6.5 is in the form of an *Internal Model Controller* (IMC) as described, for example, by Garcia and Morari (1982) and Tsypkin and Holmberg (1995).

6.1.2 Closed-loop TF for Forward Path PIP Control

The control algorithm associated with any of the above block diagrams, but most obviously obtained from Figure 6.3, is conveniently expressed in the following TF form:

$$u(k) = \frac{k_I \hat{A}(z^{-1})}{\left(G(z^{-1})\hat{A}(z^{-1}) + F(z^{-1})\hat{B}(z^{-1})\right)\Delta}(y_d(k) - y(k)) \tag{6.4}$$

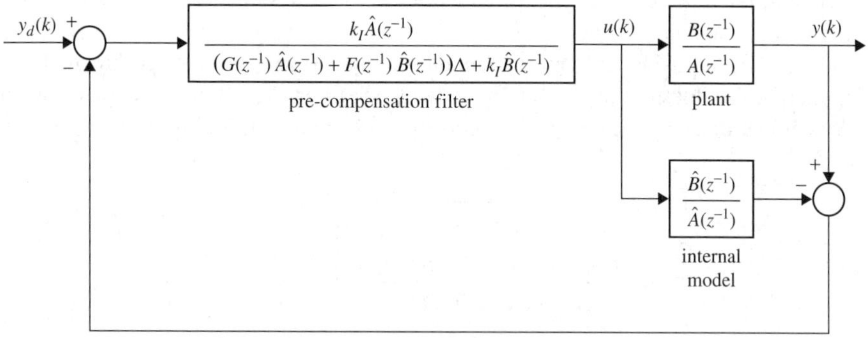

Figure 6.5 Forward path PIP control represented as an internal model controller

Retaining the distinction between the model and plant polynomials, the closed-loop TF for the forward path PIP controller is as follows:

$$y(k) = \frac{k_I B(z^{-1})\hat{A}(z^{-1})}{\left(G(z^{-1})\hat{A}(z^{-1})A(z^{-1}) + F(z^{-1})\hat{B}(z^{-1})A(z^{-1})\right)\Delta + k_I B(z^{-1})\hat{A}(z^{-1})} y_d(k) \quad (6.5)$$

In contrast to the feedback PIP closed-loop TF (6.3), equation (6.5) includes the estimated model polynomials. However, if we now assume that there is no model mismatch, i.e. $\hat{B}(z^{-1}) = B(z^{-1})$ and $\hat{A}(z^{-1}) = A(z^{-1})$, then equation (6.5) reduces to the nominal closed-loop TF (6.3). This has two main consequences: in the first place, the control gains of the forward path form are always the same as the standard feedback structure; and, secondly, the response of the two forms is identical *in this zero mismatch case*.

Finally, at steady state, $\Delta \rightarrow 0$ and the steady-state gain of equation (6.5) is unity, hence Type 1 servomechanism performance is normally ensured *even in the presence of model mismatch* (assuming closed-loop stability). However, there is one special exception to this result, which is discussed later in Example 6.3.

6.1.3 Closed-loop Behaviour and Robustness

Although a number of block diagram representations of the PIP controller have now been considered, it should be stressed that there are only two forms that have different closed-loop characteristics. In the first instance, the control structures in Figure 5.3, Figure 5.4 and Figure 6.1 are all equivalent and are considered as the standard *feedback* PIP structure. Secondly, Figure 6.2, Figure 6.3, Figure 6.4 and Figure 6.5 are similarly identical and are referred to as the *forward path* PIP structure. As we shall see, the choice of control structure has important consequences, both for the robustness of the final design to parametric uncertainty and for the disturbance rejection characteristics.

Consider, for example, the imposition of a disturbance $v(k)$ added directly to the output variable and represented as follows:

$$y(k) = \frac{B(z^{-1})}{A(z^{-1})} u(k) + v(k) \quad (6.6)$$

In many control design applications, such load disturbances are an important factor. In the control of a greenhouse environment, for example, there are a number of disturbance inputs to the system which may accumulate to a non-zero mean value (Young *et al.* 1994). Examples include solar radiation, external temperature, vents and so on. Similarly, in industrial processes it is common to encounter random steps, occurring at random times (such as changes in material quality).

In this case, the closed-loop TF for feedback PIP control becomes:

$$\begin{aligned} y(k) = {} & \frac{k_I B(z^{-1})}{\Delta\left(G(z^{-1})A(z^{-1}) + F(z^{-1})B(z^{-1})\right) + k_I B(z^{-1})} y_d(k) \\ & + \frac{\Delta G(z^{-1})A(z^{-1})}{\Delta\left(G(z^{-1})A(z^{-1}) + F(z^{-1})B(z^{-1})\right) + k_I B(z^{-1})} v(k) \end{aligned} \quad (6.7)$$

while the closed-loop TF for forward path PIP control is:

$$y(k) = \frac{k_I B(z^{-1})}{\Delta \left(G(z^{-1})A(z^{-1}) + F(z^{-1})B(z^{-1})\right) + k_I B(z^{-1})} y_d(k) + \frac{\Delta \left(G(z^{-1})A(z^{-1}) + F(z^{-1})B(z^{-1})\right)}{\Delta \left(G(z^{-1})A(z^{-1}) + F(z^{-1})B(z^{-1})\right) + k_I B(z^{-1})} v(k) \tag{6.8}$$

Equation (6.7) and equation (6.8) are determined using either straightforward (although time consuming) algebra or, more conveniently, by block diagram analysis. Further algebraic manipulation of equation (6.8) shows that:

$$y(k) = \frac{k_I B(z^{-1})}{\Delta \left(G(z^{-1})A(z^{-1}) + F(z^{-1})B(z^{-1})\right) + k_I B(z^{-1})} y_d(k) + \left(1 - \frac{k_I B(z^{-1})}{\Delta \left(G(z^{-1})A(z^{-1}) + F(z^{-1})B(z^{-1})\right) + k_I B(z^{-1})}\right) v(k) \tag{6.9}$$

Here, we see that the disturbance response in the forward path case is equal to one minus the desired closed-loop TF (6.3), thus ensuring similar disturbance response dynamics to those specified by the designer in relation to the command input response. However, in the case of the feedback PIP structure, the situation is more complex since past noisy values of the disturbed output are involved in the control signal synthesis because of the feedback filter $F(z^{-1})$. In practice, this can result in a less desirable disturbance response to that obtained with the forward path form.

Although the forward path control structure is more resilient to disturbances, the explicit cancellation of the system dynamics with an inverse model ensures that the forward path PIP controller is generally more sensitive to modelling errors than the feedback form. This is especially noticeable for marginally stable or unstable systems. These issues are illustrated in Example 6.1.

Example 6.1 Simulation Response for Feedback and Forward Path PIP Control Consider again the following non-minimum phase oscillator from earlier chapters:

$$y(k) = \frac{-z^{-2} + 2.0z^{-3}}{1 - 1.7z^{-1} + 1.0z^{-2}} u(k) \tag{6.10}$$

The present analysis utilises the pole assignment PIP controller developed in Chapter 5 (Example 5.4, Example 5.5 and Example 5.6), i.e. based on placing four closed-loop poles at 0.5 on the real axis of the complex z-plane and the fifth at the origin. This approach yields the following gain vector, for implementation using either feedback or forward path PIP control:

$$\mathbf{k}^T = [\, f_o \quad f_1 \quad g_1 \quad g_2 \quad -k_I \,] = [\, 0.176 \quad -0.464 \quad 0.700 \quad 0.928 \quad -0.063 \,] \tag{6.11}$$

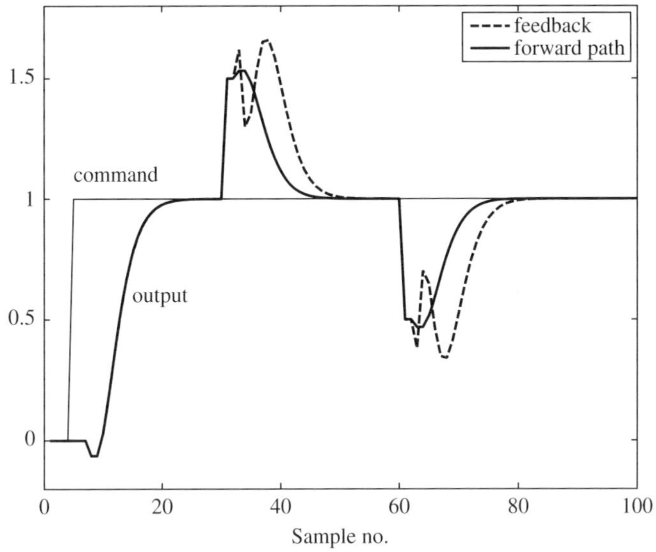

Figure 6.6 Closed-loop response for a unit step in the command level, with the feedback (dashed trace) and forward path (bold solid trace) PIP controllers of Example 6.1, when a load disturbance of magnitude 0.5 is applied at sample 30 and removed at sample 60

Figure 6.6 and Figure 6.7 show the closed-loop response for a unit step in the command level, with a constant load disturbance from samples 30 through to 60 (and zero otherwise). The nominal response for the feedback and forward path forms are identical, as shown by the first 30 samples of the simulation. However, following each step change in the disturbance, the forward path form of the PIP controller yields a smoother response than the feedback form, so requiring much reduced actuator movement (Figure 6.7) and, in practical situations, resulting in less actuator wear.

To assess the robustness of these designs to parametric uncertainty, we employ *Monte Carlo Simulation* (MCS; see Example 4.10 for another example of this approach). Here, all the parameters are considered uncertain, with the uncertainty defined by a diagonal covariance matrix $\sigma^2 \boldsymbol{I}$ with $\sigma^2 = 0.00001$ and $\boldsymbol{I}$ an identity matrix. Figure 6.8 shows the response to a unit step command input for 100 MCSs. In the case of forward path PIP control, most of responses are oscillatory and many of the realisations are, in fact, unstable, as confirmed by the closed-loop poles. These closed-loop poles for each realisation or *stochastic root loci* are plotted on the complex z-plane, which also shows the unit circle marking the stability boundary. Figure 6.8 clearly demonstrates the expected superior robustness of the feedback PIP structure when it comes to uncertainties in the estimated system dynamics.

Finally, in considering these results, it should be emphasised that, not only is the system in this case quite difficult to control [the TF model (6.10) is marginally stable], but the parametric uncertainty has been artificially scaled to facilitate the comparison. Both forms of PIP control are usually quite robust to more realistic modelling uncertainties, as demonstrated in the various practical examples considered elsewhere in this book.

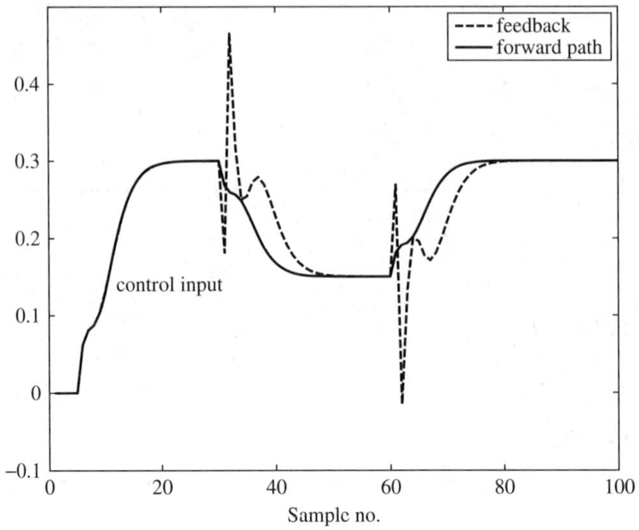

Figure 6.7 Control input signals associated with Figure 6.6

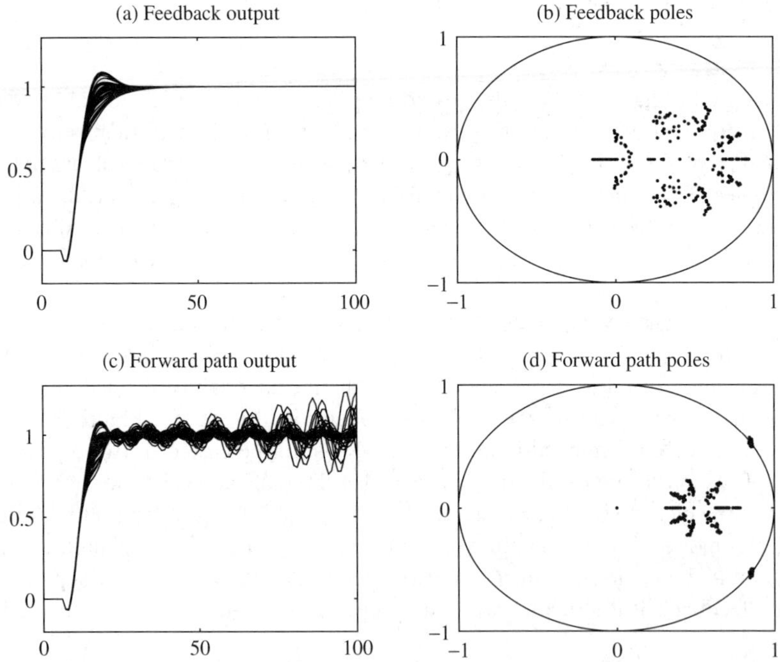

Figure 6.8 Closed-loop response for a unit step in the command level with 100 realisations, showing the output (a and c) and closed-loop poles (b and d), for the feedback (a and b) and forward path (c and d) PIP controllers of Example 6.1

6.2 Incremental Forms for Practical Implementation

The block diagrams considered above are useful for analysis and design. However, in most practical circumstances, PIP controllers should be implemented using an appropriate feedback or forward path *incremental* form, so as to avoid problems of integral 'wind-up' when the controller is subject to constraints on the actuator signal. Without the correction discussed below, a prolonged period when the input is at its practical limit and unable to keep the output at the set point, causes the integrated error signal to build up, resulting in ever larger input signals that are not achievable in practice. In such instances, when the saturation finally ends, the controller could take several samples to recover and might even drive the system into instability.

6.2.1 *Incremental Form for Feedback PIP Control*

Substituting the integral-of-error state (5.4) into the PIP control law (5.32) yields:

$$\begin{aligned} u(k) = & -f_0 y(k) - f_1 y(k-1) - \cdots - f_{n-1} y(k-n+1) \\ & - g_1 u(k-1) - \cdots - g_{m-1} u(k-m+1) + \frac{k_I}{\Delta}(y_d(k) - y(k)) \end{aligned} \tag{6.12}$$

Multiplying through by Δ yields:

$$\begin{aligned} \Delta u(k) = & -f_0 \Delta y(k) - f_1 \Delta y(k-1) - \cdots - f_{n-1} \Delta y(k-n+1) \\ & - g_1 \Delta u(k-1) - \cdots - g_{m-1} \Delta u(k-m+1) + k_I (y_d(k) - y(k)) \end{aligned} \tag{6.13}$$

Noting that $\Delta u(k) = u(k) - u(k-1)$, we obtain the following difference equation:

$$\begin{aligned} u(k) = & \, u(k-1) - f_0 (y(k) - y(k-1)) - f_1 (y(k-1) - y(k-2)) \\ & \cdots - f_{n-1} (y(k-n+1) - y(k-n)) - g_1 (u(k-1) - u(k-2)) \\ & \cdots - g_{m-1} (u(k-m+1) - u(k-m)) + k_I (y_d(k) - y(k)) \end{aligned} \tag{6.14}$$

The PIP feedback control law given by equation (6.14) is a linear function of the system variables, together with their delayed values and control gains. It is straightforward to implement such an algorithm using standard software code. Furthermore, we can specify the following limits for the input:

$$\{\overline{u}, \underline{u}\} \rightarrow \begin{cases} \text{if} \quad u(k) \geq \overline{u} \quad \text{then} \quad u(k) = \overline{u} \\ \text{if} \quad u(k) \leq \underline{u} \quad \text{then} \quad u(k) = \underline{u} \end{cases} \tag{6.15}$$

with $\overline{u}$ and $\underline{u}$ representing the maximum and minimum realisable input signals. For brevity in the block diagrams that follow, the correction (6.15) is denoted $\{\overline{u}, \underline{u}\}$. In this manner, the delayed control input variables utilised in equation (6.14) remain within these practical limits, hence avoiding integral 'wind-up' problems.

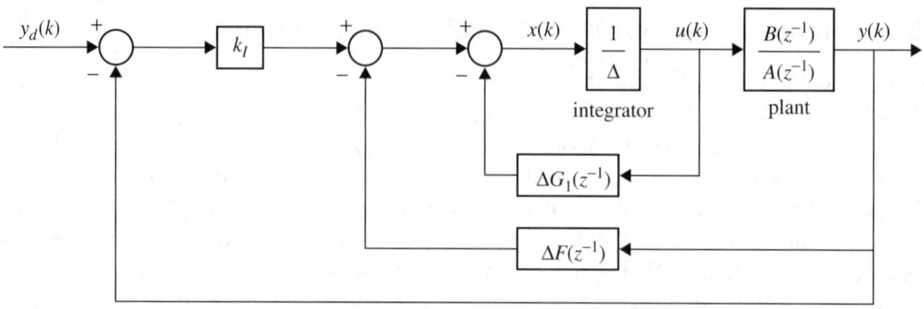

Figure 6.9 Block diagram representation of equation (6.16)

Now consider the block diagram representation of equation (6.14) and equation (6.15). Substituting the control polynomials $F(z^{-1})$ and $G_1(z^{-1}) = G(z^{-1}) - 1 = g_1 z^{-1} + \cdots + g_{m-1} z^{-m+1}$ into equation (6.13), and dividing through by Δ yields:

$$u(k) = \frac{1}{\Delta}\left[-F(z^{-1})\Delta y(k) - G_1(z^{-1})\Delta u(k) + k_I\,(y_d(k) - y(k))\right] \tag{6.16}$$

Defining $x(k) = -F(z^{-1})\Delta y(k) - G_1(z^{-1})\Delta u(k) + k_I\,(y_d(k) - y(k))$, it is clear that $u(k) = x(k)/\Delta$, as illustrated by Figure 6.9. The significance of this arrangement is that the integrator can be expanded subsequently to accommodate the correction (6.15), as illustrated by Figure 6.10. Noting that $x(k) = \Delta u(k)$, Figure 6.10 is obtained by replacing equation (6.16) with:

$$u(k) = \frac{1}{\Delta}\left(-F(z^{-1})\,\Delta y(k) - G_1(z^{-1})x(k) + k_I\,(y_d(k) - y(k))\right) \tag{6.17}$$

In Figure 6.10, the integrator $1/\Delta$ is implemented by means of a positive feedback arrangement.

The reader can verify the validity of this approach by reconsidering Figure 2.11 and equation (2.26), albeit with a sign change. Figure 6.10 represents the PIP feedback structure in block diagram form, expressed in a manner that allows for implementation of (6.15), denoted $\{\overline{u}, \underline{u}\}$.

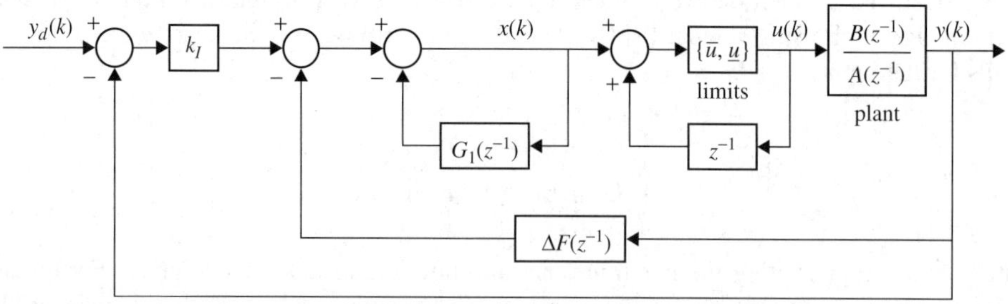

Figure 6.10 Block diagram representation of equation (6.15) and equation (6.17): the feedback PIP control system represented in incremental form with input limits

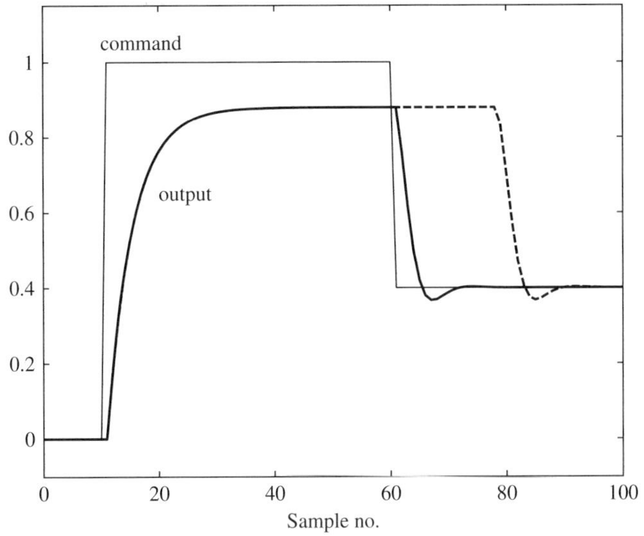

Figure 6.11 Closed-loop response with a steady-state error because of limits on the input, showing the command level (sequence of steps of varying magnitude), together with the output obtained using a controller with (bold solid trace) and without (dashed trace) the correction in equations (6.15)

This block diagram is particularly useful when using iconographic simulation or real-time control packages to represent problems with control input limits, as in Example 6.2.

Example 6.2 Simulation Experiment with Integral 'Wind-Up' Problems Consider again the pole assignment PI controller from Example 2.6 and Example 2.7, with $a_1 = -0.8$, $b_1 = 1.6$, $f_0 = 0.2188$ and $k_I = 0.1563$. Figure 6.11 and Figure 6.12 show the output and input variables respectively, in response to a sequence of changes in the command level. Here, the solid traces are based on the incremental form of the algorithm (2.39), utilising the correction (6.15), whilst the dashed traces show the response of the same controller without this correction. In both cases, however, the control input is arbitrarily limited to a maximum value of 0.11 before it is connected to the TF model in simulation, representing a level constraint on the actuator.

Note that a constant maximum control input of 0.11 yields a steady-state output $y(k \to \infty) = 0.11 \times 1.6/(1 - 0.8) = 0.88$, hence there is an error between the unity command level and the output in Figure 6.11 (for both controllers). Significantly, without the correction (6.15), the integrated error signal quickly builds up during the middle part of the simulation, resulting in ever larger input signals that are not achievable in practice: these are shown as the dashed trace in Figure 6.12. Integral wind-up can sometimes lead to instability although, in this example, control is maintained. However, when the saturation finally ends following a step down in the command level, the basic controller without the correction takes almost 20 samples to recover. By contrast, the incremental form of the algorithm has no such problems.

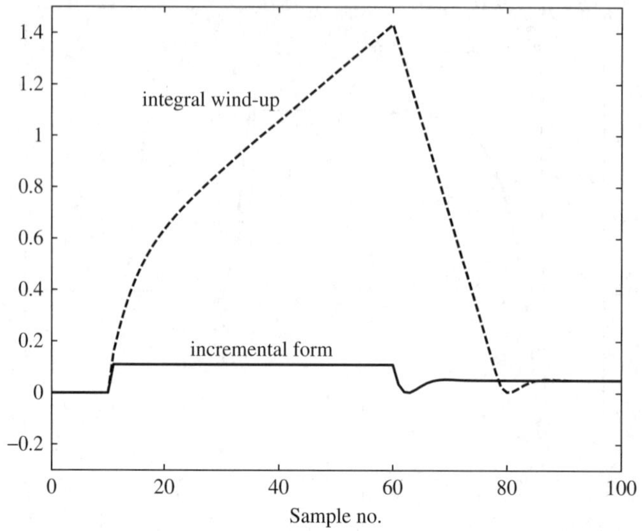

Figure 6.12 Control input signals associated with Figure 6.11

6.2.2 Incremental Form for Forward Path PIP Control

Since the integrator already appears in the denominator of the pre-compensation filter in Figure 6.3, the block diagram for the incremental form of forward path PIP control, with input limits, is straightforward to obtain, as illustrated by Figure 6.13.

The algorithm is equivalently expressed as follows, in which the correction (6.15) is again utilised to avoid integral wind-up problems:

$$u(k) = u(k-1) + \frac{k_I \hat{A}(z^{-1})}{G(z^{-1})\hat{A}(z^{-1}) + F(z^{-1})\hat{B}(z^{-1})}(y_d(k) - y(k)) \tag{6.18}$$

However, for practical implementation, equation (6.18) is usually converted into a difference equation form, as shown by Example 6.3.

Example 6.3 Incremental Form for Carbon-12 Tracer Experiments All engineering systems have constraints on the input signal or signals, including both level and rate of change limits. For these reasons, the practical examples considered in this book are all based on

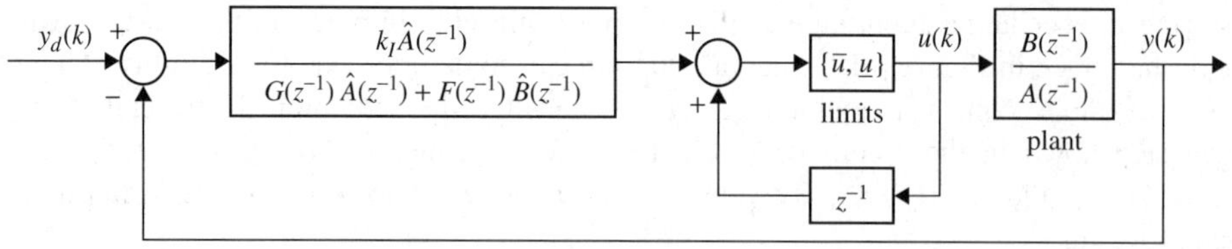

Figure 6.13 Forward path PIP control system represented in incremental form with input limits

implementing PIP control algorithms using an appropriate incremental form, as a precaution against these constraints being reached. This includes the experimental results for control of carbon dioxide ($^{12}CO_2$) level in carbon tracer experiments (Example 5.9). In fact, Figure 5.11 and Figure 5.12 were obtained as follows, using the feedback algorithm in incremental form (6.14), together with the PIP control gains (5.103):

$$\begin{aligned} u(k) = u(k-1) &- 23.00\,(y(k) - y(k-1)) - \ 0.095\,(u(k-1) - u(k-2)) \\ &- 0.099\,(u(k-2) - u(k-3)) - \ 0.104\,(u(k-3) - u(k-4)) \\ &+ 0.954\ \ (y_d(k) - y(k)) \end{aligned} \tag{6.19}$$

For comparison, Figure 6.14 illustrates the response of the system using the incremental forward path PIP algorithm. The advantage of the forward path design in this example is its ability to yield a smoother control input signal and, therefore, a more desirable constant CO2 level, without as much 'chatter' as that occurring in the feedback case: compare the control input shown in Figure 5.12 with Figure 6.14b.

However, examination of the estimated TF model (5.101) reveals a potential problem with the forward path structure: the estimated $\hat{A}(z^{-1}) = 1 - z^{-1}$ polynomial is an exact integrator in this case, which would cancel with the integral action of the forward path PIP controller, as shown in Figure 6.3 or equation (6.4). Unfortunately, the real system is not an *exact* integrator

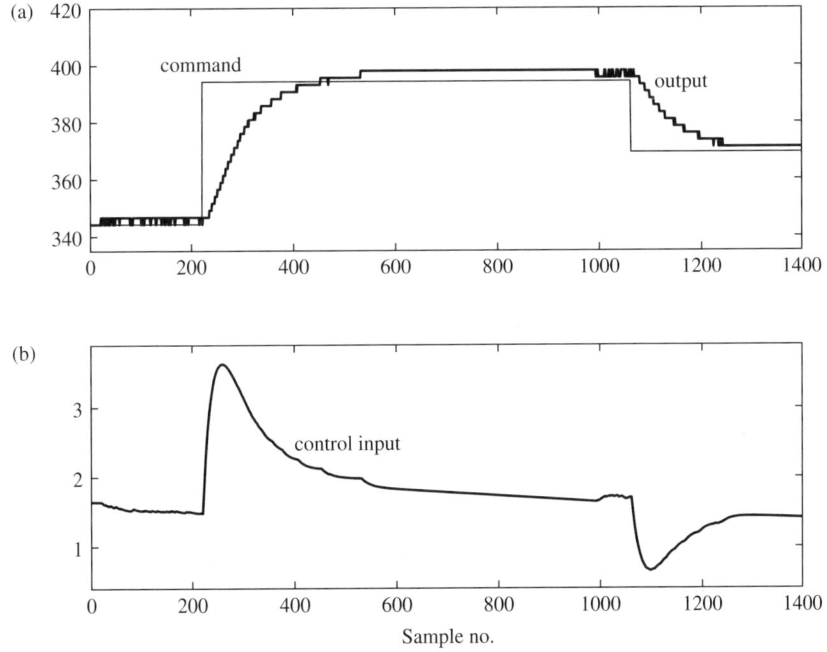

Figure 6.14 Closed-loop response using the forward path PIP controller of Example 6.3. (a) The command input (sequence of steps of varying magnitude), together with the measured CO2 concentration (ppm). (b) The control input (seconds of CO2 injection)

and, in practice, there is a small net loss of CO_2 from the leaf chamber over time; as a result, the forward path structure yields an undesirable steady-state error.

By contrast, if the CO2 uptake rate is particularly high because there is a large amount of leaf in the chamber, then the decay time constant may become significant and the estimated model will have a lag coefficient significantly different from unity, as in the following example:

$$y(k) = \frac{b_4 z^{-4}}{1 + a_1 z^{-1}} u(k) = \frac{0.0047 z^{-4}}{1 - 0.9788 z^{-1}} u(k) \tag{6.20}$$

where $y(k)$ is the CO2 concentration (ppm), $u(k)$ is the CO2 injection time (seconds) and the sampling period is 5 s. In fact, the experimental results shown by Figure 6.14 are based on this non-conservative model, together with *Linear Quadratic* (LQ) weights (5.78) obtained by trial and error experimentation in order to obtain a satisfactory closed-loop response; namely $q_y = 1$, $q_u = 1/4$ and $q_e = 1/100$. These settings yield the control gain vector:

$$\boldsymbol{k} = [\, f_o \quad g_1 \quad g_2 \quad g_3 \quad -k_I \,]^T = [\, 3.464 \quad 0.0164 \quad 0.0165 \quad 0.0166 \quad -0.0992 \,]^T \tag{6.21}$$

Evaluating equation (6.4), we obtain the incremental forward path PIP algorithm:

$$u(k) = \frac{0.0992 - 0.0971 z^{-1}}{1 - 1.9624 z^{-1} + 0.9629 z^{-2} - 0.0005 z^{-4}} (y_d(k) - y(k)) \tag{6.22}$$

Expressing this as a difference equation, yields a forward path PIP algorithm that is particularly straightforward to implement using standard software code:

$$\begin{aligned} u(k) = u(k-1) &+ 0.0992\,(y_d(k) - y(k)) - 0.0971\,(y_d(k-1) - y(k-1)) \\ &+ 0.9624 u(k-1) - 0.9629 u(k-2) + 0.0005 u(k-4) \end{aligned} \tag{6.23}$$

Note that the coefficient associated with $u(k-3)$ in equation (6.22) and equation (6.23) is zero.

The TF model (6.20) is still relatively close to being an integrator, hence the small 4 ppm steady-state error seen in Figure 6.14. In practical terms, this arises because the two $y_d(k) - y(k)$ components of equation (6.23) almost sum to zero. Hence, the integral action is acting rather slowly and, although the error is eliminated eventually (beyond the time frame of the graph), the response time is very slow. Chapter 9 introduces an alternative, δ-operator approach to the control of such systems, i.e. for models with poles that are close to the unit circle.

Nonetheless, in general terms, these results are instructive: they illustrate important differences between the two main PIP implementation structures; and act as a timely reminder that we must always consider the effect of parametric uncertainty in our designs. However, the example raises another interesting issue, one that requires detailed comment over the next few pages. The numerator polynomial of the TF model (6.20) includes a time delay of four control samples, representing 20 s pure time delay. This is sometimes referred to as a transport delay or dead time: it is a common feature of many physical systems and is a major concern in control system design.

When the time delay is relatively short in comparison with the dominant time constants of the system it can be handled fairly easily. In the case of discrete—time, sampled data control systems, where the time delay can be approximated by an integral number of sampling intervals, NMSS design methods are able to absorb the delay directly into the system model, as shown in Chapter 4 and Chapter 5. In this case, the PIP controller automatically handles the time delay by simply feeding back sufficient past values of the input variable by means of the $G(z^{-1})$ filter.

However, for systems with a particularly long time delay, and where the choice of a coarser sampling interval is not possible as a means of reducing the dimension of the NMSS model, this may require an unacceptably large number of gains in the $G(z^{-1})$ filter. Indeed, it is noteworthy that the present example, for which the open-loop TF model (6.20) is first order, yields a fifth order PIP control system (6.21) including a third order input filter, i.e. $G(z^{-1}) = 1 + 0.0164z^{-1} + 0.0165z^{-2} + 0.0166z^{-3}$. The following section considers one method for reducing the order of the PIP control system in such cases.

6.3 The Smith Predictor and its Relationship with PIP Design

Over many years, the *Smith Predictor* (SP) has proven to be one of the most popular approaches to the design of controllers for time-delay systems. Although initially formulated for continuous-time systems (Smith 1957, 1959), the SP can be implemented in any discrete-time control scheme, such as digital PIP control. In the latter case, a unit time-delay model of the system provides an estimate of the output variable, which is utilised by the feedback control filters as shown in Figure 6.15 (Taylor *et al.* 1998a). Here, $\tau > 0$ is the pure time delay of the plant in sampling intervals of Δt time units.

When the time delay is much larger than the dominant time constants of the system, it seems more efficient and parsimonious to employ a SP in this way, rather than to excessively extend the non-minimal state vector. In the present section, however, we demonstrate

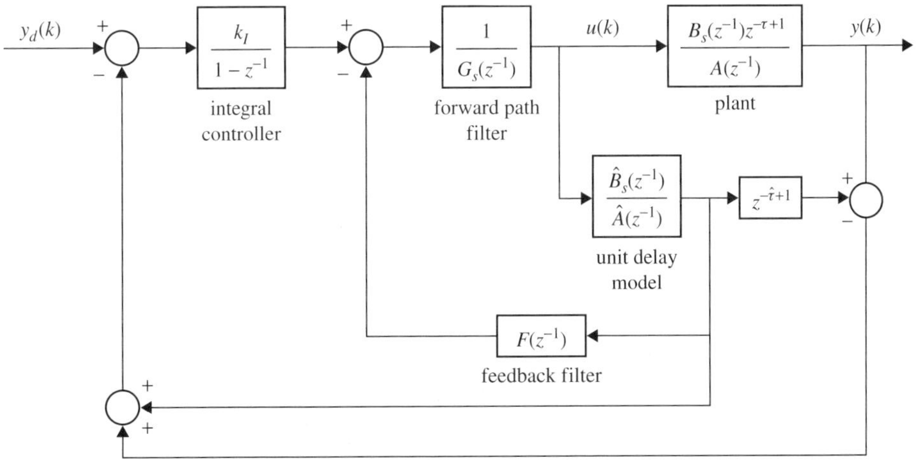

Figure 6.15 PIP control with the Smith Predictor

how, under certain non-restrictive pole assignment conditions, the forward PIP controller is *exactly* equivalent to the digital SP controller but has much greater inherent flexibility in design terms.

Discrete-time control system design assumes a delay of at least one sampling interval. In the SP-PIP controller, time delays greater than unity are external to the control loop (Figure 6.15), and so the specification of system performance is obtained by defining the following modified TF control model:

$$y(k) = \frac{B_s(z^{-1})}{A(z^{-1})}u(k-\tau+1) = \frac{b_\tau z^{-1} + b_{\tau+1}z^{-2} + \cdots + b_m z^{-m+\tau-1}}{1 + a_1 z^{-1} + \cdots + a_n z^{-n}}u(k-\tau+1) \quad (6.24)$$

This is based on equation (6.1) but $\tau - 1$ samples time delay are removed from the numerator polynomial and added to the input variable. To illustrate this, in the case of the TF model for carbon-12 tracer experiments (6.20), $m = \tau = 4$, hence $A(z^{-1}) = 1 + a_1 z^{-1}$ and $B(z^{-1}) = b_4 z^{-4}$ as usual, whilst $B_s(z^{-1}) = b_4 z^{-1}$.

The SP-NMSS form is based on equation (5.13), equation (5.14), equation (5.15) and equation (5.16), albeit with the following modifications to address the reduced order of the numerator polynomial. The $n + m - \tau + 1$ non-minimal state vector is:

$$\boldsymbol{x}(k) = [y(k) \quad \cdots \quad y(k-n+1) \quad u(k-1) \quad \cdots \quad u(k-m+\tau) \quad z(k)]^T \quad (6.25)$$

while

$$\boldsymbol{F} = \begin{bmatrix} -a_1 & -a_2 & \cdots & -a_{n-1} & -a_n & b_{\tau+1} & b_{\tau+2} & \cdots & b_{m-1} & b_m & 0 \\ 1 & 0 & \cdots & 0 & 0 & 0 & 0 & \cdots & 0 & 0 & 0 \\ 0 & 1 & \cdots & 0 & 0 & 0 & 0 & \cdots & 0 & 0 & 0 \\ \vdots & \vdots & \ddots & \vdots & \vdots & \vdots & \vdots & \ddots & \vdots & \vdots & \vdots \\ 0 & 0 & \cdots & 1 & 0 & 0 & 0 & \cdots & 0 & 0 & 0 \\ 0 & 0 & \cdots & 0 & 0 & 0 & 0 & \cdots & 0 & 0 & 0 \\ 0 & 0 & \cdots & 0 & 0 & 1 & 0 & \cdots & 0 & 0 & 0 \\ 0 & 0 & \cdots & 0 & 0 & 0 & 1 & \cdots & 0 & 0 & 0 \\ \vdots & \vdots & \ddots & \vdots & \vdots & \vdots & \vdots & \ddots & \vdots & \vdots & \vdots \\ 0 & 0 & \cdots & 0 & 0 & 0 & 0 & \cdots & 1 & 0 & 0 \\ a_1 & a_2 & \cdots & a_{n-1} & a_n & -b_{\tau+1} & -b_{\tau+2} & \cdots & -b_{m-1} & -b_m & 1 \end{bmatrix} \quad (6.26)$$

and $\boldsymbol{g} = [b_\tau \;\; 0 \;\; 0 \;\; \cdots \;\; 0 \;\; 1 \;\; 0 \;\; 0 \;\; \cdots \;\; 0 \;\; -b_\tau\,]^T$.

The $n + m - \tau + 1$ dimensional SVF control gain vector is found using pole assignment or optimal design:

$$\boldsymbol{k}_s^T = [f_0 \quad f_1 \quad \cdots \quad f_{n-1} \quad g_1 \quad \cdots \quad g_{m-\tau} \quad -k_I] \quad (6.27)$$

The numerical values of these SP-PIP control gains may differ from those of the nominal PIP controller, as discussed below. Furthermore, the order of the input filter $G_s(z^{-1})$:

$$G_s(z^{-1}) = 1 + g_1 z^{-1} + \cdots + g_{m-\tau} z^{-m+\tau} \tag{6.28}$$

is reduced by $\tau - 1$ degrees, in comparison with the nominal $G(z^{-1})$ polynomial (6.2). Indeed, this is the reason for introducing the SP in the first place. It is important to stress that any time delays larger than unity have been ignored in the model and associated SVF control algorithm. Therefore, to account for longer time delays in the plant, the final algorithm must be implemented using Figure 6.15.

Assuming that there is no model mismatch, i.e. $\hat{B}(z^{-1}) = B(z^{-1})$ and $\hat{A}(z^{-1}) = A(z^{-1})$, the closed-loop TF obtained from Figure 6.15 is as follows:

$$y(k) = \frac{k_I B_s(z^{-1}) z^{-\tau+1}}{\Delta \left(G_s(z^{-1}) A(z^{-1}) + F(z^{-1}) B_s(z^{-1}) \right) + k_I B_s(z^{-1})} y_d(k) \tag{6.29}$$

The characteristic polynomial of the SP-PIP system is of the order $n + m - \tau + 1$, compared with $n + m$ in the nominal PIP case.

6.3.1 Relationship between PIP and SP-PIP Control Gains

Equivalence between the two approaches may be obtained using Theorem 6.1.

Theorem 6.1 Relationship between PIP and SP-PIP Control Gains When there is no model mismatch nor any disturbance input, the closed-loop TF of the nominal PIP control system (6.3) for a *Single-Input, Single-Output* (SISO), time-delay system is identical to that of the SP-PIP closed-loop TF (6.29), if:

(i) $\tau - 1$ poles in the nominal PIP control system are assigned to the origin of the complex z–plane; and
(ii) the remaining $n + m - \tau + 1$ poles of the nominal PIP control system are constrained to be the same as those of the SP-PIP case.

Moreover, under these simple conditions, the nominal PIP gain vector $\boldsymbol{k}$ is given by:

$$\boldsymbol{k}^T = \boldsymbol{\Sigma}^{-1} \bullet \boldsymbol{\Sigma}_s \bullet \boldsymbol{k}_s^T \tag{6.30}$$

where $\boldsymbol{\Sigma}$ is the parameter matrix (5.67) employed for the design of a pole assignment controller; $\boldsymbol{\Sigma}_s$ is the parameter matrix employed for SP-PIP pole assignment (defined in a similar manner to $\boldsymbol{\Sigma}$ but based on the unit delay model); and $\boldsymbol{k}_s$ is the SP-PIP gain vector (6.27). The proof of this theorem is given in Appendix F.

6.3.2 Complete Equivalence of the SP-PIP and Forward Path PIP Controllers

To consider the more general case when there is model mismatch, the SP-PIP controller is converted into a unity feedback, forward path pre-compensation implementation:

$$u(k) = \frac{k_s \hat{A}(z^{-1})}{\left(G_s(z^{-1})\hat{A}(z^{-1}) + F(z^{-1})\hat{B}_s(z^{-1})\right)\Delta + k_s\left(\hat{B}_s(z^{-1}) - \hat{B}_s(z^{-1})z^{-\hat{\tau}+1}\right)} e(k) \quad (6.31)$$

in which $\hat{\tau}$ is the estimated time delay and $e(k) = y_d(k) - y(k)$. Using straightforward algebra, equation (6.31) is obtained from, and is exactly equivalent to, the control algorithm shown diagrammatically in Figure 6.15.

Theorem 6.1 has already established that, with appropriate selection of poles, the SP-PIP closed-loop TF must be identical to the closed-loop TF resulting from nominal PIP design, assuming a perfectly known model. Clearly, for this result to hold when both controllers are implemented in a forward path form, then the control filter in each case must be identical, i.e. once the polynomial multiplications have been resolved, equation (6.4) and equation (6.31) are identical. Moreover, this result applies even in the case of mismatch and disturbance inputs, since the control filter is invariant to such effects.

Example 6.4 SP-PIP Control of Carbon-12 Tracer Experiments Returning to the carbon-12 tracer experiments, as discussed in Example 6.3, the model encompasses a pure time delay of $\tau = 4$ samples. For this model, the LQ cost function yields a closed-loop characteristic equation with $\tau - 1 = 3$ poles at the origin of the complex z-plane, automatically satisfying the conditions of Theorem 6.1. This result implies that the SP-PIP approach is the optimal solution (in the LQ sense) for this system. These poles are obtained by substituting the TF model (6.20) and control gains (6.21) into the closed-loop TF (6.3). The roots of the associated characteristic equation are $p_1 = p_2 = p_3 = 0$, $p_4 = 0.981 + 0.011j$ and $p_5 = 0.981 - 0.011j$. Using the following closed-loop design polynomial:

$$D(z) = (z - 0.981 + 0.011j)(z - 0.981 - 0.011j) = z^2 - 1.962 + 0.963 = 0 \quad (6.32)$$

and $B_s(z^{-1}) = 0.0047z^{-1}$, SP-PIP pole assignment yields $F(z^{-1}) = f_0 = 3.39$, $G_s(z^{-1}) = 1$ and $k_I = 0.099$. The controller is implemented as illustrated by Figure 6.15. In this example, we see that the SP-PIP control algorithm yields two control gains, i.e. f_0 and k_I. However, noting that $B_s(z^{-1}) - B_s(z^{-1})z^{-\tau+1} = 0.0047z^{-1} - 0.0047z^{-4}$ and substituting into equation (6.31), yields exactly the same algorithm as the forward path PIP controller (6.22) or (6.23). Therefore, the closed-loop response of the SP-PIP system is identical to the forward path PIP controller, and has the same advantages and disadvantages.

Furthermore, the final control algorithm effectively has five control gains, i.e. the same as the nominal PIP controller based on the time-delay model! This is because Figure 6.15 includes the estimated time delay in the feedback loop, and this subsequently appears in the denominator of the control algorithm (6.22), as indicated above. In other words, the reduced order of the SP-PIP controller only applies at the design stage, but the overall control algorithm (and hence closed-loop system) is unchanged. It is clear from this example that, whilst the SP-PIP approach yields a well-known control structure (and hence is considered in this chapter for tutorial purposes), it can always be exactly replicated by forward path PIP control and hence does not offer any advantages over such PIP design.

Example 6.5 SP-PIP Control of Non-Minimum Phase Oscillator Consider the TF model (6.10) with $\tau = 2$ samples. The unit delay numerator polynomial is $B_s(z^{-1}) = -z^{-1} + 2.0z^{-2}$. In order to satisfy the conditions of Theorem 6.1, we are free to assign four poles anywhere on the complex z-plane, as long as the fifth is at the origin. Once again, this is the design deliberately employed earlier (Example 6.1). For example, the Monte Carlo realisations in Figure 6.8 illustrate the expected poorer robustness properties of the SP-PIP and equivalent forward path algorithms to model mismatch.

Within the context of the present discussion, however, it is interesting to examine the situation when there is uncertainty in the time delay. For this type of model uncertainty, we find that the forward path approach is typically *more* robust than feedback PIP design. The results in Figure 6.16 are for a controller design based on $\hat{\tau} = 2$ as above, but with the simulated time delay changed to $\tau = 3$. This represents a significant error in the original identification of the control model, but is useful here since it exaggerates the difference between the feedback and SP-PIP structures. We can see that while the former yields an unstable closed-loop response, with growing oscillations in the output, the forward path implementation remains stable, albeit with a small overshoot of the command level.

In this regard, it is important to emphasise that PIP pole assignment design of this type, in which the closed-loop poles associated with the time delay are assigned to the origin of the complex z-plane, is not necessarily the most robust solution. In the present example, if we specify all five closed-loop poles to lie at 0.5 on the real axis, then the feedback PIP controller

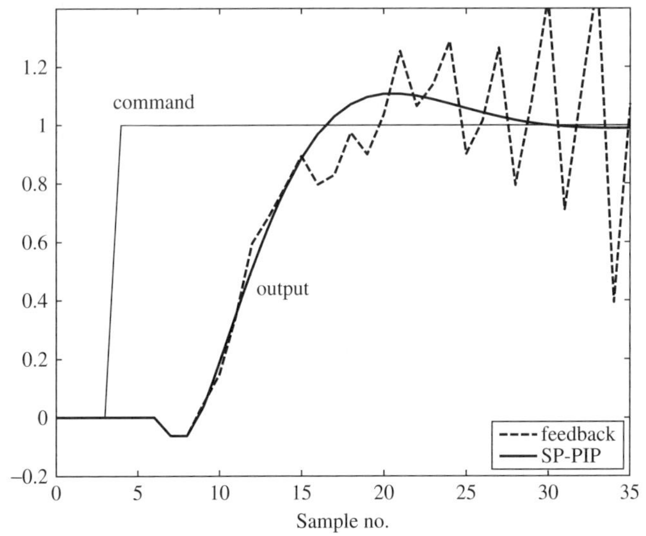

Figure 6.16 Closed-loop response for a unit step in the command level, with the feedback (dashed trace) and SP-PIP (bold solid trace) controllers of Example 6.5, when there are time-delay modelling errors

yields a stable closed-loop response similar to the bold solid trace in Figure 6.16. Here, we see the benefit of having complete freedom to assign the $n + m$ poles anywhere on the complex z-plane (including those associated with the time delay).

6.4 Stochastic Optimal PIP Design

The discussion so far has concentrated on a deterministic NMSS model of the system and associated PIP pole assignment or LQ design. However, since it is formulated in state space terms, the NMSS approach to control system design can be converted into stochastic form, by invoking the separation theorem and introducing an optimal Kalman Filter (Kalman 1960, 1961, 1963) for state estimation.

Here, the Kalman filter is employed as an observer to generate a surrogate state vector, which converges asymptotically to the true state vector in the presence of state and measurement noise. The separation theorem states that the SVF control algorithm and observer can be designed separately. When connected, they will function as an integrated control system in the desired manner (see e.g. Åström and Wittenmark 1984, pp. 273–275; Green and Limebeer 1995, pp. 179–208; Franklin *et al.* 2006, pp. 511–513; Dorf and Bishop 2008, pp. 773–775; Gopal 2010, pp. 413–428). In the present context, the approach yields PIP-*Linear Quadratic Gaussian* (PIP-LQG) designs that may be implemented in either state space or polynomial form, as demonstrated below.

Such a Kalman Filter formation is not strictly necessary for basic PIP design, but it is introduced here for completeness and as a useful starting point for some of the analysis later in this chapter. The objective of the stochastic approach is to reduce the detrimental effects of measurement noise, so as to yield a smoother input signal in the closed-loop response. This provides an alternative to the forward path PIP structure considered above. As we have seen, forward path PIP control is a pragmatic solution to the problem of measurement noise, obtained using straightforward block diagram analysis.

By contrast, the Kalman Filter approach is optimal (given assumptions about the noise) but more complex to design and implement in practical situations. Furthermore, like the forward path implementation of deterministic PIP control, it carries with it the potential disadvantage of decreased robustness. Therefore, in this chapter, we will concentrate on the use of an observer filter $D(z^{-1})$, since this offers a practical compromise between filtering, robust design and straightforward implementation[2].

6.4.1 *Stochastic NMSS Equations and the Kalman Filter*

We initially formulate the PIP-LQG control problem by introducing stochastic white noise disturbances into the regulator NMSS equations, as follows:

$$\left.\begin{aligned} \boldsymbol{x}(k) &= \boldsymbol{F}\boldsymbol{x}(k-1) + \boldsymbol{g}u(k-1) + \boldsymbol{w}(k) \\ y(k) &= \boldsymbol{h}\boldsymbol{x}(k) + \varepsilon(k) \end{aligned}\right\} \tag{6.33}$$

[2] Although the observer polynomial is sometimes denoted (e.g. Clarke 1994) by $C(z^{-1})$ or $T(z^{-1})$, we will use $D(z^{-1})$ for consistency with the various stochastic TF models considered in Chapter 8.

where $\boldsymbol{x}(k)$, $\boldsymbol{F}$, $\boldsymbol{g}$ and $\boldsymbol{h}$ are defined by equation (4.4), equation (4.5), equation (4.6), equation (4.7), equation (4.8) and equation (4.9), $\boldsymbol{w}(k)$ is a zero mean, discrete-time, Gaussian, vector white noise process with positive semi-definite covariance matrix $\boldsymbol{N}$ and $\varepsilon(k)$ is an independent, zero mean, white measurement noise signal with variance σ^2. The innovations, or Kalman Filter, representation of this system is:

$$\left.\begin{aligned} \hat{\boldsymbol{x}}(k+1/k) &= \boldsymbol{F}\hat{\boldsymbol{x}}(k/k-1) + \boldsymbol{g}u(k-1) + \boldsymbol{L}(k)e(k) \\ y(k) &= \boldsymbol{h}\hat{\boldsymbol{x}}(k/k-1) + e(k) \end{aligned}\right\} \tag{6.34}$$

where $\boldsymbol{L}(k)$ is the Kalman gain vector and $e(k) = y(k) - \boldsymbol{h}\hat{\boldsymbol{x}}(k/k-1)$ is the innovations sequence (see e.g. Young 2011). From equations (6.34), it is clear that the optimal state estimator has the form:

$$\hat{\boldsymbol{x}}(k+1/k) = \boldsymbol{F}\hat{\boldsymbol{x}}(k/k-1) + \boldsymbol{g}u(k-1) + \boldsymbol{L}(k)\left(y(k) - \boldsymbol{h}\hat{\boldsymbol{x}}(k/k-1)\right) \tag{6.35}$$

Here, we see that $\hat{\boldsymbol{x}}(k+1/k)$ represents a one step ahead estimate of the non-minimal state vector, conditional on the 'latest' measured output $y(k)$. Note that only the latest scalar observation of $y(k)$ is used to update the NMSS state at each sampling instant since the NMSS vector is composed of past values of $y(k)$ that have already been processed, as well as the present and past values of the input $u(k)$, that are known exactly. Subsequently, we use $\hat{y}(k) = \boldsymbol{h}\hat{\boldsymbol{x}}(k/k-1)$ to represent the optimal estimate of the output, hence,

$$e(k) = y(k) - \hat{y}(k) \tag{6.36}$$

The SVF control law is now formulated by reference to the separation theorem in the usual manner (see e.g. Åström and Wittenmark 1984, pp. 273–275; Green and Limebeer 1995 pp. 179–208; Franklin *et al.* 2006, pp. 511–513; Dorf and Bishop 2008, pp. 773–775; Gopal 2010, pp. 413–428):

$$u(k) = -\boldsymbol{k}^T\hat{\boldsymbol{x}}(k/k-1) \tag{6.37}$$

where $\boldsymbol{k}$ is the PIP control gain vector (5.31). Equation (6.37) is based on the deterministic control law (5.30) but adapted here to utilise the optimal estimate of the state vector.

Once the Kalman gain vector has converged to its steady-state value, i.e. $\boldsymbol{L}(k \to \infty) = \boldsymbol{L}$, it can be shown that $e(k)$ in equations (6.34) constitutes a zero mean white noise sequence with constant variance σ^2 (e.g. Young 2011). Here, $\boldsymbol{L}$ is determined from the steady-state solution of the discrete-time, matrix Riccati equations, involving the state transition matrix and observation vector, together with the covariances of the state disturbance and measurement noise inputs. In particular, $\boldsymbol{L}$ is obtained from equation (5.81) and equation (5.82) but replacing $\boldsymbol{F}$ with $\boldsymbol{F}^T$, $\boldsymbol{g}$ with $\boldsymbol{h}^T$, $\boldsymbol{Q}$ with the covariance matrix $\boldsymbol{N}^T$ and r with σ (see e.g. Åström and Wittenmark 1984, pp. 259 and 270), as follows:

$$\boldsymbol{L}^T = (\sigma + \boldsymbol{hPh}^T)^{-1}\boldsymbol{hPF}^T \tag{6.38}$$

where $\boldsymbol{P}$ is the steady-state solution of the matrix Riccati equation:

$$\boldsymbol{P} - \boldsymbol{F}\boldsymbol{P}\boldsymbol{F}^T + \boldsymbol{F}\boldsymbol{P}\boldsymbol{h}^T(r + \boldsymbol{h}\boldsymbol{P}\boldsymbol{h}^T)^{-1}\boldsymbol{h}\boldsymbol{P}\boldsymbol{F}^T - \boldsymbol{N} = \boldsymbol{0} \tag{6.39}$$

in which $\boldsymbol{0}$ is a matrix of zeros. For more information about this interesting duality between LQ control and optimal filtering, refer to the textbook references on linear control theory highlighted above.

For regulator NMSS design, the state covariance matrix $\boldsymbol{N}$ is chosen to represent the noise on the output states, while zeros are utilised to correspond to the exactly known past values of the calculated control input signal, i.e.

$$\boldsymbol{N} = \begin{bmatrix} \boldsymbol{N}_n & \boldsymbol{0} \\ \boldsymbol{0} & \boldsymbol{0} \end{bmatrix} \tag{6.40}$$

where $\boldsymbol{N}_n$ is an $n \times n$ disturbance covariance matrix and $\boldsymbol{0}$ is an appropriately defined matrix of zeros. Therefore, the general form of this minimal Kalman gain vector, where only the optimal estimate of the output and its past values are filtered, is as follows:

$$\boldsymbol{L}(k) = [l_1(k) \quad l_2(k) \quad \ldots \quad l_n(k) \quad 0 \quad \ldots \quad 0 \quad 0]^T \tag{6.41}$$

Finally, in the servomechanism formulation of the problem, the integral-of-error state variable $z(k)$, defined in equation (5.4), is included as normal, with either the measured output $y(k)$ or the optimal estimate of the output $\hat{y}(k)$ fed back to the integral controller, as discussed below.

6.4.2 *Polynomial Implementation of the Kalman Filter*

An important consequence of the convergence of $\boldsymbol{L}(k)$ to a steady-state value $\boldsymbol{L}$ is that the innovations representation of the system (6.34) may be expressed as the following *Auto-Regressive, Moving-Average eXogenous* variables (ARMAX) model:

$$y(k) = \frac{b_1 z^{-1} + \cdots + b_m z^{-m}}{1 + a_1 z^{-1} + \cdots + a_n z^{-n}} u(k) + \frac{1 + d_1 z^{-1} + \cdots + d_n z^{-n}}{1 + a_1 z^{-1} + \cdots + a_n z^{-n}} e(k) \tag{6.42}$$

or, in shorthand form:

$$y(k) = \frac{B(z^{-1})}{A(z^{-1})} u(k) + \frac{D(z^{-1})}{A(z^{-1})} e(k) \tag{6.43}$$

Equation (6.43) is derived by Young (1979) for a minimal stochastic state space representation of the system, i.e. based on the observable canonical form (3.37) introduced in Chapter 3. This paper also shows how a Kalman Filter can be constructed from polynomial filters that are used in the *Refined Instrumental Variable* (RIV) method of TF model estimation (see Chapter 8); and that, in this minimal state situation, the coefficients of the $D(z^{-1})$ polynomial are defined by $d_i = a_i + l_i$, where a_i are the parameters of the deterministic characteristic equation $A(z^{-1})$

and l_i are the minimal state Kalman Filter gains[3]. Again noting the duality between LQ control and optimal filtering, this result can be compared with equation (3.34), in which l_i are the SVF control gains obtained using the controllable canonical form.

In the non-minimal case, the coefficients of $D(z^{-1})$ are defined as follows (Hesketh 1992):

$$d_i = a_i + \tilde{l}_i \tag{6.44}$$

where

$$\begin{bmatrix} \tilde{l}_1 \\ \tilde{l}_2 \\ \tilde{l}_3 \\ \vdots \\ \tilde{l}_{n-1} \\ \tilde{l}_n \end{bmatrix} = \begin{bmatrix} 1 & 0 & 0 & \cdots & 0 & 0 \\ 0 & -a_2 & -a_3 & \cdots & -a_{n-1} & -a_n \\ 0 & -a_3 & -a_4 & \cdots & -a_n & 0 \\ \vdots & \vdots & \vdots & \vdots & \vdots & \vdots \\ 0 & -a_{n-1} & -a_n & \cdots & 0 & 0 \\ 0 & a_n & 0 & \cdots & 0 & 0 \end{bmatrix} \begin{bmatrix} l_1 \\ l_2 \\ l_3 \\ \vdots \\ l_{n-1} \\ l_n \end{bmatrix} \tag{6.45}$$

With this formulation, the noise polynomial $D(z^{-1})$ is either derived from $\boldsymbol{L}$ or obtained directly from observed data using the system identification tools discussed in Chapter 8. In the latter case, equation (6.44) and equation (6.45) could then be used in reverse to determine the equivalent Kalman gains from the estimated model. Alternatively, the Kalman Filter can be implemented in a number of ways that directly employ $D(z^{-1})$, as shown below.

Using equation (6.36) and equation (6.43):

$$y(k) = \frac{B(z^{-1})}{A(z^{-1})}u(k) + \frac{D(z^{-1})}{A(z^{-1})}\left(y(k) - \hat{y}(k)\right) \tag{6.46}$$

After rearranging,

$$\hat{y}(k) = \frac{D(z^{-1}) - A(z^{-1})}{D(z^{-1})}y(k) + \frac{B(z^{-1})}{D(z^{-1})}u(k) \tag{6.47}$$

In other words,

$$\begin{aligned} \hat{y}(k) = (d_1 - a_1)\left[\frac{z^{-1}}{D(z^{-1})}y(k)\right] + \cdots + (d_n - a_n)\left[\frac{z^{-n}}{D(z^{-1})}y(k)\right] \\ + b_1\left[\frac{z^{-1}}{D(z^{-1})}u(k)\right] + \cdots + b_m\left[\frac{z^{-m}}{D(z^{-1})}u(k)\right] \end{aligned} \tag{6.48}$$

so that the optimal estimate $\hat{y}(k)$ is the linear sum of the pre-filtered output and input variables, where the pre-filter is the numerator of the *Auto-Regressive, Moving-Average* (ARMA) noise model. See Chapter 8 for discussion on this pre-filter, as used for ARMA noise model parameter

[3] The multivariable version of this analysis is developed in Jakeman and Young (1979).

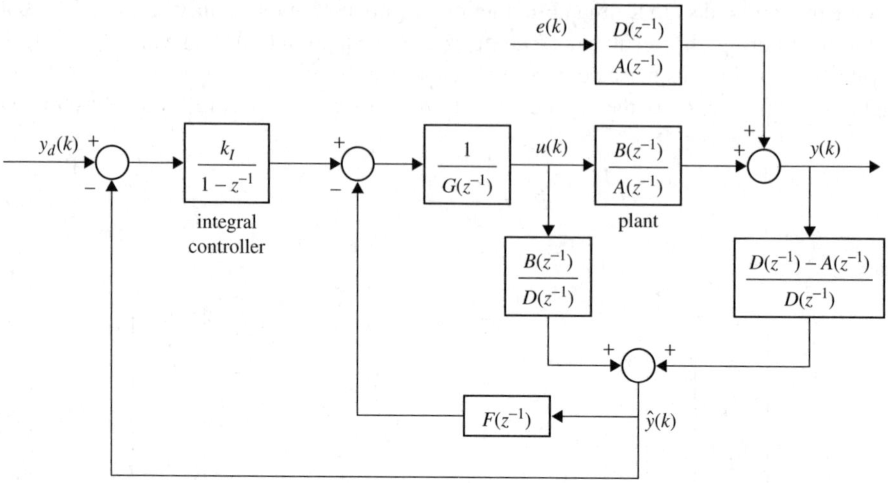

Figure 6.17 Polynomial implementation of the PIP-LQG control system with the integral controller operating on the optimal estimate of the output

estimation. On the basis of the above results, the optimal estimate $\hat{y}(k)$ can be simply fed back in place of the measured output, as shown in Figure 6.17.

In an alternative structure, the feedback filter $F(z^{-1})$ utilises $\hat{y}(k)$ but the integral controller operates on the measured unfiltered output $y(k)$, as illustrated by Figure 6.18 (cf. Figure 6.4). In the latter case, note that the scalar gain on the integral controller does not amplify the noise. The relative merits of these and other polynomial structures are discussed after the following introductory example.

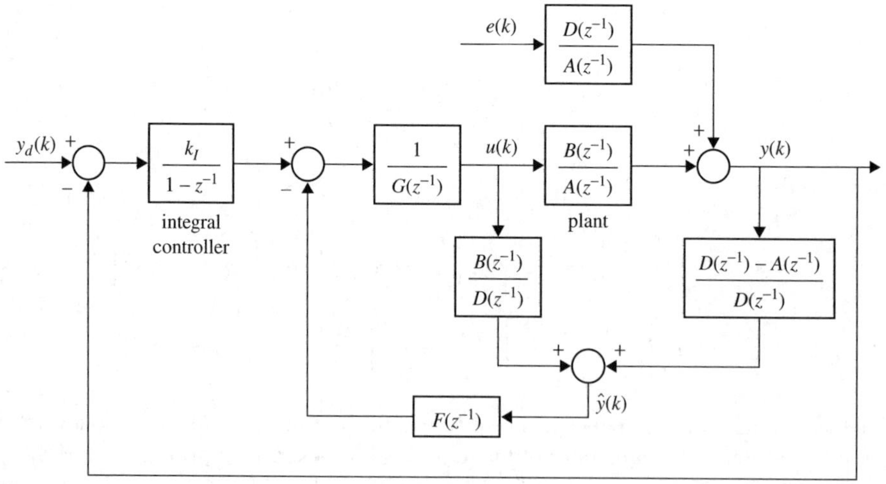

Figure 6.18 Polynomial implementation of the PIP-LQG control system with the integral controller operating on the measured output

Example 6.6 Kalman Filter Design for Noise Attenuation Consider again the non-minimum phase oscillator (6.10) and associated deterministic PIP-LQ controller developed in Chapter 5 (Example 5.8), i.e. based on the weights (5.78) with $q_y = 1/2$, $q_u = 1/3$ and $q_e = 0.1$, which yields the following control gain vector:

$$\boldsymbol{k}^T = [\, f_o \quad f_1 \quad g_1 \quad g_2 \quad -k_I \,] = [\, 0.597 \quad -1.008 \quad 1.226 \quad 2.017 \quad -0.109 \,] \tag{6.49}$$

With regard to Kalman Filter design, the system is represented using the stochastic NMSS model (6.33), a measurement noise variance $\sigma^2 = 0.1$ and the following state covariance matrix chosen to provide a reasonably balanced amount of noise filtration:

$$\boldsymbol{N} = \begin{bmatrix} 0.003 & 0 & 0 & 0 \\ 0 & 0.003 & 0 & 0 \\ 0 & 0 & 0 & 0 \\ 0 & 0 & 0 & 0 \end{bmatrix} \tag{6.50}$$

The regulator NMSS state transition matrix is:

$$\boldsymbol{F} = \begin{bmatrix} -a_1 & -a_2 & b_2 & b_3 \\ 1 & 0 & 0 & 0 \\ 0 & 0 & 0 & 0 \\ 0 & 0 & 1 & 0 \end{bmatrix} = \begin{bmatrix} 1.7 & -1 & -1 & 2 \\ 1 & 0 & 0 & 0 \\ 0 & 0 & 0 & 0 \\ 0 & 0 & 1 & 0 \end{bmatrix} \tag{6.51}$$

and the observation vector is $\boldsymbol{h} = [\,1 \;\; 0 \;\; 0 \;\; 0\,]$.

In this case, the discrete-time, matrix Riccati equation (6.38) and equation (6.39) yield the following Kalman gain vector:

$$\boldsymbol{L} = [\, 0.365 \quad 0.353 \quad 0 \quad 0 \,]^T \tag{6.52}$$

and equation (6.45) becomes

$$\begin{bmatrix} \tilde{l}_1 \\ \tilde{l}_2 \end{bmatrix} = \begin{bmatrix} 1 & 0 \\ 0 & -a_2 \end{bmatrix} \begin{bmatrix} l_1 \\ l_2 \end{bmatrix} = \begin{bmatrix} 1 & 0 \\ 0 & -1 \end{bmatrix} \begin{bmatrix} 0.365 \\ 0.353 \end{bmatrix} = \begin{bmatrix} 0.365 \\ -0.353 \end{bmatrix} \tag{6.53}$$

Hence, $D(z^{-1}) = 1 + (\tilde{l}_1 + a_1)z^{-1} + (\tilde{l}_2 + a_2)z^{-2} = 1 - 1.335z^{-1} + 0.647z^{-2}$ and the optimal estimate of the output (6.47) is determined from:

$$\hat{y}(k) = \frac{0.365z^{-1} - 0.353z^{-2}}{1 - 1.335z^{-1} + 0.647z^{-2}} y(k) + \frac{-z^{-2} + 2.0z^{-3}}{1 - 1.335z^{-1} + 0.647z^{-2}} u(k) \tag{6.54}$$

with the final control algorithm implemented as shown by Figure 6.17.

The response of the PIP-LQG and standard PIP-LQ (i.e. with and without the filtering) closed-loop system is illustrated by Figure 6.19 and Figure 6.20. The main advantage of using

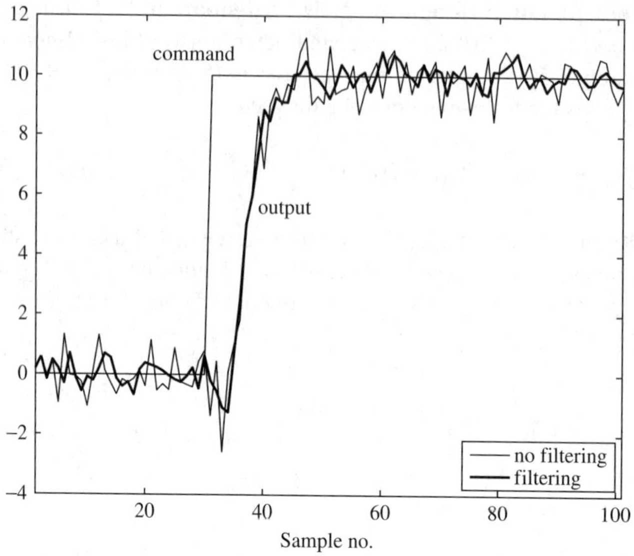

Figure 6.19 Closed-loop response for a unit step in the command level, comparing the output of the stochastic system both with (stochastic PIP-LQG design) and without (deterministic PIP-LQ design) the Kalman Filter

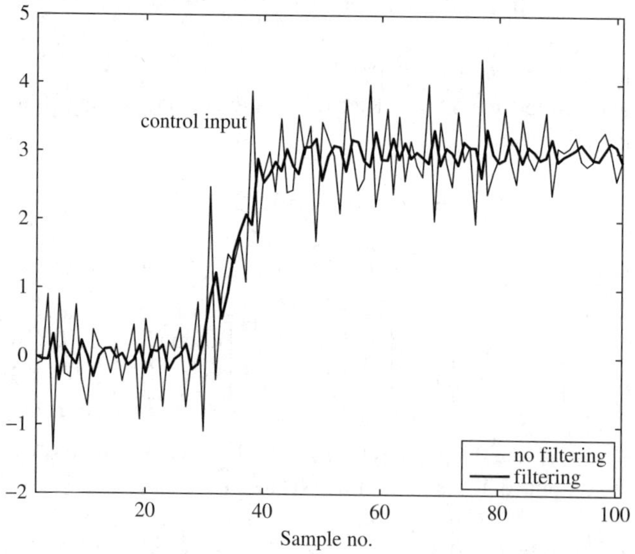

Figure 6.20 Control inputs associated with Figure 6.19

an optimally filtered output variable is that the control input is generally smoothed in some way, as demonstrated by Figure 6.20.

6.4.3 Stochastic Closed-loop System

Algebraic or block diagram analysis of Figure 6.17, reveals that the closed-loop response is given by the deterministic closed-loop TF (6.3) and a stochastic component, as follows:

$$\begin{aligned} y(k) = & \frac{k_I B(z^{-1})}{\Delta\left(G(z^{-1})A(z^{-1}) + F(z^{-1})B(z^{-1})\right) + k_I B(z^{-1})} y_d(k) \\ & + \frac{\Delta\left(G(z^{-1})D(z^{-1}) + F(z^{-1})B(z^{-1})\right) + k_I B(z^{-1})}{\Delta\left(G(z^{-1})A(z^{-1}) + F(z^{-1})B(z^{-1})\right) + k_I B(z^{-1})} e(k) \end{aligned} \tag{6.55}$$

Similarly for Figure 6.18:

$$\begin{aligned} y(k) = & \frac{k_I B(z^{-1})}{\Delta\left(G(z^{-1})A(z^{-1}) + F(z^{-1})B(z^{-1})\right) + k_I B(z^{-1})} y_d(k) \\ & + \frac{\Delta\left(G(z^{-1})D(z^{-1}) + F(z^{-1})B(z^{-1})\right)}{\Delta\left(G(z^{-1})A(z^{-1}) + F(z^{-1})B(z^{-1})\right) + k_I B(z^{-1})} e(k) \end{aligned} \tag{6.56}$$

These take a similar form to the deterministic disturbance response TF models considered earlier, i.e. equation (6.7) and equation (6.8). In the present analysis, however, the numerator of the stochastic component includes the expression $G(z^{-1})D(z^{-1})$, which contrasts with the $G(z^{-1})A(z^{-1})$ term in the denominator.

It is clear that, as usual for observer design, the $D(z^{-1})$ polynomial cancels out in the command response (Clarke 1994; Yoon and Clarke 1994). By contrast, the stochastic component in equation (6.55) relates to the disturbance vector $\boldsymbol{w}(k)$ of the stochastic NMSS model (6.33). As discussed above, the Kalman Filter is designed to attenuate the measurement noise $\varepsilon(k)$ but to pass the state disturbances $\boldsymbol{w}(k)$. The latter are regarded as 'real' disturbances that have to be handled by the control system and, for example, could be dealt with by feed-forward control (see section 6.5). Of course, Figure 6.17 and Figure 6.18 have actually been derived from the innovations representation (6.34), hence the noise is encompassed by the $D(z^{-1})/A(z^{-1})$ TF model.

In the ideal case represented by Figure 6.17 or Figure 6.18, it is straightforward to verify that $\hat{y}(k) = y(k) - e(k)$ exactly. In this context, it is interesting to also consider the special case of a white noise disturbance i.e. $D(z^{-1}) = A(z^{-1})$. Here, the open-loop TF model (6.43) is:

$$y(k) = \frac{B(z^{-1})}{A(z^{-1})} u(k) + e(k) \tag{6.57}$$

Since $D(z^{-1}) - A(z^{-1}) = 0$, the optimal estimate of the output (6.47) is determined using the deterministic open-loop TF model, as follows:

$$\hat{y}(k) = \frac{B(z^{-1})}{A(z^{-1})} u(k) \tag{6.58}$$

This is a very unsatisfactory solution in practical terms because $\hat{y}(k)$ is generated from $u(k)$, with no reference at all to the measured output, and it clearly exposes some lack of robustness in the Kalman Filter when used in the minimal state feedback context.

To further illustrate the implication of a white noise disturbance $D(z^{-1}) = A(z^{-1})$, consider Figure 6.17 or Figure 6.18 again. Combining equation (6.58) and Figure 6.18 yields the block diagram for the deterministic forward path PIP controller (Figure 6.4). Equivalently, the closed-loop response (6.56) based on Figure 6.18 reduces to the forward path load disturbance closed-loop response (6.8). This interpretation shows that, in the absence of state disturbance noise, forward path PIP design is optimal.

By contrast, with $D(z^{-1}) = A(z^{-1})$, equation (6.55) based on Figure 6.17 becomes:

$$y(k) = \frac{k_I B(z^{-1})}{\Delta \left(G(z^{-1})A(z^{-1}) + F(z^{-1})B(z^{-1}) \right) + k_I B(z^{-1})} y_d(k) + e(k) \tag{6.59}$$

In other words, when the output is contaminated by white measurement noise only, Figure 6.17 simply yields the closed-loop form of equation (6.57).

6.4.4 *Other Stochastic Control Structures*

At this juncture, it should be remembered that the NMSS approach to control system design does not need any form of state reconstruction. In essence, therefore, the Kalman Filter is only required for noise attenuation and not for state estimation per se. Furthermore, like the forward path implementation of basic PIP control, the Kalman Filter carries with it the disadvantage of decreased robustness, as revealed also by equation (6.58): in particular, the state estimation is dependent upon the input-driven model of the system and the cancellation of the observer dynamics.

This limitation has led to the concept of *Loop Transfer Recovery* (LTR: see e.g. Maciejowski 1985; Bitmead *et al.* 1990), which attempts to recover the robustness of the full state feedback properties. However, full state feedback is always achieved in the NMSS case, so LTR is not required. For this reason, it is sometimes advantageous to utilise different, nominally sub-optimal, approaches to noise attenuation which do not involve the Kalman Filter and do not necessitate complicated design manipulation such as LTR. In this regard, it should be stressed that Figure 6.17 and Figure 6.18 illustrate just two possible polynomial implementations of the PIP-LQG control system, among a wide range of other options available in the literature. For example, Dixon (1996) and Taylor (1996) consider a number of other approaches to noise attenuation that utilise either the Kalman Filter (implemented in various forms, including a special non-minimal filter), explicit pre-filtering of the non-minimal states and/or the introduction of additional filters.

One limitation of the stochastic model (6.43), is that it does not allow for non-stationary (time-varying) disturbance inputs to the system. This problem manifests itself in the form of a non-zero steady-state error, if the controller shown by Figure 6.17 is utilised on a system that is affected by such disturbances. This is verified by noting that the steady-state gain of the stochastic component of equation (6.55) is unity. To combat the problem, we could follow a similar approach to forward path PIP control, by utilising Figure 6.18, where the steady-state gain of the stochastic component of equation (6.56) is zero, as required. Arguably, however, a more satisfactory solution in the context of optimal filtering, can be achieved by explicitly representing the disturbance in the NMSS model, as shown below.

6.4.5 *Modified Kalman Filter for Non-Stationary Disturbances*

To address the problem of non-stationary disturbances, consider the following modified innovations representation, which is based on equations (6.34) but with the introduction of a random walk state variable $r(k)$ to represent the unknown disturbance:

$$\left.\begin{aligned} \begin{bmatrix} \hat{\boldsymbol{x}}(k+1/k) \\ r(k+1/k) \end{bmatrix} &= \begin{bmatrix} \boldsymbol{F} & \boldsymbol{0}_{m+n,1} \\ \boldsymbol{0}_{1,n+m} & 1 \end{bmatrix} \begin{bmatrix} \hat{\boldsymbol{x}}(k/k-1) \\ r(k/k-1) \end{bmatrix} + \begin{bmatrix} \boldsymbol{g} \\ 0 \end{bmatrix} u(k) + \begin{bmatrix} \boldsymbol{L}(k) \\ l_{n+1}(k) \end{bmatrix} e(k) \\ y(k) &= \begin{bmatrix} \boldsymbol{h} & 1 \end{bmatrix} \begin{bmatrix} \hat{\boldsymbol{x}}(k/k-1) \\ r(k/k-1) \end{bmatrix} + e(k) \end{aligned}\right\} \tag{6.60}$$

where $\boldsymbol{0}$ represents a vector of $n+m$ zeros, while $r(k) = r(k-1) + \xi(k)$, in which $\xi(k)$ is a zero mean white noise sequence with constant variance. As usual, the Kalman gain vector obtained from equations (6.60), including the gain associated with $r(k)$, is estimated using the steady-state solution of the discrete-time, matrix Riccati equations.

Following again a similar approach to Young (1979) and Hesketh (1992), the polynomial representation of the modified system is the following *Auto–Regressive, Integrated Moving-Average eXogenous* variables (ARIMAX) model[4]:

$$y(k) = \frac{B(z^{-1})}{A(z^{-1})}u(k) + \frac{D_1(z^{-1})}{\Delta A(z^{-1})}e(k) \tag{6.61}$$

where $D_1(z^{-1}) = D(z^{-1})\Delta + l_{n+1}A(z^{-1})$. Utilising equation (6.36) and rearranging (6.61) yields the optimal estimate of the output:

$$\hat{y}(k) = \frac{D_1(z^{-1}) - \Delta A(z^{-1})}{D_1(z^{-1})}y(k) + \frac{\Delta B(z^{-1})}{D_1(z^{-1})}u(k) \tag{6.62}$$

The PIP control algorithm is now implemented in either SVF form based on equation (6.35) and equation (6.37), or in polynomial form in a similar manner to Figure 6.17. In both cases, the

[4] This is sometimes called the CARIMA (Controlled Auto-Regressive Integrated Moving-Average) model.

optimal estimate $\hat{y}(k)$ now includes the difference operator $\Delta = 1 - z^{-1}$, which automatically introduces the necessary integral action into the observer.

The discussion above has utilised the stochastic NMSS model to develop PIP-LQG control systems. In section 6.4.6, we will briefly consider the introduction of a risk sensitive, *Linear Exponential-of-Quadratic Gaussian* (LEQG) cost function.

6.4.6 Stochastic PIP Control using a Risk Sensitive Criterion

Robust control design aims to minimise problems caused by uncertainty in the control system (e.g. Green and Limebeer 1995). For example, H_∞ control ensures that bounds on the H_∞-norm of certain TF models, representing the relationship between disturbance signals and particular variables of interest, are achieved. This contrasts with the more conventional H_2 approach where, in the optimal stochastic LQG case considered above, the deterministic control law involves the minimisation of the LQ criterion:

$$J = \sum_{k=0}^{N-1} \boldsymbol{x}(k)^T \boldsymbol{Q}\boldsymbol{x}(k) + r(u(k)^2) \tag{6.63}$$

in which $\boldsymbol{Q}$ is a symmetric, positive semi-definite matrix, r is a positive scalar and N is an integer number of samples (the optimisation window). The cost function is presented in this $N-1$ step ahead form for the purposes of the later discussion on predictive control (section 6.6). However, $N \to \infty$ ensures equivalence with the standard implementation of optimal PIP-LQ control introduced in Chapter 5, i.e. the steady-state or infinite time solution (5.75).

Unlike in the LQG formulation, a particular H_∞ criterion does not yield a unique solution. One approach, for example, is the minimum entropy solution of Mustafa and Glover (1990) which, in turn, is closely related to the following LEQG cost function:

$$\gamma(\theta) = -2\theta^{-1} \log_e E\left\{e^{-\theta J/2}\right\} \tag{6.64}$$

where θ a scalar 'risk-sensitive' parameter and $E\{\cdot\}$ represents the expectation (Jacobson 1973; Kumar and van Schuppen 1981; Whittle 1981, 1990). An interesting result in the present context is the relationship between LEQG optimisation and the H_∞ approach. This relationship means that the LEQG formulation has certain robustness properties that enhance its potential importance to control engineers (Glover and Doyle 1988).

To develop the PIP-LEQG control system, we first define a stochastic servomechanism NMSS model, similar to equations (6.33) but here including the integral-of-error state variable. The control gains are obtained using the following backwards Riccati recursion:

$$\left.\begin{aligned} \boldsymbol{P}(k) &= \boldsymbol{Q} + \boldsymbol{F}^T\left[\boldsymbol{g}r^{-1}\boldsymbol{g}^T + \theta\boldsymbol{N} + \boldsymbol{P}(k+1)^{-1}\right]^{-1}\boldsymbol{F} \\ \boldsymbol{k}^T(k) &= r^{-1}\boldsymbol{g}^T\left[\boldsymbol{g}r^{-1}\boldsymbol{g}^T + \theta\boldsymbol{N} + \boldsymbol{P}(k+1)^{-1}\right]^{-1}\boldsymbol{F} \end{aligned}\right\} \tag{6.65}$$

where $\boldsymbol{k}(k)$ is the state variable feedback control gain matrix (Whittle 1981). In contrast to the minimal approach, the PIP-LEQG algorithm may be implemented in entirely deterministic form without filtering if desired, i.e. the NMSS equations employed here ensure that it is possible to implement what is referred to, in this context, as the 'complete observations' solution. In this case, LEQG design does not include any consideration of the measurement noise.

In deterministic PIP-LEQG design, solving (6.65) with $\theta > 0$ generally moves the closed-loop pole positions towards the origin of the complex z-plane and so increases the speed of response in comparison with the PIP-LQ solution (Taylor *et al.* 1996). By contrast, with $\theta < 0$, the control input signal is generally smoother than in the LQ case, but at the expense of a slower output response. In other words, θ is essentially an additional tuning parameter for optimal PIP control design. For this reason, it is best combined with a multi-objective optimisation approach to control system design, as discussed in Chapter 7.

6.5 Generalised NMSS Design

One advantage of the NMSS model is that the state vector is readily extended to account for the availability of additional measured or estimated information. As we have seen, such additional states can be utilised to develop more sophisticated algorithms that address specific control problems. This concept is illustrated by the modified Kalman Filter for non-stationary disturbances (6.60). In the present section, the flexibility of the NMSS model is further illustrated using two new examples, namely feed-forward control design and command input anticipation.

6.5.1 Feed-forward PIP Control based on an Extended Servomechanism NMSS Model

Here, we aim to improve the response of the controller to measured system disturbances (see e.g. Figure 6.6), by utilising these measurements in a *feed-forward* structure. Therefore, the following multi-input, single-output TF model is estimated using the system identification tools in Chapter 8. The model takes a similar form to equation (6.43), but the unknown noise signal is replaced with the measured disturbance $v(k)$, i.e.

$$y(k) = \frac{B(z^{-1})}{A(z^{-1})}u(k) + \frac{D(z^{-1})}{A(z^{-1})}v(k) \tag{6.66}$$

Although equation (6.66) suffices for the present illustrative discussion, generalised versions of this model are also possible, including multiple disturbance signals with independent denominator polynomials (Young *et al.* 1994). In conventional feed-forward design, we derive a control structure algebraically to exactly cancel the disturbance signal in the zero model mismatch case (e.g. Ogata 2001). This cancellation approach can also be utilised for PIP design (Cross *et al.* 2011), but an alternative is to extend the NMSS model, as follows:

$$\left.\begin{aligned} \boldsymbol{x}(k) &= \tilde{\boldsymbol{F}}\boldsymbol{x}(k-1) + \tilde{\boldsymbol{g}}u(k-1) + \tilde{\boldsymbol{d}}\, y_d(k) + \tilde{\boldsymbol{b}}\, v(k) \\ y(k) &= \tilde{\boldsymbol{h}}\boldsymbol{x}(k) \end{aligned}\right\} \tag{6.67}$$

where the non-minimal state vector is, for example

$$\boldsymbol{x}(k) = [y(k) \quad \cdots \quad y(k-n) \;\; u(k-1) \cdots u(k-m+1) \;\; z(k) \;\; v(k-N+1) \cdots v(k)]^{\mathrm{T}} \tag{6.68}$$

in which N is the order of the disturbance polynomial $D(z^{-1})$. Here, $\tilde{\boldsymbol{F}}, \tilde{\boldsymbol{g}}, \tilde{\boldsymbol{d}}, \tilde{\boldsymbol{b}}$ and $\tilde{\boldsymbol{h}}$ are defined in the obvious way, following a similar approach to the nominal servomechanism NMSS model of Chapter 5, but here extended to represent equation (6.66) in difference equation form. The SVF control gain vector:

$$\boldsymbol{k}^T = [f_0 \quad f_1 \quad \cdots \quad f_{n-1} \quad g_1 \quad \cdots \quad g_{m-1} \quad -k_I \quad -p_N \quad \cdots \quad -p_1 \quad -p_0] \tag{6.69}$$

is obtained by minimising the LQ cost function (5.75) or (6.63) in the usual manner[5]. In block diagram terms, an additional control filter operates on the measured disturbance signal:

$$P(z^{-1}) = p_0 + p_1 z^{-1} + \cdots + p_{N-1} z^{-(N-1)} \tag{6.70}$$

Young *et al.* (1994), for example, utilise a feed-forward PIP approach based on these equations to control temperature in a greenhouse, significantly reducing the effects of the solar radiation disturbance. More recently, Cross *et al.* (2011) developed feed-forward PIP control systems for regulating the power take-off of a nonlinear wave energy convertor simulation. In this case, the feed-forward approach addresses the influence of the wave force disturbance, ultimately improving power capture.

6.5.2 *Command Anticipation based on an Extended Servomechanism NMSS Model*

In many applications of control, the command level is either constant or any adjustments are known well in advance. In the case of robotic manipulators, for example, the joint angle trajectory is often determined by solving the kinematics offline. This is the approach taken for the laboratory excavator discussed in Chapter 5 (Example 5.2), in which the PIP controlled manipulator joint angles are planned at the start of each manoeuvre. In this situation, the NMSS model can be extended to handle N future known commands by simply appending the state vector with these values (Taylor *et al.* 2000a):

$$\boldsymbol{x}(k) = [y(k) \quad \cdots \quad y(k-n) \quad u(k-1) \quad \cdots \quad u(k-m+1) \quad z(k) \quad \boldsymbol{y}_d(k)^T]^T \tag{6.71}$$

[5] For consistency with the later discussion on command anticipation and model predictive control, the order of the disturbance polynomial is given by N, which is also the number of iterations of the finite-horizon LQ cost function (6.63). However, this connection is not a general requirement of the feed-forward PIP formulation. Indeed, the practical implementation results discussed by Young *et al.* (1994) are based on the infinite horizon cost function (5.75), with the order of the disturbance polynomial obtained independently using the system identification algorithms discussed in Chapter 8.

where

$$\mathbf{y}_d(k) = [y_d(k+1) \quad y_d(k+2) \; \ldots \; y_d(k+N)]^T \tag{6.72}$$

The command input NMSS form with these modifications is:

$$\left.\begin{aligned} \mathbf{x}(k) &= \tilde{\mathbf{F}}\mathbf{x}(k-1) + \tilde{\mathbf{g}}u(k-1) + \tilde{\mathbf{d}}y_d(k+N) \\ y(k) &= \tilde{\mathbf{h}}\mathbf{x}(k) \end{aligned}\right\} \tag{6.73}$$

where

$\tilde{\mathbf{g}} = [\mathbf{g} \quad 0 \quad \cdots \quad 0 \quad 0]^T$, $\tilde{\mathbf{d}} = [0 \quad 0 \quad \cdots \quad 0 \quad \mathbf{d}]^T$, $\tilde{\mathbf{h}} = [\mathbf{h} \quad 0 \quad \cdots \quad 0 \quad 0]$ and

$$\tilde{\mathbf{F}} = \begin{bmatrix} \mathbf{F} & \mathbf{d} & \mathbf{O}_{n+m,N-1} \\ \mathbf{O}_{N-1,n+m} & \mathbf{O}_{N-1,1} & \mathbf{I}_{N-1} \\ 0 & \ldots & 0 \end{bmatrix} \tag{6.74}$$

in which $\mathbf{F}$, $\mathbf{g}$ and $\mathbf{d}$ are from the servomechanism NMSS model, i.e. they are defined by equation (5.15) and equation (5.16). Finally, $\mathbf{O}$ is an appropriately defined matrix of zeros and $\mathbf{I}_{N-1}$ an identity matrix.

The extended PIP control algorithm is implemented in either SVF or incremental form, with the SVF control gain vector (6.69) again obtained by minimising the infinite-horizon LQ cost function (5.75). Because of the integral-of-error state variable, and in contrast to the minimal approach (see e.g. Bitmead *et al.* 1990), no special LQ state weighting arrangements are required to ensure steady-state tracking. Therefore $\mathbf{Q}$ is usually formed as a diagonal matrix similar to equation (5.77), with the additional elements set to zero:

$$\mathbf{Q} = diag\left(q_y \quad \cdots \quad q_y \quad q_u \quad \cdots \quad q_u \quad q_e \quad 0 \quad \cdots \quad 0\right) \tag{6.75}$$

In fact, the command input is clearly not controllable, so the choice of associated weights in the $\mathbf{Q}$ matrix has no effect on the final PIP gains. Furthermore, for a given model and LQ weights, the gain associated with each future command input is always the same, regardless of the value of N. Hence, above a certain level, the closed-loop response is not particularly sensitive to the value of N and selecting a high value (say greater than 10) has little effect, since the gains trail off to zero.

In block diagram terms, the controller can be implemented as shown in Figure 6.21, in which $P(z^{-1})$ is given by equation (6.70).

Note that $P(z^{-1})$ operates on the N-step ahead command input $y_d(k+N)$, whilst that the integral-of-error state variable $z(k)$ is defined using the 'current' command $y_d(k)$. Inspection of Figure 6.21 shows:

$$u(k) = \frac{1}{G(z^{-1})}\left(-F(z^{-1})y(k) + P(z^{-1})y_d(k+N) + \frac{k_I}{\Delta}\left(z^{-N}y_d(k+N) - y(k)\right)\right) \tag{6.76}$$

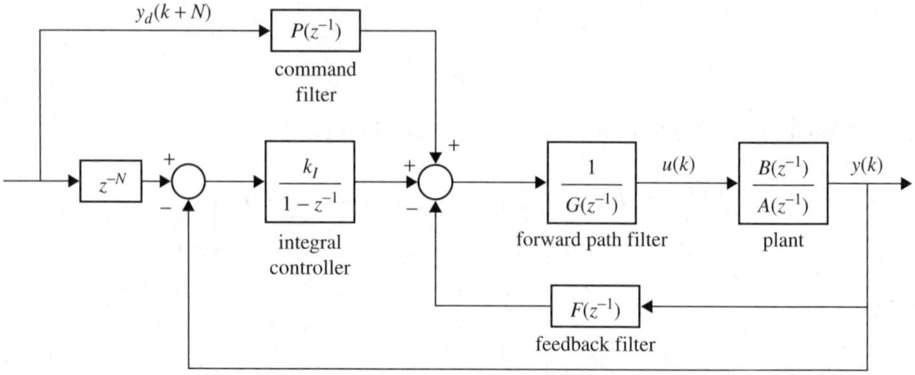

Figure 6.21 Feedback PIP control with command input anticipation

Rearranging gives:

$$G(z^{-1})\Delta u(k) = -\left(F(z^{-1})\Delta + k_I\right) y(k) \;+\; \left(k_I z^{-N} + P(z^{-1})\Delta\right) y_d(k+N) \tag{6.77}$$

Further algebraic or block diagram analysis yields:

$$y(k) = \frac{B(z^{-1})\left(k_I z^{-N} + P(z^{-1})\Delta\right)}{\Delta\left(G(z^{-1})A(z^{-1}) + F(z^{-1})B(z^{-1})\right) + k_I B(z^{-1})} y_d(k+N) \tag{6.78}$$

Here, only the numerator polynomial of the closed-loop TF is modified in comparison with equation (6.3), hence LQ design yields the same closed-loop poles as for the nominal case. Furthermore, since $\Delta = 1 - z^{-1} \rightarrow 0$ for $k \rightarrow \infty$, the steady-state gain of equation (6.78) is unity and Type 1 servomechanism performance is obtained, as required.

Other NMSS implementations of command anticipation are also possible, including forward path and stochastic solutions. Finally, as Hesketh (1992) points out, the z^{-N} time delay employed in Figure 6.21 could be replaced by a TF model, in order to induce a model following response in the closed-loop.

Example 6.7 Command Input Anticipation Design Example There are two major advantages of introducing the future command input states: in the first place, the closed-loop system anticipates changes and reacts to the command input sooner; and, secondly, simulation studies show that the control input is generally smoothed in some manner. An attractive consequence of the latter point can be seen in the control of the familiar model (6.10), where the open-loop non-minimum phase characteristics have been completely eliminated, as illustrated by Figure 6.22.

To emphasise the difference between the command anticipation and nominal responses, the PIP-LQ controller in Figure 6.22 is based on an increased integral-of-error weighting

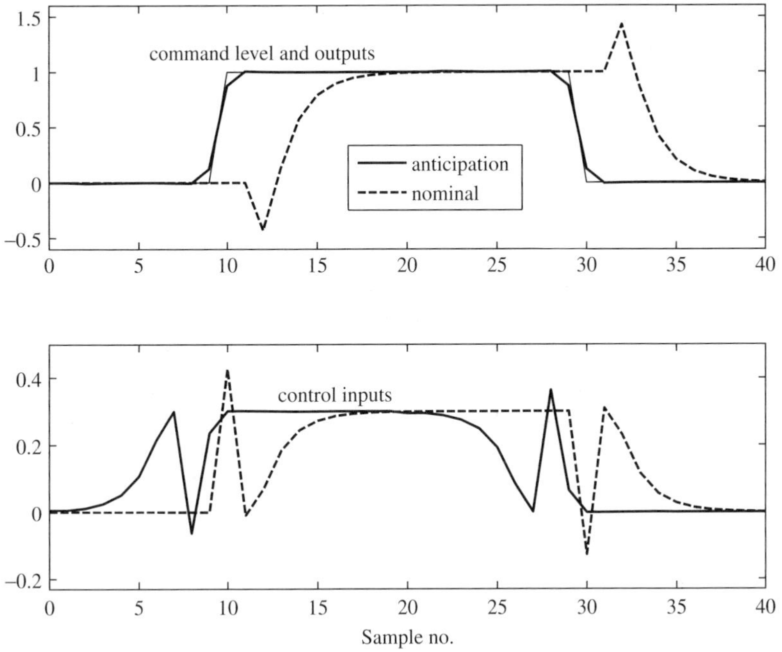

Figure 6.22 Closed-loop response for positive and negative unit steps in the command level, comparing the output of the PIP-LQ optimal control system with (bold solid trace) and without (dashed trace) a command anticipation filter

compared with Example 6.1, hence a faster closed-loop response, i.e. based on the weights (5.78) with $q_y = 1/2$, $q_u = 1/3$ and $q_e = 10$, which yields:

$$\mathbf{k}^T = [\, f_0 \quad f_1 \quad g_1 \quad g_2 \quad -k_I \,] = [\, 2.440 \quad -2.877 \quad 2.023 \quad 5.753 \quad -0.428 \,] \tag{6.79}$$

With command anticipation based on $N = 10$, the above SVF gains remain unchanged, whilst the additional gains (6.70) are $p_0 = 0.006$, $p_1 = 0.012$, $p_2 = 0.023$, $p_3 = 0.045$, $p_4 = 0.089$, $p_5 = 0.182$, $p_6 = 0.369$, $p_7 = 0.611$, $p_8 = 0.428$ and $p_9 = 0.428$. Another example of PIP command anticipation, in which the NMSS model is stated in full for tutorial purposes, is discussed later (Example 6.8).

6.6 Model Predictive Control

Model Predictive Control (MPC) is a popular approach to control system design with many successful applications, particularly in the chemical and process engineering industries (Morari and Lee 1999). An attractive feature of MPC is the ability to handle constraints on the system variables, such as equations (6.15), as an inherent part of the design process, albeit at the cost of increased complexity (Maciejowski 2002; Rossiter 2003). MPC algorithms based on minimal state space models typically utilise an observer and exploit the separation theorem in a similar manner to equation (6.37). Hence, much research effort has been directed towards increasing the robustness of MPC design and to reducing the online computational load (e.g. Bemporad *et al.* 2002).

6.6.1 Model Predictive Control based on NMSS Models

As we have seen, PIP control systems are generally implemented using an incremental form, in order to avoid integral 'wind-up' problems. To address more general system constraint problems, such as limits on the output signal and other measured variables, we will consider later the use of multi-objective optimisation techniques where the PIP-LQ weights are optimised offline using computer simulation, as described in Chapter 7. However, in recent years, various deterministic NMSS model structures have also been utilised for MPC design (see e.g. Taylor *et al.* 2000a; Wang and Young 2006; Exadaktylos *et al.* 2009; Gonzales *et al.* 2009). The latter citations all define a modified form of the regulator NMSS model, in which differenced values of the control input signal $\Delta u(k)$ are utilised as state variables. By contrast, the relative merits of various integral action schemes for MPC have been considered by Pannocchia and Rawlings (2003), and these results are exploited by Exadaktylos and Taylor (2010) to develop MPC systems using the servomechanism NMSS model of Chapter 5.

The MPC cost function is generally solved online to address the problem of system constraints. In particular, the future constrained control input variables, $u(k)$, $u(k+1)$, $u(k+2)$ through to $u(k+N)$, are determined by solving a quadratic optimisation problem at every sampling instant k. Following standard MPC practice, a receding-horizon approach is subsequently employed, with only the first of these actually utilised at each control sample. By contrast, the focus of the present book is on *linear control theory* and on the development of fixed gain controllers that can be analysed in block diagram terms. In this context, an important step in the evolution of MPC is the *Generalised Predictive Control* (GPC) approach of Clarke *et al.* (1987). Hence, the final part of this chapter focuses on the relationship between GPC and PIP control.

6.6.2 Generalised Predictive Control

Unlike PIP control, GPC was not originally formulated overtly in an optimal state space context. Rather, it is normally presented in an alternative receding-horizon optimal setting, defined by the following cost function:

$$J = E\left\{\sum_{j=1}^{N}(y(k+j) - y_d(k+j))^2 + \sum_{j=1}^{NU}\lambda\left(\Delta u(k+j-1)\right)^2\right\} \tag{6.80}$$

where N and NU are the output and input costing horizons, respectively, and $E\{\cdot\}$ represents the expectation.

Here, the control strategy is posed in terms of the differenced input signal, i.e. $\Delta u(k) = u(k) - u(k-1)$. The approach utilises the ARIMAX model (6.61), which can be rearranged and restated as follows:

$$\tilde{A}(z^{-1})y(k) = B(z^{-1})\Delta u(k) + D(z^{-1})\,e(k) \tag{6.81}$$

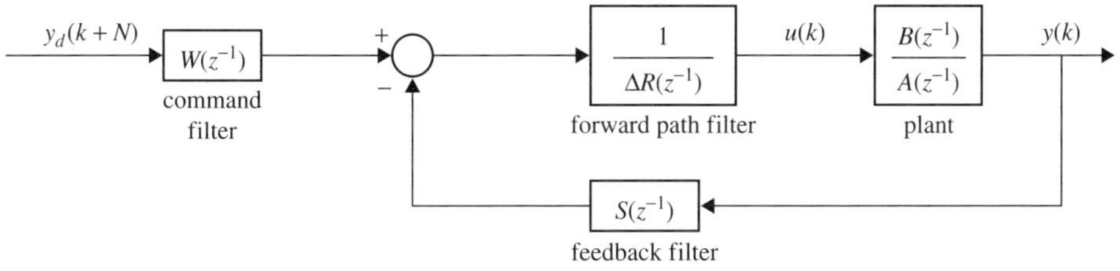

Figure 6.23 The NMSS servomechanism system in predictive form

where $\tilde{A}(z^{-1}) = \Delta A(z^{-1}) = 1 + \tilde{a}_1 z^{-1} + \cdots + \tilde{a}_{n+1} z^{-(n+1)}$, $A(z^{-1})$ and $B(z^{-1})$ are defined by equation (6.1), and the generalised noise polynomial $D(z^{-1}) = 1 + d_1 z^{-1} + \cdots + d_p z^{-p}$.

The simplest form of GPC is based on the ARIMAX system model (6.81) with the noise polynomial set to unity, i.e. $D(z^{-1}) = 1$. We will consider the general case using the noise polynomial for filtering purposes later on. Minimising the cost function (6.80) yields:

$$\Delta u(k) = \boldsymbol{g}^T(\boldsymbol{y}_d - \boldsymbol{f}) \tag{6.82}$$

in which $\boldsymbol{g}^T = [g_1 \ g_2 \ \ldots \ g_N]$ is a vector of gain coefficients, $\boldsymbol{y_d}$ is a vector of future command inputs (6.72) and $\boldsymbol{f} = [f(k+1) \ \ f(k+2) \ \ \ldots \ \ f(k+N)]^T$ is called the model free response prediction. As shown, for example, by Clarke *et al.* (1987) and Bitmead *et al.* (1990), evaluation of equation (6.82) yields a fixed gain control algorithm that can be conveniently expressed in block diagram form, as illustrated by Figure 6.23.

The control polynomials in Figure 6.23 are defined:

$$\left.\begin{aligned} S(z^{-1}) &= s_0 + s_1 z^{-1} + \cdots + s_n z^{-n} \\ R(z^{-1}) &= 1 + r_1 z^{-1} + \cdots + r_{m-1} z^{-(m-1)} \\ W(z^{-1}) &= w_0 + w_1 z^{-1} + \cdots + w_{N-1} z^{-(N-1)} \end{aligned}\right\} \tag{6.83}$$

Using Figure 6.23, the GPC algorithm can be expressed:

$$R(z^{-1})\Delta u(k) \ = \ -S(z^{-1})y(k) \ + W(z^{-1})y_d(k+N) \tag{6.84}$$

Further algebraic or block diagram analysis yields the closed-loop TF:

$$y(k) = \frac{W(z^{-1})B(z^{-1})}{\Delta R(z^{-1})A(z^{-1}) + S(z^{-1})B(z^{-1})} y_d(k+N) \tag{6.85}$$

Note that $s_0 + s_1 + \cdots + s_n = w_0 + w_1 + \cdots + w_{N-1}$, hence the steady-state gain of closed-loop system (6.85) is unity. The integral action implicit in the model (6.81), ensures an integrator in the forward path control filter and so yields Type 1 servomechanism performance.

6.6.3 *Equivalence Between GPC and PIP Control*

Returning now to the command anticipation form of PIP control, Figure 6.21 can be algebraically transformed so that it conforms to the same structure as GPC in Figure 6.23 (Taylor *et al.* 1994, 2000a). In fact, comparing equation (6.77) and equation (6.84), it is straightforward to determine the equivalent terms, as follows:

$$\left.\begin{aligned} S(z^{-1}) &\equiv \left(F(z^{-1})\Delta + k_I\right) = F^*(z^{-1}) \\ R(z^{-1}) &\equiv G(z^{-1}) \\ W(z^{-1}) &\equiv \left(k_I z^{-N} + P(z^{-1})\Delta\right) = P^*(z^{-1}) \end{aligned}\right\} \tag{6.86}$$

where $F^*(z^{-1})$ and $P^*(z^{-1})$ can be considered as modified versions of the $F(z^{-1})$ and $P(z^{-1})$ polynomials in the command anticipation PIP controller.

Furthermore, it can be seen from their definitions in equation (6.2) and equation (6.83), that the GPC polynomials $S(z^{-1})$, $R(z^{-1})$ and $W(z^{-1})$ are of the same order as their PIP equivalents, $F^*(z^{-1})$, $G(z^{-1})$ and $P^*(z^{-1})$. Note, however, that the control gains and corresponding closed-loop response will only match if, for example, PIP pole assignment is employed to ensure the same pole positions or, as shown by Theorem 6.2, appropriate LQ conditions are specified.

The relationship between GPC and minimal LQG control is very well known (see e.g. Clarke *et al.* 1987; Bitmead *et al.* 1990). For NMSS design, the control gains are usually obtained from the infinite time solution of (6.63), with N set to infinity. However, in order to demonstrate full equivalence with the finite horizon cost function employed in GPC, Theorem 6.2 (Taylor *et al.* 2000a) instead selects suitable initial conditions for the matrix Riccati equations and only completes $N - 1$ recursions over the selected horizon.

Theorem 6.2 Equivalence Between GPC and (Constrained) PIP-LQ The command anticipation PIP-LQ control law, based on the NMSS model (6.73), is identical to that of the GPC algorithm (6.82) when $N = NU$ (for simplicity), if the following three conditions all hold:

(i) $\boldsymbol{Q} = \boldsymbol{h}_1^T \boldsymbol{h}_1$
(ii) $\boldsymbol{P}_0 = \boldsymbol{h}_1^T \boldsymbol{h}_1$
(iii) $r = \lambda$

Here, $\boldsymbol{Q}$ is the state weighting matrix and r is the scalar weight on the input in the LQ cost function (6.63), $\boldsymbol{P}_0$ is the initial value for the Riccati equation $\boldsymbol{P}$ matrix in equation (5.82), whilst $\boldsymbol{h}_1 = [0 \ldots 0\ 1\ 0 \ldots 0]$ is a specially defined vector, of dimension $n + m + N$, chosen to ensure that the LQ cost function weights apply only to the integral-of-error state.

The details of this theorem are similar to those in Bitmead *et al.* (1990), since the NMSS form is just another equivalent state space representation of the model (6.81). In particular, note that the GPC cost function (6.80) involves differencing the control input, which is equivalent to integrating the error term. Therefore, the basis for the equivalence is that the only state assigned a positive weighting in the $\boldsymbol{Q}$ matrix is the one equivalent to the GPC predicted error.

As in the case of Theorem 4.2 (for the relationship between minimal and non-minimal forms), it should be stressed that, while *any* GPC design can be exactly replicated with an equivalent PIP controller, the reverse is not true: the relationship only holds under the constraining conditions above. Finally, in this example, N in the LQ criterion (6.63) takes the same value as the GPC forecasting horizon. However, it is clear that this is not a general requirement of the command anticipation PIP formulation and, in practice, N is a free design parameter, independent of the number of iterations of the Riccati equation. To illustrate this, the PIP control gains (6.79) are based on the infinite solution of (6.63), while the order of the command anticipation filter, $N = 10$.

Example 6.8 Generalised Predictive Control and Command Anticipation PIP Control System Design In order to illustrate the discussion above, the example considered by Bitmead *et al.* (1990, p. 78) is extended to the non-minimal case. The system is defined by (6.81), with $D(z^{-1}) = 1$, $A(z^{-1}) = 1 - 0.7z^{-1}$, $\tilde{A}(z^{-1}) = 1 - 1.7z^{-1} + 0.7z^{-2}$ and $B(z^{-1}) = 0.9z^{-1} - 0.6z^{-2}$. Hence, $n = 1$ and $m = 2$, while the difference equation representation of the model is:

$$y(k) = 1.7y(k-1) - 0.7y(k-2) + 0.9\Delta u(k-1) - 0.6\Delta u(k-2) + e(k) \tag{6.87}$$

Selecting $N = NU = 3$ and $\lambda = 0.1$ in the GPC cost function (6.80) and evaluating equation (6.82) in the manner of Clarke *et al.* (1987), the GPC polynomials (6.83) for implementation using Figure 6.23 are:

$$\left.\begin{aligned} S(z^{-1}) &= 1.748 - 0.751z^{-1} \\ R(z^{-1}) &= 1 - 0.644z^{-1} \\ W(z^{-1}) &= 0.010 + 0.093z^{-1} + 0.895z^{-2} \end{aligned}\right\} \tag{6.88}$$

The non-minimal state vector (6.71) for PIP design is:

$$\boldsymbol{x}(k) = [\, y(k) \quad u(k-1) \quad z(k) \quad y_d(k+1) \quad y_d(k+2) \quad y_d(k+3)\,]^T \tag{6.89}$$

The command anticipation servomechanism NMSS model (6.73) is defined by:

$$\tilde{\boldsymbol{F}} = \begin{bmatrix} 0.7 & -0.6 & 0 & 0 & 0 & 0 \\ 0 & 0 & 0 & 0 & 0 & 0 \\ -0.7 & 0.6 & 1 & 1 & 0 & 0 \\ 0 & 0 & 0 & 0 & 1 & 0 \\ 0 & 0 & 0 & 0 & 0 & 1 \\ 0 & 0 & 0 & 0 & 0 & 0 \end{bmatrix}; \quad \tilde{\boldsymbol{g}} = \begin{bmatrix} 0.9 \\ 1 \\ -0.9 \\ 0 \\ 0 \\ 0 \end{bmatrix}; \quad \tilde{\boldsymbol{d}} = \begin{bmatrix} 0 \\ 0 \\ 0 \\ 0 \\ 0 \\ 1 \end{bmatrix}; \quad \tilde{\boldsymbol{h}}^T = \begin{bmatrix} 1 \\ 0 \\ 0 \\ 0 \\ 0 \\ 0 \end{bmatrix} \tag{6.90}$$

Using Theorem 6.2 to obtain equivalence with GPC, the LQ cost function weights are $r = 0.1$ and $\boldsymbol{Q} = \boldsymbol{P}_0 = \text{diag}[0 \;\; 0 \;\; 1 \;\; 0 \;\; 0 \;\; 0]$. Two iterations of the Riccati equation yield $k_I = 0.997$ and

$$\left.\begin{aligned} F(z^{-1}) &= f_0 = 0.751 \\ G(z^{-1}) &= 1 + g_1 z^{-1} = 1 - 0.644z^{-1} \\ P(z^{-1}) &= p_0 + p_1 z^{-1} + p_2 z^{-2} = 0.010 + 0.102z^{-1} + 0.997z^{-2} \end{aligned}\right\} \tag{6.91}$$

The PIP control system is implemented as shown by Figure 6.21. Finally, utilising equations (6.86), the following modified control polynomials are obtained:

$$\left.\begin{aligned} F^*(z^{-1}) &= (0.751\Delta + 0.997) = 1.748 - 0.751z^{-1} \\ P^*(z^{-1}) &= \left(0.997z^{-3} + P(z^{-1})\Delta\right) = 0.010 + 0.093z^{-1} + 0.895z^{-2} \end{aligned}\right\} \tag{6.92}$$

It is clear that $F^*(z^{-1}) = S(z^{-1})$, $P^*(z^{-1}) = W(z^{-1})$ and $G(z^{-1}) = R(z^{-1})$ as expected.

6.6.4 Observer Filters

For brevity, the discussion above has assumed that $D(z^{-1}) = 1$ in the model (6.81). However, we have already seen in section 6.4 that such an observer polynomial can be introduced into PIP design: the optimally filtered output is obtained using the asymptotic gain, innovations representation of the Kalman Filter, converted into TF form (6.62). Subsequently, the stochastic PIP control law can be implemented using a similar approach to Figure 6.17 or Figure 6.18 (modified to use the non-stationary disturbances version of the model).

Alternatively, the observer polynomial can be treated as a design 'parameter' [sometimes denoted by $T(z^{-1})$ to highlight its generic form]. For GPC, it is chosen by the designer to improve disturbance rejection or to improve the robustness of the closed-loop system to mismatch between the model and the real plant (Clarke 1994; Yoon and Clarke 1994). In order to mimic this GPC approach for PIP design, using equation (6.77) [here replacing $y(k)$ with the optimal estimate $\hat{y}(k)$] and equation (6.86),

$$\Delta u(k) = -\frac{F^*(z^{-1})}{G(z^{-1})}\hat{y}(k) + \frac{P^*(z^{-1})}{G(z^{-1})}y_d(k+N) \tag{6.93}$$

In this case, substituting from equation (6.62) [here based on the generalized noise polynomial defined below equation (6.81)] yields:

$$\Delta u(k) = -\frac{F^*(z^{-1})}{G(z^{-1})}\left\{\frac{D(z^{-1}) - \Delta A(z^{-1})}{D(z^{-1})}y(k) + \frac{\Delta B(z^{-1})}{D(z^{-1})}u(k)\right\} + \frac{P^*(z^{-1})}{G(z^{-1})}y_d(k+N) \tag{6.94}$$

which can be reduced to the following form:

$$\Delta u(k) = -\frac{F^a(z^{-1})}{D(z^{-1})}\frac{D(z^{-1})}{G^a(z^{-1})}y(k) + \frac{D(z^{-1})P^*(z^{-1})}{G^a(z^{-1})}y_d(k+N) \tag{6.95}$$

where

$$\left.\begin{aligned} F^a(z^{-1}) &= F^*(z^{-1})\left(D(z^{-1}) - \Delta A(z^{-1})\right) \\ G^a(z^{-1}) &= G(z^{-1})D(z^{-1}) + B(z^{-1})F^*(z^{-1}) \end{aligned}\right\} \tag{6.96}$$

The block diagram of the complete closed-loop stochastic PIP system using this control structure is shown in Figure 6.24, which is equivalent to the usual GPC formulation. Of course, as in the GPC case, it is possible to replace $D(z^{-1})$ by the more general filter $T(z^{-1})$ and select it either heuristically or optimally to satisfy other requirements.

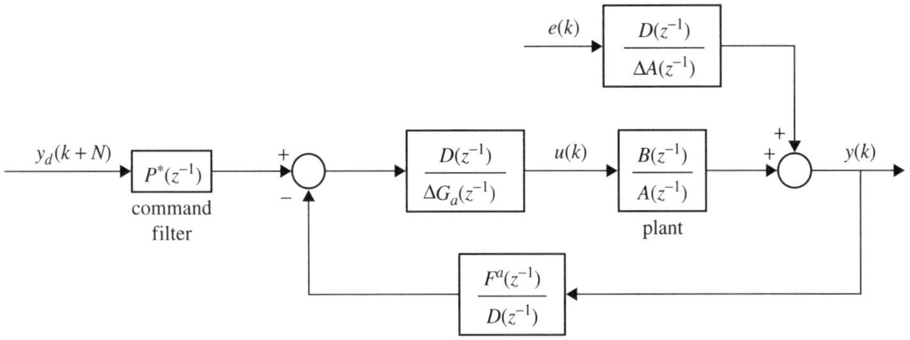

Figure 6.24 Stochastic PIP-LQG control conformed to show its relationship with the equivalent GPC system

6.7 Concluding Remarks

In this chapter, we have examined the two main structures for implementing PIP control systems, namely the incremental *feedback* and *forward path* forms. We have seen how the latter utilises an internal deterministic model to generate a noise-free estimate of the output. For this reason, the feedback form is inherently more robust to parametric uncertainty, especially for open-loop unstable and marginally stable systems. Practical examples of feedback PIP control are described by Chotai *et al.* (1991), Young *et al.* (1994), Seward *et al.* (1997), Taylor *et al.* (2000b, 2004a, b), Gu *et al.* (2004) and Shaban *et al.* (2008), among others.

However, many systems in industry and the environment are inherently open-loop stable, so that the reduced robustness of the forward path PIP controller is often less important. In such circumstances, the improved disturbance rejection and input characteristics of the forward path structure can prove advantageous. Example applications of forward path PIP control are given by Lees *et al.* (1996, 1998) and Taylor and Shaban (2006).

We have also considered various ways of either extending or modifying the basic NMSS form to include stochastic, disturbance and command input elements. An interesting aspect of such generalised NMSS models and the associated PIP control systems is the manner in which they may be readily compared with other approaches to control system design. Figure 5.3 and Appendix D show how the PIP controller may be considered as one extension to the classical Proportional-Integral-Derivative algorithm. In the present chapter, we have seen how the SP for time-delay systems and the GPC may similarly be treated as special constrained cases of PIP design. In demonstrating this ability to mimic exactly other control approaches, we see again the power and flexibility of the PIP algorithm.

It is important to emphasise that PIP pole assignment or optimal design of this type, in which the closed-loop poles or optimisation weights, are constrained in some manner to ensure equivalence with the other control system design methodologies, is not necessarily the most robust solution to the design problem. For example, the feedback PIP approach, which effectively subsumes the SP in the time-delay situation, will normally yield better and more robust closed-loop behaviour, making it more desirable for practical applications. The advantage of the PIP approach is that full order pole assignment or optimal control can be handled in these various forms, with the final implementation structure chosen according to the particular control objectives and plant in question.

Finally, it is noteworthy that the discussion so far has been entirely limited to SISO systems. This is largely for reasons of brevity and for tutorial purposes. Therefore, before we can claim that NMSS methods provide a unified approach to digital control system design, we first need to consider, in Chapter 7, the more general multivariable (*Multi-Input, Multi-Output* or MIMO) case.

References

Åström, K.J. and Wittenmark, B. (1984) *Computer-Controlled Systems Theory and Design*, Prentice Hall, Englewood Cliffs, NJ.

Bemporad, A., Morari, M., Dua, V. and Pistikopoulos, E. (2002) The explicit linear quadratic regulator for constrained systems, *Automatica*, **38**, pp. 3–20.

Bitmead, R.R., Gevers, M. and Wertz, V. (1990) *Adaptive Optimal Control: The Thinking Man's GPC*, Prentice Hall, Harlow.

Chotai, A., Young, P.C. and Behzadi, M.A. (1991) Self-adaptive design of a non-linear temperature control system, *IEE Proceedings: Control Theory and Applications*, **38**, pp. 41–49.

Clarke, D.W. (1994) Advances in model-based predictive control. In D.W. Clarke (Ed.), *Advances in Model-Based Predictive Control*, Oxford University Press, Oxford, pp. 3–21.

Clarke, D.W., Mohtadi, C. and Tuffs, P.S. (1987) Generalised Predictive Control *Automatica*, **23**, Part I The basic algorithm, pp. 137–148, Part II Extensions and interpretations, pp. 149–160.

Cross, P., Taylor, C.J. and Aggidis, G.A. (2011) State dependent feed-forward control of a wave energy converter model, *9th European Wave and Tidal Energy Conference* (EWTEC-11), 5–9 September 2011, University of Southampton, UK, Paper 61.

Dixon, R. (1996) *Design and Application of Proportional-Integral-Plus (PIP) Control Systems*, PhD thesis, Environmental Science Department, Lancaster University.

Dorf, R.C. and Bishop, R.H. (2008) *Modern Control Systems*, Eleventh Edition, Pearson Prentice Hall, Upper Saddle River, NJ.

Exadaktylos, V. and Taylor, C.J. (2010) Multi-objective performance optimisation for model predictive control by goal attainment, *International Journal of Control*, **83**, pp. 1374–1386.

Exadaktylos, V., Taylor, C.J., Wang, L. and Young, P.C. (2009) Forward path model predictive control using a non-minimal state space form, *IMECHE Proceedings: Systems and Control Engineering*, **223**, pp. 353–369.

Franklin, G.F., Powell, J.D. and Emami-Naeini, A. (2006) *Feedback Control of Dynamic Systems*, Fifth Edition, Pearson Prentice Hall, Upper Saddle River, NJ.

Garcia, C.E. and Morari, M. (1982) Internal model control. A unifying review and some new results, *Industrial & Engineering Chemistry Process Design and Development*, **21**, pp. 308–323.

Gawthrop, P.J., Wang, L. and Young, P.C. (2007) Continuous-time non-minimal state-space design, *International Journal of Control*, **80**, pp. 1690–1697.

Glover, K. and Doyle, J.C. (1988) State-space formulae for all stabilising controllers that satisfy an H-norm bound and relations to risk sensitivity, *System and Control Letters*, **11**, 167–172.

Gonzalez, A., Perez, J. and Odloak, D. (2009) Infinite horizon with non-minimal state space feedback, *Journal of Process Control*, **19**, pp. 473–481.

Gopal, M. (2010) *Digital Control and State Variable Methods*, McGraw-Hill International Edition, McGraw-Hill, Singapore.

Green, M. and Limebeer, D.J.N. (1995) *Linear Robust Control*, Prentice Hall, Englewood Cliffs, NJ.

Gu, J., Taylor, J. and Seward, D. (2004) Proportional-Integral-Plus control of an intelligent excavator, *Journal of Computer-Aided Civil and Infrastructure Engineering*, **19**, pp. 16–27.

Hesketh, T. (1992) Linear Quadratic methods for adaptive control – A tutorial, *Control Systems Centre Report 765*, UMIST, Manchester.

Jakeman, A.J. and Young, P.C. (1979) Joint parameter/state estimation, *Electronics Letters*, **15**, 19, pp. 582–583.

Jacobson, D.H. (1973) Optimal stochastic linear systems with exponential performance criteria and their relation to deterministic differential games, *IEEE Transactions on Automatic Control*, **AC-18**, pp. 124–131.

Kalman, R.E. (1960) A new approach to linear filtering and prediction problems, *ASME Journal of Basic Engineering*, **82**, pp. 34–45.

Kalman, R.E. (1961) On the general theory of control systems, *Proceedings 1st IFAC Congress*, **1**, pp. 481–491.

Kalman, R.E. (1963) Mathematical description of linear dynamical systems, *SIAM Journal of Control*, **1**, pp. 152–192.

Kumar, P.R. and van Schuppen, J.H. (1981) On the optimal control of stochastic systems with an exponential of integral performance index, *Journal of Mathematical Analysis and Applications*, **80**, pp. 312–332.

Lees, M.J. (1996) *Multivariable Modelling and Proportional-Integral-Plus Control of Greenhouse Climate*, PhD thesis, Environmental Science Department, Lancaster University.

Lees, M.J., Taylor, J., Chotai, A., Young, P.C. and Chalabi, Z.S. (1996) Design and implementation of a Proportional-Integral-Plus (PIP) control system for temperature, humidity, and carbon dioxide in a glasshouse, *Acta Horticulturae*, **406**, pp. 115–123.

Lees, M.J., Taylor, C.J., Young, P.C. and Chotai, A. (1998) Modelling and PIP control design for open top chambers, *Control Engineering Practice*, **6**, pp. 1209–1216.

Maciejowski, J. (2002) *Predictive Control with Constraints*, Prentice Hall, Harlow, UK.

Maciejowski, J.M. (1985) Asymptotic recovery for discrete-time systems, *IEEE Transactions on Automatic Control*, **30**, pp. 602–605.

Morari, M. and Lee, J.H. (1999) Model Predictive Control: past, present and future, *Computers and Chemical Engineering*, **23**, 5, pp. 667–682.

Mustafa, D. and Glover, K. (1990) *Minimum Entropy H_∞ Control*, Springer-Verlag, Berlin.

Ogata, K. (2001) *Modern Control Engineering*, Prentice Hall, New York.

Pannocchia, G. and Rawlings, J. (2003) Disturbance models for offset-free model predictive control, *AIChE Journal*, **39**, pp. 262–287.

Rossiter, J.A. (2003) *Model Based Predictive Control: A Practical Approach*, CRC Press, Boca Raton, FL.

Seward, D.W., Scott, J.N., Dixon, R., Findlay, J.D. and Kinniburgh, H. (1997) The automation of piling rig positioning using satellite GPS, *Automation in Construction*, **6**, pp. 229–240.

Shaban, E.M., Ako, S., Taylor, C.J. and Seward, D.W. (2008) Development of an automated verticality alignment system for a vibro-lance, *Automation in Construction*, **17**, pp. 645–655.

Smith, O.J. (1959) A controller to overcome dead time, *ISA Journal*, **6**, pp. 28–33.

Smith, O.J.M. (1957) Closer control of loops with dead time, *Chemical Engineering Progress*, **53**, pp. 217–219.

Taylor, C.J. (1996) *Generalised Proportional-Integral-Plus Control*, PhD thesis, Environmental Science Department, Lancaster University.

Taylor, C.J., Chotai, A. and Cross, P. (2012) Non-minimal state variable feedback decoupling control for multivariable continuous-time systems, *International Journal of Control*, **85**, pp. 722–734.

Taylor, C.J., Chotai, A. and Young, P.C. (1998a) Proportional-Integral-Plus (PIP) control of time-delay systems, *IMECHE Proceedings: Systems and Control Engineering*, **212**, Part I, pp. 37–48.

Taylor, C.J., Chotai, A. and Young, P.C. (1998b) Continuous time Proportional-Integral-Plus (PIP) control with filtering polynomials, *UKACC International Conference* (Control-98), 1–4 September, University of Wales, Swansea, UK, Institution of Electrical Engineers Conference Publication No. 455, Vol. 2, pp. 1391–1396.

Taylor, C.J., Chotai, A. and Young, P.C. (2000a) State space control system design based on non-minimal state-variable feedback: further generalisation and unification results, *International Journal of Control*, **73**, pp. 1329–1345.

Taylor, C.J., Leigh, P.A., Chotai, A., Young, P.C., Vranken, E. and Berckmans, D. (2004a) Cost effective combined axial fan and throttling valve control of ventilation rate, *IEE Proceedings: Control Theory and Applications*, **151**, pp. 577–584.

Taylor, C.J., Leigh, P., Price, L., Young, P.C., Berckmans, D., Janssens, K., Vranken, E. and Gevers, R. (2004b) Proportional-Integral-Plus (PIP) control of ventilation rate in agricultural buildings, *Control Engineering Practice*, **12**, pp. 225–233.

Taylor, C.J. and Shaban, E.M. (2006) Multivariable Proportional-Integral-Plus (PIP) control of the ALSTOM nonlinear gasifier simulation, *IEE Proceedings: Control Theory and Applications*, **153**, pp. 277–285.

Taylor, J., Young, P.C. and Chotai, A. (1994) On the relationship between GPC and PIP control. In D.W. Clarke (Ed.), *Advances in Model-Based Predictive Control*, Oxford University Press, Oxford, pp. 53–68.

Taylor, C.J., Young, P.C. and Chotai, A. (1996) PIP optimal control with a risk sensitive criterion, *UKACC International Conference*, 2–5 September, University of Exeter, UK, pp. 959–964.

Taylor, C.J., Young, P.C., Chotai, A., McLeod, A.R. and Glasock, A.R. (2000b) Modelling and proportional-integral-plus control design for free air carbon dioxide enrichment systems, *Journal of Agricultural Engineering Research*, **75**, pp. 365–374.

Tsypkin, Y.A.Z. and Holmberg, U. (1995) Robust stochastic control using the internal model principle and internal model control, *International Journal of Control*, **61**, pp. 809–822.

Wang, L. and Young, P. (2006) An improved structure for model predictive control using non-minimal state space realisation, *Journal of Process Control*, **16**, pp. 355–371.

Whittle, P. (1981) Risk sensitive Linear/Quadratic/Gaussian control, *Advances in Applied Probability*, **13**, pp. 764–777.

Whittle, P. (1990) *Risk-Sensitive Optimal Control*, John Wiley & Sons, Ltd, Chichester.

Yoon, T.-W. and Clarke, D.W. (1994) Towards robust adaptive predictive control. In D.W. Clarke (Ed.), *Advances in Model-Based Predictive Control*, Oxford University Press, Oxford, pp. 402–414.

Young, P.C. (1979) Self-Adaptive Kalman filter, *Electronics Letters*, **15**, pp. 358–360.

Young, P.C. (2011) *Recursive Estimation and Time-Series Analysis: An Introduction for the Student and Practitioner*, Springer-Verlag, Berlin.

Young, P.C., Behzadi, M.A., Wang, C.L. and Chotai, A. (1987) Direct Digital and Adaptive Control by input–output, state variable feedback pole assignment, *International Journal of Control*, **46**, pp. 1867–1881.

Young, P.C., Chotai, A., McKenna, P.G. and Tych, W. (1998) Proportional-Integral-Plus (PIP) design for delta operator systems: Part 1, SISO systems, *International Journal of Control*, **70**, pp. 149–168.

Young, P.C., Lees, M., Chotai, A., Tych, W. and Chalabi, Z.S. (1994) Modelling and PIP control of a glasshouse micro-climate, *Control Engineering Practice*, **2**, pp. 591–604.

7

True Digital Control for Multivariable Systems

The general state space formulation of the linear *Single-Input, Single-Output* (SISO) *Proportional-Integral-Plus* (PIP) control system, introduced in Chapter 5, facilitates straightforward extension to the multivariable case. This chapter describes the *True Digital Control* (TDC) design philosophy for such *Multi-Input, Multi-Output* (MIMO) systems. Here, the system is characterised by multiple control inputs that affect the state and output variables in a potentially complicated and cross-coupled manner (Wolovich 1974; Kailath 1980; Albertos and Sala 2004; Skogestad and Postlethwaite 2005).

Multivariable PIP design should take account of this natural cross-coupling and generate control inputs which ameliorate any unsatisfactory aspects of the system behaviour that arise from it. In particular, it is often an advantage if the control system is able to *decouple* the effects of different command inputs and their related outputs. In this case, each command signal leads only to the required changes in the specified output variable, without disturbing the other output variables. This can be a rather difficult task, which is achieved by manipulating the control inputs for all of the input–output channels simultaneously, so that the effects of the natural coupling are neutralised.

We start with the *Transfer Function Matrix* (TFM) model and use it to derive a left *Matrix Fraction Description* (MFD) of the multivariable system. In the TDC approach to control system design, these models are obtained from either experimental or simulated data using multiple SISO or *Multi-Input, Single-Output* (MISO) versions of the system identification tools described in Chapter 8. The control approach is subsequently based on the definition of a suitable multivariable *Non-Minimal State Space* (NMSS) form constructed from these multiple SISO or MISO models. As in the univariate case, the NMSS setting of the control problem allows for the use of full *State Variable Feedback* (SVF), involving only the measured input and output variables and their past values, so avoiding the need for an explicit state reconstruction filter or observer (section 7.1).

True Digital Control: Statistical Modelling and Non-Minimal State Space Design, First Edition.
C. James Taylor, Peter C. Young and Arun Chotai.

The resulting multivariable PIP control system is, therefore, relatively straightforward to design and implement in practical applications (section 7.2). For example, the weighting matrices required for optimal *Linear Quadratic* (LQ) design are directly related to the input and output state variables, and so can be manually tuned in a similar manner to univariate systems (section 7.3).

In addition to multivariable decoupling control, we often require different response characteristics between each command input and the associated output variable, so that the control system has to satisfy these and other conflicting requirements: in other words, we need to simultaneously satisfy multiple objectives. In this regard, the PIP-LQ control system is ideal for incorporation within a multi-objective optimisation framework. Often, it is possible to obtain a satisfactory compromise between conflicting objectives such as robustness, system constraints, overshoot, rise times and multivariable decoupling, by concurrent numerical optimisation of the diagonal and off-diagonal elements of the weighting matrices in the cost function (section 7.4).

By contrast, one approach to multivariable PIP *pole assignment* is based on a transformation of the NMSS model into the Luenberger controller canonical form (Luenberger 1971). This method, while achieving the desired pole assignment, does not address the problem of ensuring good transient cross-coupling characteristics. However, it is well known that assignment of the closed-loop poles of the multivariable system does not, in itself, uniquely specify the SVF gain matrix (see e.g. Luenberger 1967; Gopinath 1971; Young and Willems 1972; Munro 1979; Kailath 1980). Consequently, this extra design freedom can be used to instead develop a PIP algorithm that provides combined multivariable decoupling *and* pole assignment (section 7.5).

7.1 The Multivariable NMSS (Servomechanism) Representation

One approach for the control of multivariable systems is to utilise a number of independent SISO controllers. However, an important advantage of model-based multivariable design over such multiple-loop SISO controllers is the ability of the former to dynamically decouple the control channels. This is achieved by exploiting information about the interactions contained in the control model. Example 7.1 illustrates how a matrix of *Transfer Function* (TF) models may be employed to describe these interactions.

Example 7.1 Multivariable TF Representation of a Two-Input, Two-Output System

Consider the following pair of two-input, single-output models, which have been arbitrarily chosen here for tutorial purposes:

$$
\begin{aligned}
y_1(k) &= \frac{0.9z^{-1}}{1-0.5z^{-1}}u_1(k) + \frac{2z^{-1}}{1-0.5z^{-1}}u_2(k) \\
y_2(k) &= \frac{1.1z^{-1}}{1-0.8z^{-1}}u_1(k) + \frac{z^{-1}}{1-0.8z^{-1}}u_2(k)
\end{aligned}
\tag{7.1}
$$

Here, the dynamic response of the two output variables, $y_1(k)$ and $y_2(k)$, are both determined by the two control input signals, $u_1(k)$ and $u_2(k)$. Such a coupled two-input, two-output system can be represented equivalently in the following TF matrix or TFM form:

$$\begin{bmatrix} y_1(k) \\ y_2(k) \end{bmatrix} = \begin{bmatrix} \dfrac{0.9z^{-1}}{1-0.5z^{-1}} & \dfrac{2z^{-1}}{1-0.5z^{-1}} \\ \dfrac{1.1z^{-1}}{1-0.8z^{-1}} & \dfrac{z^{-1}}{1-0.8z^{-1}} \end{bmatrix} \begin{bmatrix} u_1(k) \\ u_2(k) \end{bmatrix} \tag{7.2}$$

Here, the open-loop system consists of four first order TF models, with common denominator polynomials, unity time delay in each case, and no zeros. Despite the apparent simplicity in this model, Figure 7.1 shows clear evidence of cross-coupling, with each input variable influencing both outputs, as expected from the above system definition.

Similar figures will be employed on several further occasions in the present chapter, so it is worthwhile clarifying exactly what is presented here: Figure 7.1 shows the response of two separate open-loop simulation experiments: it illustrates the open-loop response of $y_1(k)$ (Figure 7.1a) and $y_2(k)$ (Figure 7.1b) to a unit step change in $u_1(k)$ when $u_2(k) = 0$; and the response of $y_1(k)$ (Figure 7.1c) and $y_2(k)$ (Figure 7.1d) to a unit step change in $u_2(k)$ when $u_1(k) = 0$. Later, a similar arrangement will be utilised to illustrate the closed-loop response to step changes in each command input.

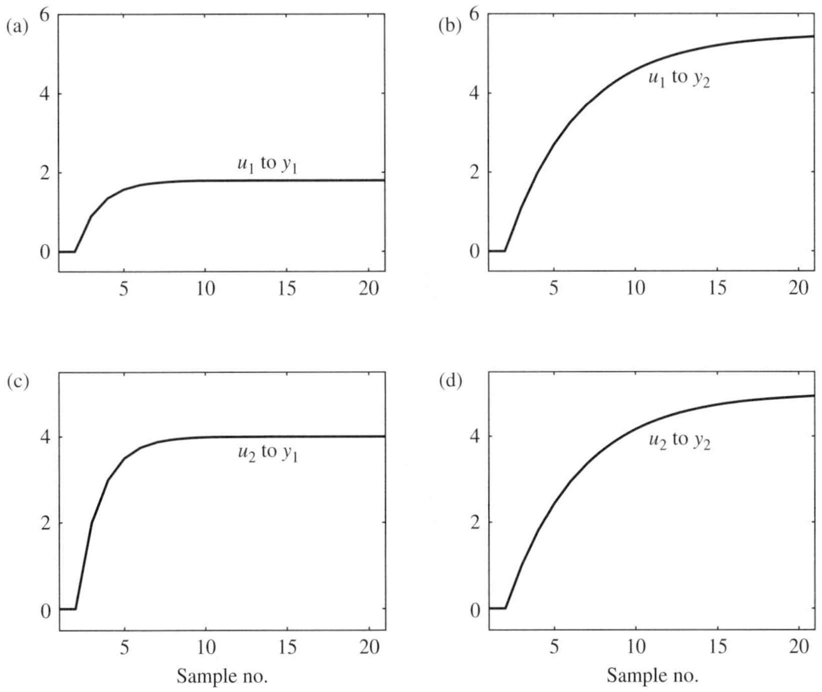

Figure 7.1 Open-loop unit step response of the multivariable system in Example 7.1

For analogy with the SISO case described in earlier chapters, we can write equation (7.2) in the following form:

$$\begin{bmatrix} y_1(k) \\ y_2(k) \end{bmatrix} = (\boldsymbol{A}(z^{-1}))^{-1}\boldsymbol{B}(z^{-1})\begin{bmatrix} u_1(k) \\ u_2(k) \end{bmatrix} \tag{7.3}$$

Alternatively,

$$\boldsymbol{A}(z^{-1})\begin{bmatrix} y_1(k) \\ y_2(k) \end{bmatrix} = \boldsymbol{B}(z^{-1})\begin{bmatrix} u_1(k) \\ u_2(k) \end{bmatrix} \tag{7.4}$$

where, for this example,

$$\boldsymbol{A}(z^{-1}) = \begin{bmatrix} 1-0.5z^{-1} & 0 \\ 0 & 1-0.8z^{-1} \end{bmatrix} = \begin{bmatrix} 1 & 0 \\ 0 & 1 \end{bmatrix} + \begin{bmatrix} -0.5 & 0 \\ 0 & -0.8 \end{bmatrix} z^{-1} \tag{7.5}$$

and

$$\boldsymbol{B}(z^{-1}) = \begin{bmatrix} 0.9z^{-1} & 2z^{-1} \\ 1.1z^{-1} & z^{-1} \end{bmatrix} = \begin{bmatrix} 0.9 & 2 \\ 1.1 & 1 \end{bmatrix} z^{-1} \tag{7.6}$$

Compare these matrix equations with their scalar equivalents (5.12) in Chapter 5. Hence, following a similar approach to the state equation in equations (5.13), we can define the following servomechanism NMSS form:

$$\begin{bmatrix} y_1(k) \\ y_2(k) \\ z_1(k) \\ z_2(k) \end{bmatrix} = \begin{bmatrix} 0.5 & 0 & 0 & 0 \\ 0 & 0.8 & 0 & 0 \\ -0.5 & 0 & 1 & 0 \\ 0 & -0.8 & 0 & 1 \end{bmatrix} \begin{bmatrix} y_1(k-1) \\ y_2(k-1) \\ z_1(k-1) \\ z_2(k-1) \end{bmatrix} + \begin{bmatrix} 0.9 & 2 \\ 1.1 & 1 \\ -0.9 & -2 \\ -1.1 & -1 \end{bmatrix} \boldsymbol{u}(k) + \begin{bmatrix} 0 & 0 \\ 0 & 0 \\ 1 & 0 \\ 0 & 1 \end{bmatrix} \boldsymbol{y}_d(k) \tag{7.7}$$

where $\boldsymbol{u}(k) = [\, u_1(k-1) \quad u_2(k-1) \,]^T$ and $\boldsymbol{y}_d(k) = [\, y_{d1}(k) \quad y_{d2}(k) \,]^T$, in which $y_{d1}(k)$ and $y_{d2}(k)$ are the command levels associated with $y_1(k)$ and $y_2(k)$, respectively. As usual, the integral-of-error state variables are based on equation (5.3), i.e.

$$z_1(k) = z_1(k-1) + (y_{d1}(k) - y_1(k)); \quad z_2(k) = z_2(k-1) + (y_{d2}(k) - y_2(k)) \tag{7.8}$$

Note that no input states are required for this simplest example. However, a higher order system requiring input state variables is described later (Example 7.3).

7.1.1 *The General Multivariable System Description*

Generalising from Example 7.1, consider a multivariable discrete-time system described by a set of ordinary difference equations, represented in the following TFM form:

$$\boldsymbol{y}(k) = \boldsymbol{G}(z^{-1})\boldsymbol{u}(k) \tag{7.9}$$

where $\boldsymbol{u}(k)$ and $\boldsymbol{y}(k)$ are the following vectors of r input and p output variables, respectively:

$$\begin{aligned}\boldsymbol{y}(k) &= \begin{bmatrix} y_1(k) & y_2(k) & \cdots & y_p(k) \end{bmatrix}^T \\ \boldsymbol{u}(k) &= \begin{bmatrix} u_1(k) & u_2(k) & \cdots & u_r(k) \end{bmatrix}^T\end{aligned} \tag{7.10}$$

and $\boldsymbol{G}(z^{-1})$ is a matrix of TF models, such as (7.2). Here, the TF model for each input–output pathway takes the same general univariate form as employed in earlier chapters: see e.g. equation (5.12). In general, no prior assumptions are made about the nature of this TF matrix, which may be marginally stable, unstable, or possess non-minimum phase characteristics. The presentation here is based on the general case when the number of input and output variables may differ. However, in practice, Type 1 servomechanism performance requires at least the same number of input variables as outputs.

The TFM (7.9) can be transformed into the following left MFD form (Wolovich 1974; Kailath 1980), which represents a generalisation of equation (7.3):

$$\boldsymbol{y}(k) = (\boldsymbol{A}(z^{-1}))^{-1}\boldsymbol{B}(z^{-1})\boldsymbol{u}(k) \tag{7.11}$$

Alternatively,

$$\boldsymbol{A}(z^{-1})\boldsymbol{y}(k) = \boldsymbol{B}(z^{-1})\boldsymbol{u}(k) \tag{7.12}$$

where

$$\begin{aligned}\boldsymbol{A}(z^{-1}) &= \boldsymbol{I} + \boldsymbol{A}_1 z^{-1} + \cdots + \boldsymbol{A}_n z^{-n} && (\boldsymbol{A}_n \neq \boldsymbol{0}) \\ \boldsymbol{B}(z^{-1}) &= \boldsymbol{B}_1 z^{-1} + \boldsymbol{B}_2 z^{-2} + \cdots + \boldsymbol{B}_m z^{-m} && (\boldsymbol{B}_m \neq \boldsymbol{0})\end{aligned} \tag{7.13}$$

Here, $\boldsymbol{A}_i$ $(i = 1, 2, \ldots, n)$ are $p \times p$ and $\boldsymbol{B}_i (i = 1, 2, \ldots, m)$ are $p \times r$ matrices, while $\boldsymbol{I}$ is an $p \times p$ identity matrix. In a similar manner to that described in Chapter 5 for the univariate case, some of the initial $\boldsymbol{B}(z^{-1})$ terms could take null values to accommodate pure time delays in the system. Note that, in the context of the system identification and parameter estimation literature, the left MFD model (7.11) is known as a vector TF model and, in the stochastic case, the associated, additive, multivariable noise model is known as the vector *Auto-Regressive, Moving Average* (ARMA) model. Finally, its relationship with the nominal minimal state space representation of a multivariable system is well established (see e.g. Wolovich 1974; Kailath 1980; Gevers and Wertz 1984).

7.1.2 Multivariable NMSS Form

The multivariable non-minimal state vector is defined as follows[1]:

$$\boldsymbol{x}(k) = \begin{bmatrix}\boldsymbol{y}(k)^T & \cdots & \boldsymbol{y}(k-n+1)^T & \boldsymbol{u}(k-1)^T & \cdots & \boldsymbol{u}(k-m+1)^T & \boldsymbol{z}(k)^T\end{bmatrix}^T \tag{7.14}$$

[1] In this book, the multivariable non-minimal state vector (7.14) is assigned the same notation as for the SISO case; see e.g. equation (5.14). Of course, the dimension of the multivariable state vector (7.14) will generally be higher, since it includes present and past values of *all* the system variables.

where $\boldsymbol{x}(k)$ is composed of vectors of sampled past and present system outputs and past control inputs, while the integral-of-error state vector $\boldsymbol{z}(k)$ is defined:

$$\boldsymbol{z}(k) = \frac{\left(\boldsymbol{y}_d(k) - \boldsymbol{y}(k)\right)}{1 - z^{-1}} \tag{7.15}$$

Alternatively,

$$\boldsymbol{z}(k) = \boldsymbol{z}(k-1) + \left(\boldsymbol{y}_d(k) - \boldsymbol{y}(k)\right) \tag{7.16}$$

in which $\boldsymbol{y}_d(k)$ is the reference or command input vector:

$$\boldsymbol{y}_d(k) = \begin{bmatrix} y_{d1}(k) & y_{d2}(k) & \cdots & y_{dp}(k) \end{bmatrix}^T \tag{7.17}$$

and, finally, $\boldsymbol{z}(k)$ similarly consists of p elements:

$$\boldsymbol{z}(k) = \begin{bmatrix} z_1(k) & z_2(k) & \cdots & z_p(k) \end{bmatrix}^T \tag{7.18}$$

Type 1 servomechanism performance is automatically accommodated by the introduction of this integral-of-error vector $\boldsymbol{z}(k)$, in which the definitions of each element follow a similar approach to equations (7.8). Consequently, provided the closed-loop system is stable, then steady state decoupling is inherent in the basic design, i.e. each output variable $y_i(k)$ will asymptotically converge to its associated command level $y_{di}(k)$: in other words, there is the automatic *steady-state* decoupling in the sense that although, in general, any maintained step command input signal will cause perturbations in all the output variables, it will lead eventually to a steady-state change only in its associated output.

It should be noticed that the NMSS state vector (7.14) has an order of $p(n+1) + r(m-1)$, which is usually higher than that of the original system representation, as expected for any *non-minimal* formulation of the problem.

With the above definitions, the NMSS representation associated with the MFD model (7.11) is defined directly in terms of the following discrete-time state equations:

$$\boldsymbol{x}(k) = \boldsymbol{F}\boldsymbol{x}(k-1) + \boldsymbol{G}\boldsymbol{u}(k-1) + \boldsymbol{D}\boldsymbol{y}_d(k) \tag{7.19}$$

where

$$\boldsymbol{F} = \begin{bmatrix}
-\boldsymbol{A}_1 & -\boldsymbol{A}_2 & \cdots & -\boldsymbol{A}_{n-1} & -\boldsymbol{A}_n & \boldsymbol{B}_2 & \boldsymbol{B}_3 & \cdots & \boldsymbol{B}_{m-1} & \boldsymbol{B}_m & \boldsymbol{0} \\
\boldsymbol{I}_p & \boldsymbol{0} & \cdots & \boldsymbol{0} & \boldsymbol{0} & \boldsymbol{0} & \boldsymbol{0} & \cdots & \boldsymbol{0} & \boldsymbol{0} & \boldsymbol{0} \\
\boldsymbol{0} & \boldsymbol{I}_p & \cdots & \boldsymbol{0} & \boldsymbol{0} & \boldsymbol{0} & \boldsymbol{0} & \cdots & \boldsymbol{0} & \boldsymbol{0} & \boldsymbol{0} \\
\vdots & \vdots & \ddots & \vdots & \vdots & \vdots & \vdots & \ddots & \vdots & \vdots & \vdots \\
\boldsymbol{0} & \boldsymbol{0} & \cdots & \boldsymbol{I}_p & \boldsymbol{0} & \boldsymbol{0} & \boldsymbol{0} & \cdots & \boldsymbol{0} & \boldsymbol{0} & \boldsymbol{0} \\
\boldsymbol{0} & \boldsymbol{0} & \cdots & \boldsymbol{0} & \boldsymbol{0} & \boldsymbol{0} & \boldsymbol{0} & \cdots & \boldsymbol{0} & \boldsymbol{0} & \boldsymbol{0} \\
\boldsymbol{0} & \boldsymbol{0} & \cdots & \boldsymbol{0} & \boldsymbol{0} & \boldsymbol{I}_r & \boldsymbol{0} & \cdots & \boldsymbol{0} & \boldsymbol{0} & \boldsymbol{0} \\
\boldsymbol{0} & \boldsymbol{0} & \cdots & \boldsymbol{0} & \boldsymbol{0} & \boldsymbol{0} & \boldsymbol{I}_r & \cdots & \boldsymbol{0} & \boldsymbol{0} & \boldsymbol{0} \\
\vdots & \vdots & \ddots & \vdots & \vdots & \vdots & \vdots & \ddots & \vdots & \vdots & \vdots \\
\boldsymbol{0} & \boldsymbol{0} & \cdots & \boldsymbol{0} & \boldsymbol{0} & \boldsymbol{0} & \boldsymbol{0} & \cdots & \boldsymbol{I}_r & \boldsymbol{0} & \boldsymbol{0} \\
\boldsymbol{A}_1 & \boldsymbol{A}_2 & \cdots & \boldsymbol{A}_{n-1} & \boldsymbol{A}_n & -\boldsymbol{B}_2 & -\boldsymbol{B}_3 & \cdots & -\boldsymbol{B}_{m-1} & -\boldsymbol{B}_m & \boldsymbol{I}_p
\end{bmatrix} \tag{7.20}$$

and

$$G = \begin{bmatrix} B_1 \\ 0 \\ 0 \\ \vdots \\ 0 \\ I_r \\ 0 \\ 0 \\ \vdots \\ 0 \\ -B_1 \end{bmatrix}; \quad D = \begin{bmatrix} 0 \\ 0 \\ 0 \\ \vdots \\ 0 \\ 0 \\ 0 \\ 0 \\ \vdots \\ 0 \\ I_p \end{bmatrix} \tag{7.21}$$

Here, the block matrices I_p and I_r denote $p \times p$ and $p \times r$ identity matrices, respectively.

The left MFD model is particularly convenient to use here, because it allows for the formation of an appropriate NMSS model in a very similar manner to that described for univariate systems in Chapter 5. Although the multivariable case naturally requires a vector-matrix notation, the analogy with equation (5.12), equation (5.13), equation (5.14), equation (5.15) and equation (5.16) is clear.

7.1.3 The Characteristic Polynomial of the Multivariable NMSS Model

The open-loop characteristic polynomial $S(\lambda)$ of the multivariable NMSS representation (7.19) is given by:

$$S(\lambda) = |I\lambda - F| = (\lambda - 1)^p \lambda^{r(m-1)} \det(A^*(\lambda)) \tag{7.22}$$

where $\det(A^*(\lambda))$ is the determinant of $A^*(\lambda)$ and

$$A^*(\lambda) = I\lambda^n + A_1\lambda^{n-1} + \cdots + A_n \tag{7.23}$$

Equation (7.22) is obtained from the transition matrix of the state space model, as shown in earlier chapters; see e.g. equation (3.64). It shows that $S(\lambda)$ is the product of three components: $(\lambda - 1)^p$ are associated with the integral-of-error state vector; $\lambda^{r(m-1)}$ are due to the input variables, i.e. $u(k-1)$, $u(k-2), \ldots, u(k-m+1)$; and $\det(A^*(\lambda))$ is composed of the modes of the original open-loop representation (7.9), together with extra modes at $\lambda = 0$. The latter are called 'transmission zeros' and are introduced by the delayed output variables when the p row degrees $n_1, n_2, \ldots, n_p$ of the polynomial matrix $A(z^{-1})$ are not equal (this occurs when the TF models associated with each input–output pathway are of different orders: see Example 7.4).

Chapter 3 introduced the concept of controllability. To obtain a satisfactory SVF control law, the NMSS model must be controllable. In this case, all of the $p(n+1) + r(m-1)$ eigenvalues may be assigned to any desired positions in the complex z-plane, subject only to the normal restriction that complex eigenvalues appear in conjugate pairs. If the NMSS model is not completely controllable, then the controllable eigenvalues can be assigned arbitrarily, while the uncontrollable ones remain invariant (as in the case of the coupled drives system considered in Example 7.4). The conditions for controllability are listed below.

Theorem 7.1 Controllability of the Multivariable NMSS Model Given a multivariable linear discrete-time system represented in the left MFD form (7.11), the NMSS model (7.19), as described by the pair $[\boldsymbol{F}, \boldsymbol{G}]$, is completely controllable if and only if the following conditions are all satisfied:

(i) the two polynomial matrices $\boldsymbol{A}(z^{-1})$ and $\boldsymbol{B}(z^{-1})$ are left coprime;
(ii) the row degrees of $\boldsymbol{A}(z^{-1})$ and $\boldsymbol{B}(z^{-1})$ satisfy the following identity,

$$(n - n_i)\,(m - m_i) = 0 \text{ for } i = 1,\ 2, \ldots,\ p$$

where n_i and m_i are the ith row degrees of $\boldsymbol{A}(z^{-1})$ and $\boldsymbol{B}(z^{-1})$, respectively;
(iii) when $z = 1$, Rank $\left(\boldsymbol{B}(z^{-1})\right) = p$.

The proof of Theorem 7.1 is a straightforward extension of the regulator SISO case described by Theorem 4.1 (see also Appendix C) and so, for brevity, is omitted here.

The first controllability condition simply ensures that there are no pole-zero cancellations in the original system, i.e. the system representation should be derived from a controllable and observable state space representation. The second condition requires that $\boldsymbol{A}(z^{-1})$ and $\boldsymbol{B}(z^{-1})$ should possess no lower degrees in the same row than their highest row degrees n and m, respectively. When this condition is not satisfied, redundant delayed output variables are introduced into the non-minimal state vector and these redundant variables add uncontrollable modes at the origin ($z = 0$). However, such uncontrollable modes cause no problem since they are all at the origin and are, therefore, stable. In fact, they are cancelled out in the TFM and have no effect upon the closed-loop system response. Note that the extra $r(m - 1)$ modes at $z = 0$, which are introduced into the NMSS model by $\boldsymbol{u}(k - 1), \boldsymbol{u}(k - 2), \ldots, \boldsymbol{u}(k - m + 1)$ are always inherently controllable, since $\boldsymbol{u}(k)$ represents the vector of control input signals.

The third condition of Theorem 7.1 requires that, as for all multivariable SVF servomechanism systems, there must be at least the same number of independent input variables as there are independent output variables. Hence, for simplicity, we usually define the control system such that $p = r$ in equations (7.10). This condition also avoids the presence of zeros at $z = 1$ on the complex z-plane, which would otherwise cancel with the poles associated with the integral-of-error states.

In a similar manner to the SISO case, the controllability conditions of Theorem 7.1 are equivalent to the normal requirement that the following controllability matrix:

$$\boldsymbol{S}_1 = \begin{bmatrix} \boldsymbol{G} & \boldsymbol{F}\boldsymbol{G} & \boldsymbol{F}^2\boldsymbol{G} & \cdots & \boldsymbol{F}^{p(n+1)+r(m-1)-1}\boldsymbol{G} \end{bmatrix} \tag{7.24}$$

has full rank $p(n + 1) + r(m - 1)$. Equation (7.24) takes a similar form to equation (3.79), here revised to account for the order of the multivariable NMSS model.

7.2 Multivariable PIP Control

The SVF control law associated with the NMSS model (7.19) takes the usual form:

$$\boldsymbol{u}(k) = -\boldsymbol{K}\boldsymbol{x}(k) \tag{7.25}$$

where the control gain matrix:

$$\boldsymbol{K} = \begin{bmatrix} \boldsymbol{L}_0 & \boldsymbol{L}_1 & \cdots & \boldsymbol{L}_{n-1} & \boldsymbol{M}_1 & \cdots & \boldsymbol{M}_{m-1} & -\boldsymbol{K}_I \end{bmatrix} \tag{7.26}$$

Expanding the terms in (7.25) yields:

$$\begin{aligned} \boldsymbol{u}(k) = -\boldsymbol{L}_0\boldsymbol{y}(k) - \boldsymbol{L}_1\boldsymbol{y}(k-1) - \cdots - \boldsymbol{L}_{n-1}\boldsymbol{y}(k-n+1) \\ -\boldsymbol{M}_1\boldsymbol{u}(k-1) - \cdots - \boldsymbol{M}_{m-1}\boldsymbol{u}(k-m+1) + \boldsymbol{K}_I\boldsymbol{z}(k) \end{aligned} \tag{7.27}$$

The gain matrix $\boldsymbol{K}$ is usually decomposed into the following output feedback and input feedback polynomial matrices $\boldsymbol{L}(z^{-1})$ and $\boldsymbol{M}(z^{-1})$, respectively, together with the integral-of-error gain matrix $\boldsymbol{K}_I$, where

$$\begin{aligned} \boldsymbol{L}(z^{-1}) &= \boldsymbol{L}_0 + \boldsymbol{L}_1 z^{-1} + \cdots + \boldsymbol{L}_{n-1} z^{-n+1} \\ \boldsymbol{M}(z^{-1}) &= \boldsymbol{M}_1 z^{-1} + \boldsymbol{M}_2 z^{-2} + \cdots + \boldsymbol{M}_{m-1} z^{-m+1} \end{aligned} \tag{7.28}$$

From equation (7.27) and equation (7.28), the control law becomes:

$$\boldsymbol{u}(k) = -\boldsymbol{L}(z^{-1})\boldsymbol{y}(k) - \boldsymbol{M}(z^{-1})\boldsymbol{u}(k) + \boldsymbol{K}_I\boldsymbol{z}(k) \tag{7.29}$$

The block diagram for such a PIP control system is illustrated by Figure 7.2, which reveals the structural similarity between this approach and multivariable *Proportional-Integral* (PI) and *Proportional-Integral-Derivative* (PID) design. As for the SISO case, the negative sign associated with $\boldsymbol{K}_I$ in (7.26) is introduced to allow the integral states to take on the same structural form as multivariable PI and PID control. In this manner, the PIP approach can be interpreted as a logical extension of these standard industrial controllers,

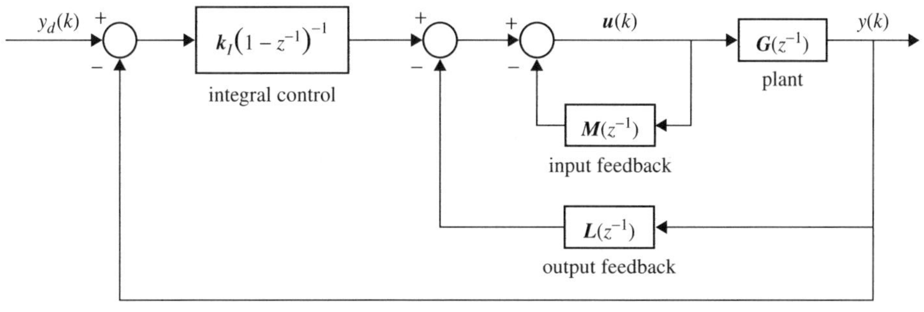

Figure 7.2 Multivariable PIP control in feedback form

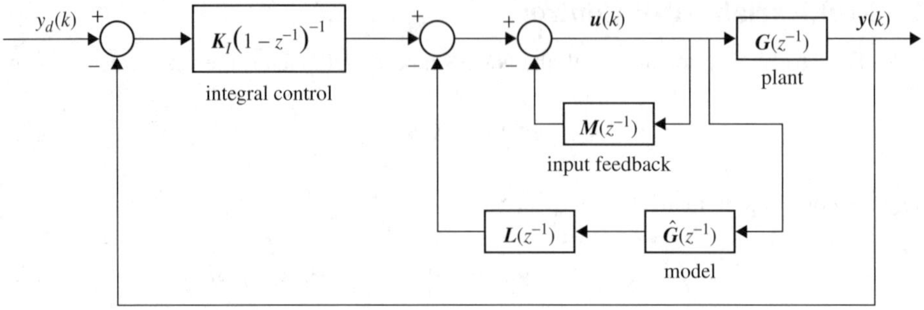

Figure 7.3 Multivariable PIP control in forward path form

with additional dynamic feedback and input compensators introduced automatically when the process has second order or higher dynamics; or more than a single sample pure time delay.

As we have already seen in the univariate case, the PIP formulation is inherently much more flexible than conventional PI or PID designs, allowing for the implementation of well known SVF control strategies such as closed-loop pole assignment, with complete (or partial) decoupling control; or deterministic LQ optimal design. It also facilitates consideration of other control strategies based on state space concepts, such as the risk averse designs proposed by Whittle (1981), the related robust H_∞ design (e.g. Green and Limebeer 1995) and predictive control (e.g. Rossiter 2003). These approaches are readily extended from the SISO solutions discussed in Chapter 6. Examples of multivariable NMSS based control in various contexts are given by Dixon and Lees (1994), Lees *et al.* (1994), Young *et al.* (1994), Dixon *et al.* (1997), Chotai *et al.* (1998), Taylor *et al.* (1998), Taylor and Shaban (2006) and Exadaktylos and Taylor (2010), among others.

The multivariable PIP controller can be implemented in an alternative forward path form, as illustrated in Figure 7.3 (cf. Figure 6.4). Here, the disturbance response and relative parametric sensitivity of the two control structures are similar to the SISO case described in Chapter 6. Finally, in practice PIP controllers are generally implemented in an equivalent incremental form to avoid problems associated with integral wind-up. Again, these are simply the vector-matrix form of the scalar examples discussed in section 6.2.

For example, the incremental feedback algorithm is as follows:

$$\boldsymbol{u}(k) = \boldsymbol{u}(k-1) - \boldsymbol{L}(z^{-1})\,\Delta\boldsymbol{y}(k) - \boldsymbol{M}(z^{-1})\,\Delta\boldsymbol{u}(k) + \boldsymbol{K}_I\left(\boldsymbol{y}_d(k) - \boldsymbol{y}(k)\right) \tag{7.30}$$

where $\Delta = 1 - z^{-1}$ is the difference operator, and the correction:

$$\left\{\bar{\boldsymbol{u}}, \underline{\boldsymbol{u}}\right\} \rightarrow \left\{ \begin{array}{llll} \text{if} & \boldsymbol{u}(k) \geq \bar{\boldsymbol{u}} & \text{then} & \boldsymbol{u}(k) = \bar{\boldsymbol{u}} \\ \text{if} & \boldsymbol{u}(k) \leq \underline{\boldsymbol{u}} & \text{then} & \boldsymbol{u}(k) = \underline{\boldsymbol{u}} \end{array} \right. \tag{7.31}$$

in which $\bar{\boldsymbol{u}}$ and $\underline{\boldsymbol{u}}$ are the specified constraints on the input control signals $\boldsymbol{u}(k)$.

7.3 Optimal Design for Multivariable PIP Control

For optimal design, the aim is to design a feedback gain matrix $\boldsymbol{K}$ that minimises the following LQ cost function:

$$J = \frac{1}{2}\sum_{k=0}^{\infty} \boldsymbol{x}(k)^T \boldsymbol{Q}\boldsymbol{x}(k) + \boldsymbol{u}(k)^T \boldsymbol{R}\boldsymbol{u}(k) \tag{7.32}$$

Here $\boldsymbol{Q} = \boldsymbol{q}^T\boldsymbol{q}$ is a positive semi-definite symmetric state weighting matrix and $\boldsymbol{R} = \boldsymbol{r}^T\boldsymbol{r}$ is a positive definite symmetric input weighting matrix, where $\boldsymbol{q}$ and $\boldsymbol{r}$ are the associated Choleski factors. Equation (7.32) is the multivariable equivalent of the univariate cost function (5.75) in Chapter 5. It is the standard formulation of the infinite time, LQ optimal servomechanism cost function for a multivariable system. The optimal control law that minimises the performance measure (7.32) is given by:

$$\boldsymbol{u}(k) = -\left(\boldsymbol{R} + \boldsymbol{G}^T\boldsymbol{P}\boldsymbol{G}\right)^{-1}\boldsymbol{G}^T\boldsymbol{P}\boldsymbol{F}\boldsymbol{x}(k) \tag{7.33}$$

where the matrix $\boldsymbol{P}$ is the steady-state solution of the following discrete-time, algebraic Riccati equation (e.g. Kuo 1997):

$$\boldsymbol{P} - \boldsymbol{F}^T\boldsymbol{P}\boldsymbol{F} + \boldsymbol{F}^T\boldsymbol{P}\boldsymbol{G}(\boldsymbol{R} + \boldsymbol{G}^T\boldsymbol{P}\boldsymbol{G})^{-1}\boldsymbol{G}^T\boldsymbol{P}\boldsymbol{F} - \boldsymbol{Q} = \boldsymbol{O} \tag{7.34}$$

Using equation (7.26), the optimal feedback gain matrix $\boldsymbol{K}$ is given by:

$$\boldsymbol{K} = \left(\boldsymbol{R} + \boldsymbol{G}^T\boldsymbol{P}\boldsymbol{G}\right)^{-1}\boldsymbol{G}^T\boldsymbol{P}\boldsymbol{F} = \begin{bmatrix} \boldsymbol{L}_0 & \boldsymbol{L}_1 & \cdots & \boldsymbol{L}_{n-1} & \boldsymbol{M}_1 & \cdots & \boldsymbol{M}_{m-1} & -\boldsymbol{K}_I \end{bmatrix} \tag{7.35}$$

It is important to note that, with a perfect model, the control law (7.33) guarantees the asymptotic stability of the resulting closed-loop system, as long as the NMSS model is either completely controllable or at least stabilisable. The closed-loop control system in this case is given by:

$$\boldsymbol{x}(k) = \left(\boldsymbol{F} - \boldsymbol{G}\left(\boldsymbol{R} + \boldsymbol{G}^T\boldsymbol{P}\boldsymbol{G}\right)^{-1}\boldsymbol{G}^T\boldsymbol{P}\boldsymbol{F}\right)\boldsymbol{x}(k-1) + \boldsymbol{D}\boldsymbol{y}_d(k) \tag{7.36}$$

and the associated closed-loop poles or eigenvalues can be obtained from the following characteristic equation:

$$\det\left\{\lambda\boldsymbol{I} - \boldsymbol{F} + \boldsymbol{G}\left(\boldsymbol{R} + \boldsymbol{G}^T\boldsymbol{P}\boldsymbol{G}\right)^{-1}\boldsymbol{G}^T\boldsymbol{P}\boldsymbol{F}\right\} = 0 \tag{7.37}$$

It is clear that, due to the special structure of the non-minimal state vector, the elements of the LQ weighting matrices have a particularly simple interpretation, since the diagonal elements directly define the weights assigned to the measured input and output variables, together with the integral-of-error states. In this regard, the following convention is sometimes employed for the choice of the multivariable weighting matrices $\boldsymbol{Q}$ and $\boldsymbol{R}$ (Taylor *et al.* 2000; Taylor and Shaban 2006):

$$\boldsymbol{Q} = \text{diag}[\bar{y}_1 \ldots \bar{y}_n \; \bar{u}_1 \ldots \bar{u}_{m-1} \; \bar{z}] \tag{7.38}$$

where $\bar{y}_i(i = 1 \ldots n)$, $\bar{u}_i(i = 1 \ldots m-1)$ and $\bar{z}$ are defined as follows:

$$\begin{aligned} \bar{y}_i \;\; (i = 1 \;\ldots\; n) &= \begin{bmatrix} \dfrac{y_1^w}{n} & \ldots & \dfrac{y_p^w}{n} \end{bmatrix} \\ \bar{u}_i \;\; (i = 1 \;\ldots\; m-1) &= \begin{bmatrix} \dfrac{u_1^w}{m} & \ldots & \dfrac{u_r^w}{m} \end{bmatrix} \\ \bar{z} &= \begin{bmatrix} z_1^w & \ldots & z_p^w \end{bmatrix} \end{aligned} \tag{7.39}$$

in which, $y_1^w \ldots y_p^w$, $u_1^w \ldots u_r^w$ and $z_1^w \ldots z_p^w$ are the user selected weighting parameters.

In this case, the corresponding input weighting matrix takes the following form:

$$\boldsymbol{R} = \text{diag}\begin{bmatrix} \dfrac{u_1^w}{m} \ldots \dfrac{u_r^w}{m} \end{bmatrix} \tag{7.40}$$

Although convoluted in description, the purpose of equation (7.38), equation (7.39) and equation (7.40) is to simplify the choice of the LQ weightings, so that the designer selects only a *total* weight associated with all the present and past values of each input and output signal, together with each integral-of-error state. This formulation is the multivariable equivalent of equation (5.78) and, in a similar manner, the 'default' weightings are obtained by setting each of the user selected parameters to unity.

Example 7.2 Multivariable PIP-LQ control of a Two-Input, Two-Output System Consider again the first order multivariable system described in Example 7.1, i.e. the TFM model (7.2) and NMSS form (7.7). Since $p = r = 2$ and $n = m = 1$:

$$\boldsymbol{Q} = \text{diag}\begin{bmatrix} y_1^w \; y_2^w \; z_1^w \; z_2^w \end{bmatrix}; \quad \boldsymbol{R} = \text{diag}\begin{bmatrix} u_1^w \; u_2^w \end{bmatrix} \tag{7.41}$$

In this case, the default state weighting matrix $\boldsymbol{Q}$ and input weighting matrix $\boldsymbol{R}$, are 4×4 and 2×2 identity matrices, respectively. Solving the LQ cost function (7.32), using (7.34) and (7.35) yields:

$$\boldsymbol{K} = \begin{bmatrix} -0.2079 & 0.7891 & 0.2347 & -0.4838 \\ 0.3005 & -0.2807 & -0.3569 & 0.1469 \end{bmatrix} \tag{7.42}$$

Hence, the control gain matrices associated with Figure 7.2 or Figure 7.3 are:

$$\boldsymbol{L}(z^{-1}) = \boldsymbol{L}_0 = \begin{bmatrix} -0.2079 & 0.7891 \\ 0.3005 & -0.2807 \end{bmatrix}; \quad \boldsymbol{K}_I = \begin{bmatrix} -0.2347 & 0.4838 \\ 0.3569 & -0.1469 \end{bmatrix} \tag{7.43}$$

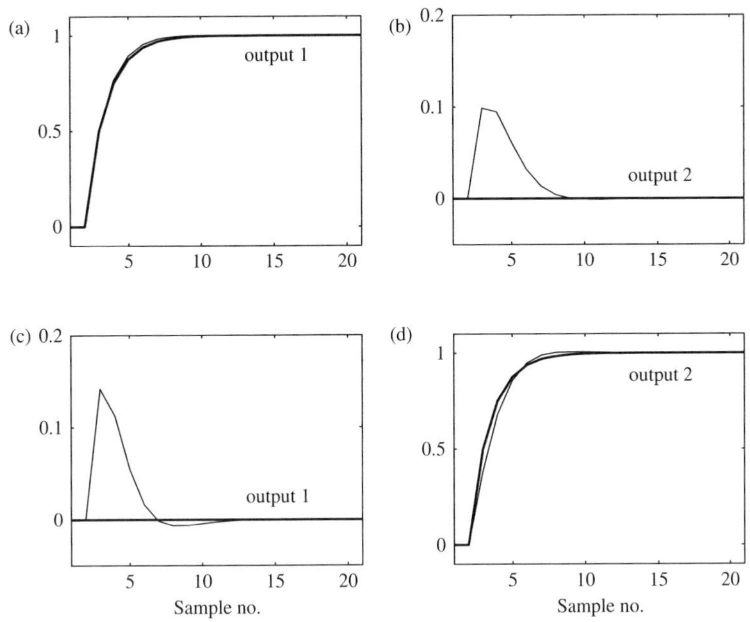

Figure 7.4 Closed-loop response of the multivariable system in Example 7.2, comparing optimal design using diagonal weights (thin traces) and algebraic decoupling by pole assignment (thick traces: see Example 7.6)

The various closed-loop responses are illustrated by the thin traces in Figure 7.4, which shows the response of $y_1(k)$ (Figure 7.4a) and $y_2(k)$ (Figure 7.4b) to a unit step change in $y_{d1}(k)$ when $y_{d2}(k) = 0$; and the response of $y_1(k)$ (Figure 7.4c) and $y_2(k)$ (Figure 7.4d) to a unit step change in $y_{d2}(k)$ when $y_{d1}(k) = 0$. In other words, the diagonal subplots show the response of each output variable to a unit step in the associated command level, whilst the off-diagonals show the cross-coupling dynamics.

For this example with these default weights, the closed-loop cross-coupling dynamics are relatively small (10–15%), as shown by the off-diagonal subplots, noting the differing scales on the axis. Of course, improved responses may be obtained by utilising either analytical decoupling by pole assignment or numerical optimisation of the LQ weighting matrices, as discussed later. For example, a pole assignment solution with analytical decoupling is shown by the thick traces in Figure 7.4 (see Example 7.6 for details).

Example 7.3 Multivariable PIP-LQ control of an Unstable System Consider another two-input, two-output system represented in TFM form (7.9):

$$\boldsymbol{G}(z^{-1}) = \begin{bmatrix} \dfrac{z^{-1}+z^{-2}}{1-2z^{-1}+z^{-2}} & \dfrac{z^{-1}}{1-z^{-1}} \\ \dfrac{2z^{-1}}{1-2z^{-1}} & \dfrac{3z^{-1}}{1-0.5z^{-1}} \end{bmatrix} \tag{7.44}$$

where it will be noted that $1 - 2z^{-1} + z^{-2} = \left(1 - z^{-1}\right)\left(1 - z^{-1}\right)$.

For stability analysis, the TFM is usually expressed in terms of the forward shift operator (see Example 2.3), as follows:

$$\mathbf{G}(z) = \begin{bmatrix} \dfrac{z+1}{z^2-2z+1} & \dfrac{1}{z-1} \\ \dfrac{2}{z-2} & \dfrac{3}{z-0.5} \end{bmatrix} = \begin{bmatrix} \dfrac{z+1}{(z-1)(z-1)} & \dfrac{1}{z-1} \\ \dfrac{2}{z-2} & \dfrac{3}{z-0.5} \end{bmatrix} \tag{7.45}$$

Hence, the open-loop system is rather badly behaved with the TF models for the first output variable consisting of a double integrator and a single integrator; while those associated with the second output variable are both first order systems, one of which is unstable. To develop the MFD form, we first determine common denominators for each output variable, as follows:

$$\begin{aligned} y_1(k) &= \frac{z^{-1}+z^{-2}}{1-2z^{-1}+z^{-2}}u_1(k) + \frac{z^{-1}\left(1-z^{-1}\right)}{\left(1-z^{-1}\right)\left(1-z^{-1}\right)}u_2(k) \\ y_2(k) &= \frac{2z^{-1}\left(1-0.5z^{-1}\right)}{\left(1-2z^{-1}\right)\left(1-0.5z^{-1}\right)}u_1(k) + \frac{3z^{-1}\left(1-2z^{-1}\right)}{\left(1-2z^{-1}\right)\left(1-0.5z^{-1}\right)}u_2(k) \end{aligned} \tag{7.46}$$

where $y_1(k)$, $y_2(k)$, $u_1(k)$ and $u_2(k)$ are the output and input variables associated with the TFM model (7.44). Hence,

$$\mathbf{G}(z^{-1}) = \begin{bmatrix} \dfrac{z^{-1}+z^{-2}}{1-2z^{-1}+z^{-2}} & \dfrac{z^{-1}-z^{-2}}{1-2z^{-1}+z^{-2}} \\ \dfrac{2z^{-1}-z^{-2}}{1-2.5z^{-1}+z^{-2}} & \dfrac{3z^{-1}-6z^{-2}}{1-2.5z^{-1}+z^{-2}} \end{bmatrix} \tag{7.47}$$

The system representation is next transformed into the following left MFD description:

$$\mathbf{A}(z^{-1})\begin{bmatrix} y_1(k) \\ y_2(k) \end{bmatrix} = \mathbf{B}(z^{-1})\begin{bmatrix} u_1(k) \\ u_2(k) \end{bmatrix} \tag{7.48}$$

in which

$$\mathbf{A}(z^{-1}) = \begin{bmatrix} 1-2z^{-1}+z^{-2} & 0 \\ 0 & 1-2.5z^{-1}+z^{-2} \end{bmatrix} = \begin{bmatrix} 1 & 0 \\ 0 & 1 \end{bmatrix} + \begin{bmatrix} -2 & 0 \\ 0 & -2.5 \end{bmatrix}z^{-1} + \begin{bmatrix} 1 & 0 \\ 0 & 1 \end{bmatrix}z^{-2} \tag{7.49}$$

and

$$\mathbf{B}(z^{-1}) = \begin{bmatrix} z^{-1}+z^{-2} & z^{-1}-z^{-2} \\ 2z^{-1}-z^{-2} & 3z^{-1}-6z^{-2} \end{bmatrix} = \begin{bmatrix} 1 & 1 \\ 2 & 3 \end{bmatrix}z^{-1} + \begin{bmatrix} 1 & -1 \\ -1 & -6 \end{bmatrix}z^{-2} \tag{7.50}$$

Finally, the NMSS model is:

$$\begin{bmatrix} y_1(k) \\ y_2(k) \\ y_1(k-1) \\ y_2(k-1) \\ u_1(k-1) \\ u_2(k-1) \\ z_1(k) \\ z_2(k) \end{bmatrix} = \begin{bmatrix} 2 & 0 & -1 & 0 & 1 & -1 & 0 & 0 \\ 0 & 2.5 & 0 & -1 & -1 & -6 & 0 & 0 \\ 1 & 0 & 0 & 0 & 0 & 0 & 0 & 0 \\ 0 & 1 & 0 & 0 & 0 & 0 & 0 & 0 \\ 0 & 0 & 0 & 0 & 0 & 0 & 0 & 0 \\ 0 & 0 & 0 & 0 & 0 & 0 & 0 & 0 \\ -2 & 0 & 1 & 0 & -1 & 1 & 1 & 0 \\ 0 & -2.5 & 0 & 1 & 1 & 6 & 0 & 1 \end{bmatrix} \begin{bmatrix} y_1(k-1) \\ y_2(k-1) \\ y_1(k-2) \\ y_2(k-2) \\ u_1(k-2) \\ u_2(k-2) \\ z_1(k-1) \\ z_2(k-1) \end{bmatrix}$$
$$+ \begin{bmatrix} 1 & 1 \\ 2 & 3 \\ 0 & 0 \\ 0 & 0 \\ 1 & 0 \\ 0 & 1 \\ -1 & -1 \\ -2 & -3 \end{bmatrix} \begin{bmatrix} u_1(k-1) \\ u_2(k-1) \end{bmatrix} + \begin{bmatrix} 0 & 0 \\ 0 & 0 \\ 0 & 0 \\ 0 & 0 \\ 0 & 0 \\ 0 & 0 \\ 1 & 0 \\ 0 & 1 \end{bmatrix} \begin{bmatrix} y_{d1}(k) \\ y_{d2}(k) \end{bmatrix} \tag{7.51}$$

where $z_1(k)$ and $z_2(k)$ are the integral-of-error states (7.8). The weighting matrices in the quadratic cost function are selected as follows:

$$\boldsymbol{Q} = \text{diag}\begin{bmatrix} 0.5 & 0.5 & 0.5 & 0.5 & 0.5 & 0.5 & 0.1 & 0.1 \end{bmatrix}; \boldsymbol{R} = \text{diag}\begin{bmatrix} 0.5 & 0.5 \end{bmatrix} \tag{7.52}$$

These particular $\boldsymbol{Q}$ and $\boldsymbol{R}$ matrices were tuned by trial and error to obtain a desirable closed-loop response, using (7.39) with $y_1^w = 1$, $y_2^w = 1$, $u_1^w = 1$, $u_2^w = 1$, $z_1^w = 0.1$ and $z_2^w = 0.1$. In contrast to the default solution, we have simply reduced the integral-of-error weightings by a factor of 10, in order to slow down the speed of response. Solving the LQ cost function (7.32), using (7.34) and (7.35) yields:

$$\boldsymbol{K} = \begin{bmatrix} \boldsymbol{L}_0 & \boldsymbol{L}_1 & \boldsymbol{M}_1 & -\boldsymbol{K}_I \end{bmatrix} \tag{7.53}$$

where the control gain matrices associated with Figure 7.2 or Figure 7.3 are:

$$\begin{aligned} \boldsymbol{L}_0 &= \begin{bmatrix} -1.3568 & 3.6790 \\ 1.4314 & -2.2330 \end{bmatrix}; \quad \boldsymbol{L}_1 = \begin{bmatrix} 1.0537 & -1.831 \\ -1.0324 & 1.1610 \end{bmatrix} \\ \boldsymbol{M}_1 &= \begin{bmatrix} -2.8669 & -9.8251 \\ 2.1934 & 5.9339 \end{bmatrix}; \quad \boldsymbol{K}_I = \begin{bmatrix} -0.0402 & 0.0397 \\ 0.0716 & 0.0392 \end{bmatrix} \end{aligned} \tag{7.54}$$

The resulting closed-loop step responses are shown in Figure 7.5a–d, with the corresponding control input signals shown in Figure 7.6a–d.

As usual, (a) and (b) show the response to a unit step change in $y_{d1}(k)$ when $y_{d2}(k) = 0$, whilst (c) and (d) are for a unit step change in $y_{d2}(k)$ when $y_{d1}(k) = 0$. Although the output variables shown in Figure 7.5 are not completely decoupled, the closed-loop response of this

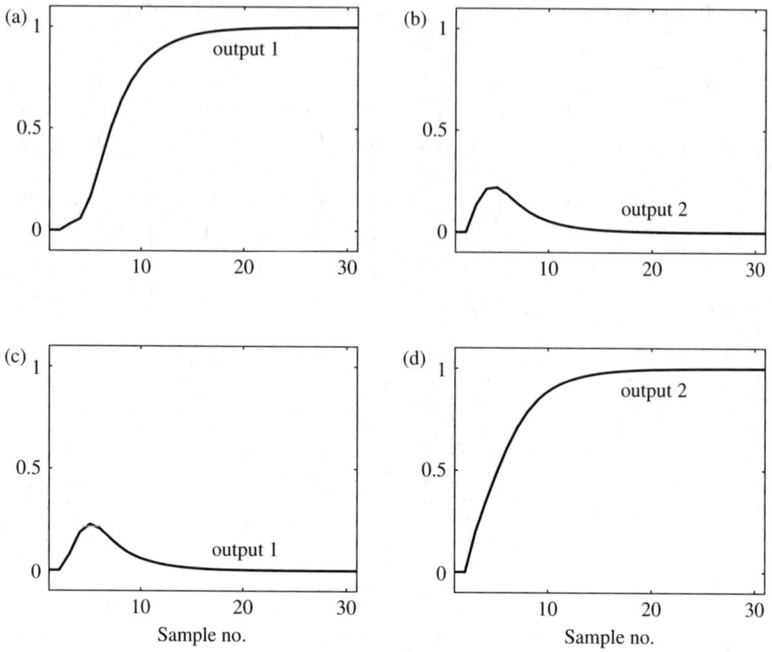

Figure 7.5 Closed-loop output response of the multivariable system in Example 7.3

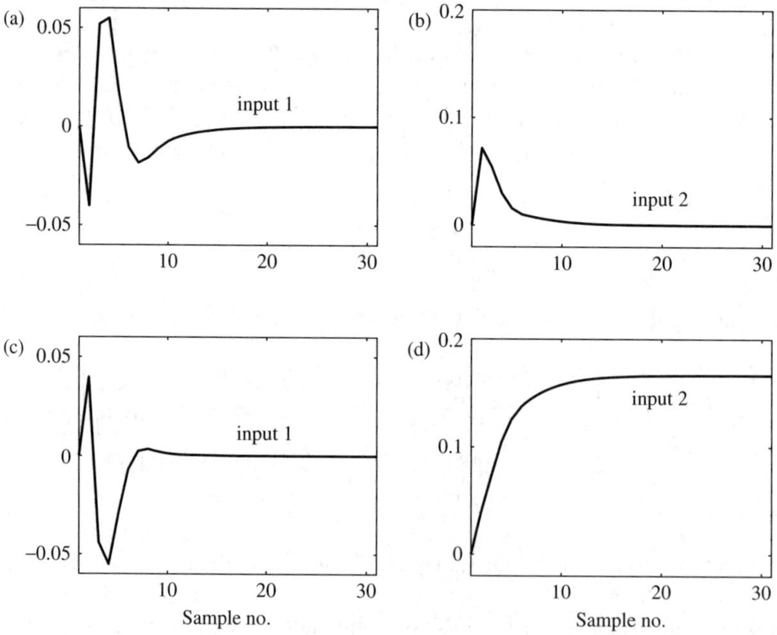

Figure 7.6 Control input variables associated with Figure 7.5

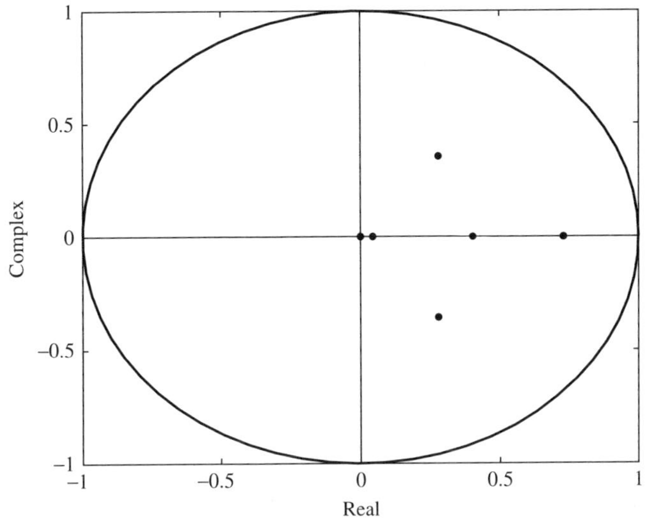

Figure 7.7 Closed-loop pole positions associated with Figure 7.5

open-loop unstable system is quite satisfactory, i.e. stable and with relatively small interaction effects.

Finally, the closed-loop pole locations obtained from equation (7.37), including a complex conjugate pair and two poles at the origin are: $p_1 = 0.0443$, $p_2 = 0.4036$, $p_3 = 0.7301$, $p_4 = 0.7320$, $p_5 = p_6 = 0$, $p_7 = 0.2805 + 0.3553j$ and $p_8 = 0.2805 - 0.3553j$. These are plotted on the complex z-plane in Figure 7.7.

Example 7.4 Multivariable PIP-LQ Control of a Coupled Drive System The coupled drive system illustrated in Figure 7.8 is described, for example, by Dixon and Lees (1994), Dixon (1996), Young *et al.* (1998) and Chotai *et al.* (1998). It is a two-input, two-output,

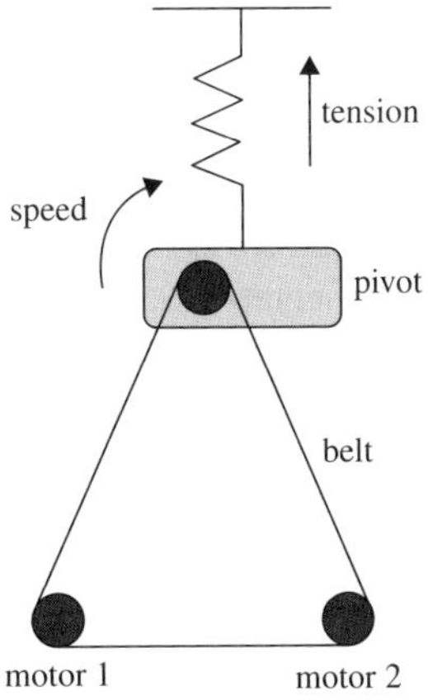

Figure 7.8 Schematic diagram of the coupled drives apparatus

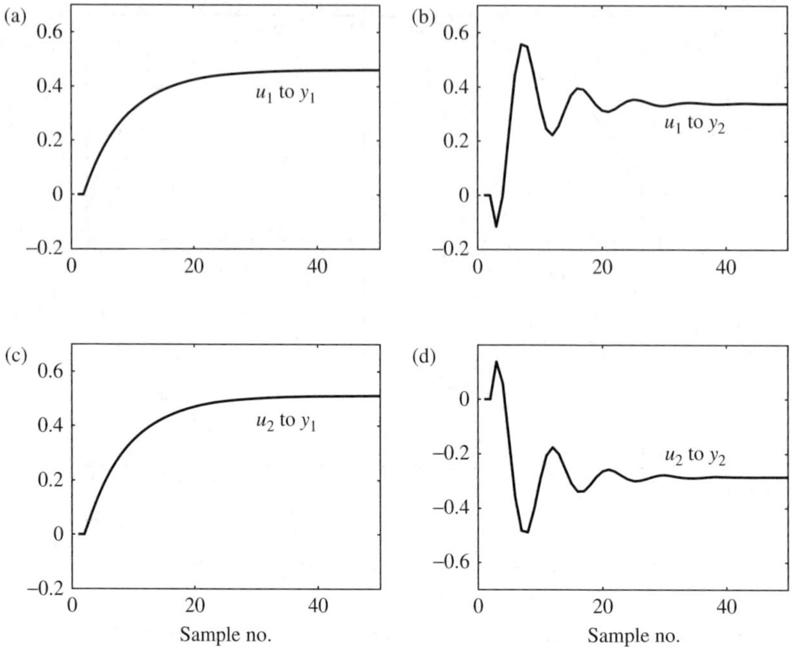

Figure 7.9 Open-loop unit step response of the coupled drives model in Example 7.4

laboratory scale apparatus, analogous in its dynamic behaviour to systems employed in the manufacturing of strip metal, wire, paper and textile fibres. The rig consists of three pulley wheels connected by a single rubber band. Two of these wheels are in fixed positions and connected to independently driven motors. The third is mounted upon a pivoting arm that, in turn, is attached to the main body of the rig by a spring.

There is a high degree of dynamical coupling between the tension and the speed of the rig, as shown in Figure 7.9a–d. Furthermore, the tension variable exhibits oscillatory dynamic behaviour. The multivariable design problem is to independently control both the speed of the third pulley and the tension in the spring, thus decoupling these two control channels.

Hence, in the equations that follow, $y_1(k)$ and $y_2(k)$ are the measured speed and tension, respectively; whilst the control inputs $u_1(k)$ and $u_2(k)$ relate to the motor 1 and motor 2 applied toque, respectively. These variables are all scaled for the purposes of the present example, so no units are given; details of the instrumentation set-up are given by Dixon (1996).

Statistical analysis of data obtained from open-loop experiments with a sampling rate of 40 Hz (0.025 s), utilising the *Simplified Refined Instrumental Variable* (SRIV) algorithm described in Chapter 8, yields the following TFM (7.9):

$$\boldsymbol{G}(z^{-1}) = \begin{bmatrix} \dfrac{0.0602z^{-1}}{1 - 0.8696z^{-1}} & \dfrac{0.0666z^{-1}}{1 - 0.8696z^{-1}} \\ \dfrac{-0.1150z^{-1} + 0.2603z^{-2}}{1 - 1.3077z^{-1} + 0.7374z^{-2}} & \dfrac{0.1385z^{-1} - 0.2610z^{-2}}{1 - 1.3077z^{-1} + 0.7374z^{-2}} \end{bmatrix} \qquad (7.55)$$

The TFM is converted into MFD form (7.11) with:

$$A(z^{-1}) = \begin{bmatrix} 1 & 0 \\ 0 & 1 \end{bmatrix} + \begin{bmatrix} -0.8696 & 0 \\ 0 & -1.3077 \end{bmatrix} z^{-1} + \begin{bmatrix} 0 & 0 \\ 0 & 0.7374 \end{bmatrix} z^{-2} \tag{7.56}$$

and

$$B(z^{1}) = \begin{bmatrix} 0.0602 & 0.0666 \\ -0.1150 & 0.1385 \end{bmatrix} z^{-1} + \begin{bmatrix} 0 & 0 \\ 0.2603 & -0.261 \end{bmatrix} z^{-2} \tag{7.57}$$

The associated multivariate NMSS model (7.19) is defined by:

$$F = \begin{bmatrix} 0.8696 & 0 & 0 & 0 & 0 & 0 & 0 & 0 \\ 0 & 1.3077 & 0 & -0.7374 & 0.2603 & -0.2610 & 0 & 0 \\ 1 & 0 & 0 & 0 & 0 & 0 & 0 & 0 \\ 0 & 1 & 0 & 0 & 0 & 0 & 0 & 0 \\ 0 & 0 & 0 & 0 & 0 & 0 & 0 & 0 \\ 0 & 0 & 0 & 0 & 0 & 0 & 0 & 0 \\ -0.8696 & 0 & 0 & 0 & 0 & 0 & 1 & 0 \\ 0 & -1.3077 & 0 & 0.7374 & -0.2603 & 0.2610 & 0 & 1 \end{bmatrix}$$

$$G = \begin{bmatrix} 0.0602 & 0.0666 \\ -0.1150 & 0.1385 \\ 0 & 0 \\ 0 & 0 \\ 1 & 0 \\ 0 & 1 \\ -0.0602 & -0.0666 \\ 0.1150 & -0.1385 \end{bmatrix}; \quad D = \begin{bmatrix} 0 & 0 \\ 0 & 0 \\ 0 & 0 \\ 0 & 0 \\ 0 & 0 \\ 0 & 0 \\ 1 & 0 \\ 0 & 1 \end{bmatrix} \tag{7.58}$$

Here, the non-minimal state vector is:

$$x(k) = \begin{bmatrix} y_1(k) & y_2(k) & y_1(k-1) & y_2(k-1) & u_1(k-1) & u_2(k-1) & z_1(k) & z_2(k) \end{bmatrix}^T \tag{7.59}$$

where $z_1(k)$ and $z_2(k)$ are the integral-of-error states for the speed and tension signals, respectively, defined in the same way as (7.8). Trial and error tuning, using both simulation and the laboratory apparatus, yields the following weighting matrices for a satisfactory closed-loop response:

$$\begin{aligned} Q &= \mathrm{diag}\begin{bmatrix} 0.25 & 0.25 & 0.25 & 0.25 & 0.5 & 0.5 & 0.01 & 0.01 \end{bmatrix} \\ R &= \mathrm{diag}\begin{bmatrix} 1.0 & 1.0 \end{bmatrix} \end{aligned} \tag{7.60}$$

In this case, solving the LQ cost function (7.32), using (7.34) and (7.35) yields:

$$K = \begin{bmatrix} 0.3452 & 0.0636 & 0 & -0.1936 & 0.0683 & -0.0685 & -0.0519 & -0.0579 \\ 0.3990 & -0.0249 & 0 & 0.1545 & -0.0545 & 0.0547 & -0.0600 & 0.0497 \end{bmatrix} \tag{7.61}$$

Consequently, the control gain matrices associated with Figure 7.2 or Figure 7.3 are:

$$L(z^{-1}) = \begin{bmatrix} 0.3452 & 0.0636 \\ 0.3990 & -0.0249 \end{bmatrix} + \begin{bmatrix} 0 & -0.1936 \\ 0 & 0.1545 \end{bmatrix} z^{-1} \tag{7.62}$$

$$M(z^{-1}) = \begin{bmatrix} 0.0683 & -0.0685 \\ -0.0545 & 0.0547 \end{bmatrix} z^{-1}; \quad K_I = \begin{bmatrix} 0.0519 & 0.0579 \\ 0.0600 & -0.0497 \end{bmatrix} \tag{7.63}$$

For this coupled drive system, both polynomial matrices $A(z^{-1})$ and $B(z^{-1})$ are left coprime, hence condition (i) of Theorem 7.1 is fulfilled. However, the two row degrees of $A(z^{-1})$ are different so that condition (ii) is not satisfied. As a result, there is one uncontrollable mode at the origin in the NMSS model. Furthermore, both terms in the output feedback polynomial matrix (7.62) corresponding to $y_1(k-1)$ are zero, i.e. although $y_1(k-1)$ is introduced into the NMSS representation, it does not appear in the optimal control law and hence this mode has no effect upon the closed-loop response.

Nonetheless, laboratory experiments reveal that the PIP–LQ control system effectively removes almost all the coupling between the output variables, as illustrated in Figure 7.10a–d for simulated data. The subplots in Figure 7.10 are arranged in the same manner as for the earlier examples. Similar results based on closed-loop experimental data are illustrated in the references given earlier.

7.4 Multi-Objective Optimisation for PIP Control

In many cases, satisfactory closed-loop performance of the optimal PIP-LQ control system is obtained by straightforward manual tuning of the diagonal weights, as discussed in Example 7.2, Example 7.3 and Example 7.4 and, for univariate controllers, in Chapter 5 and Chapter 6. In more difficult situations, the PIP approach is ideal for incorporation within a multi-objective optimisation framework. Here satisfactory compromise can be obtained between conflicting objectives such as robustness, overshoot, settling times, rise times, control input level and rate constraints, frequency domain band-pass and, for multivariable systems, decoupling control. Examples of such multi-objective optimisation can be found in Fleming and Pashkevich (1986), Dixon and Pike (2006), Simm and Liu (2006), Wojsznis *et al.* (2007) and Exadaktylos and Taylor (2010), among others. In the case of PIP-LQ design, this is achieved by concurrent optimisation of the diagonal and off-diagonal elements of the weighting matrices in the LQ cost function (Chotai *et al.* 1998).

Such multi-objective optimisation is a convenient and practical approach to mapping the technical characteristics into elements of the LQ weighting matrices. It is straightforwardly

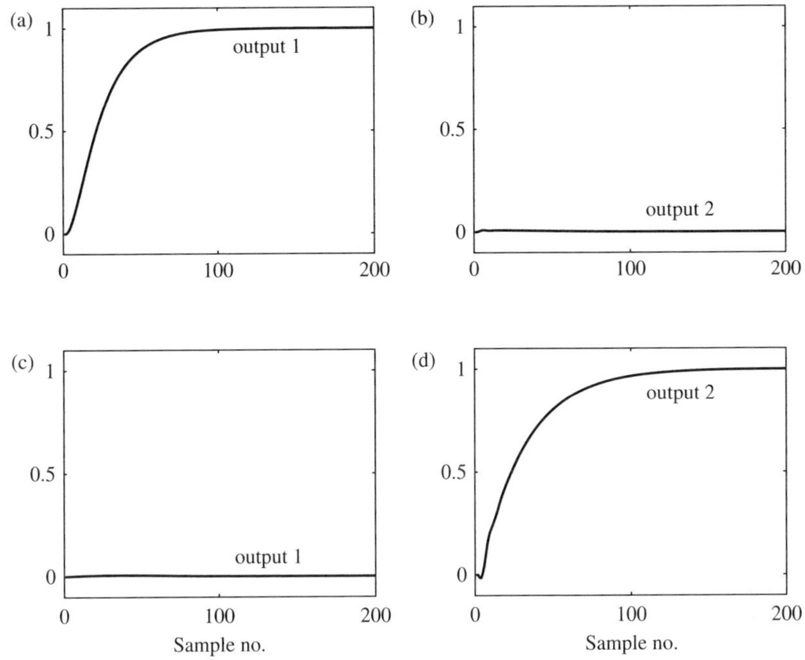

Figure 7.10 Closed-loop output of the coupled drives model in Example 7.4

implemented by exploitation of goal-attainment functions in readily available numerical optimisation packages, such as MATLAB®[2]. Here, our novel approach is to numerically optimise the Cholesky factors of the weighting matrices in order to satisfy the user-defined requirements, while ensuring the positive definiteness of the LQ weighting matrices (Tych and Taylor 1996; Chotai *et al.* 1998). The approach differs from related methods that involve direct optimisation of the control gains themselves (e.g. Fleming and Pashkevitch 1986). The optimisation of the LQ *weights* has the advantage of generating only guaranteed stable optimal solutions (in the case of no model mismatch), thus allowing for better (smoother) defined optimisation problems.

7.4.1 Goal Attainment

In order to meet the multiple objectives, a solution is sought within a subspace of controller parameters which fulfil both the LQ optimality criteria (7.32) and the technical conditions, such as multivariable decoupling. The goal attainment method involves expressing a set of design objectives or goals $\boldsymbol{F}^*$ associated with a set of objective functions $\boldsymbol{F}(\boldsymbol{f})$, where $\boldsymbol{f}$ is a vector of optimisation parameters. The problem formulation permits the objectives to be over- or under-achieved, allowing for very optimistic goals to be defined without leading to

[2] MATLAB®, The MathWorks Inc., Natick, MA, USA. Various commercial and freely available tools address generic optimisation problems. The examples in this book all utilise the fgoalattain function in the Optimisation Toolbox for MATLAB®.

an infeasible problem. A relative weighting $\boldsymbol{\omega}$ among the objectives is also specified, which enables the designer to include every desirable objective but to assign a relative importance to each. With a set of h goals, we define the goal, objective function and weighting vectors as follows:

$$\begin{aligned} \boldsymbol{F}^* &= \begin{bmatrix} F_1^* & F_2^* & \cdots & F_h^* \end{bmatrix} \\ \boldsymbol{F}(\boldsymbol{f}) &= [F_1(\boldsymbol{f}) \quad F_2(\boldsymbol{f}) \quad \cdots \quad F_h(\boldsymbol{f})] \\ \boldsymbol{\omega} &= [\omega_1 \quad \omega_2 \quad \cdots \quad \omega_h] \end{aligned} \tag{7.64}$$

The optimisation problem is then to minimise the real scalar variable λ such that:

$$\boldsymbol{F}(\boldsymbol{f}) - \boldsymbol{\omega}\lambda \leq \boldsymbol{F}^* \tag{7.65}$$

In the case that an element ω_i of the weighting vector is zero, the associated objective is intended as 'hard' and (if possible) the solution will explicitly satisfy the inequality above. Depending on their nature, it can be necessary to constrain the optimisation parameters $\boldsymbol{f}$ to a specified region (linear or nonlinear) imposed by process-related factors. In the present context, therefore, $\boldsymbol{f}$ is a vector containing the elements of the lower triangular matrices $\boldsymbol{q}$ and $\boldsymbol{r}$, where $\boldsymbol{q}$ and $\boldsymbol{r}$ are the Choleski factors of the weighting matrices in the LQ cost function (7.32).

In fact, it is not always necessary to exploit the full dimension of the weighting matrices. In practice, excellent results are often achieved by simply optimising the combined diagonal weighting parameters $y_1^w \ldots y_p^w$, $u_1^w \ldots u_p^w$ and $z_1^w \ldots z_p^w$ defined by equations (7.39).

However, in the case of multivariable decoupling, practical experience reveals that the off-diagonals associated with the $z(k)$ states can be very important. As discussed above, these states are explicitly introduced into the PIP control system to ensure steady-state tracking and, in the multivariable case, static decoupling. Example 7.5 shows how optimisation of the diagonal elements of $\boldsymbol{Q}$ and $\boldsymbol{R}$, together with the off-diagonals associated with the $z(k)$ states only, yields almost complete dynamic decoupling of a three-input, three-output system.

Example 7.5 PIP-LQ control of the Shell Heavy Oil Fractionator Simulation The Shell Heavy Oil Fractionator simulation was introduced by Shell in 1987. It is a highly coupled continuous-time linear model that illustrates some of the control problems associated with a typical multivariable industrial plant (Sandoz *et al.* 2000). The full benchmark simulation, with 7 outputs and 5 inputs, is based on 35 differential equations with various time delays. However, four of the outputs are auxiliary variables that do not have to be controlled, while two of the input variables are uncontrollable disturbances. Hence, in the simplest design terms considered here, the system is limited to 3 control inputs and 3 measured outputs.

For convenience, the time units employed in this study are minutes. Clearly, however, any consistent set of units may be applied to the simulation, which is illustrative of a typical industrial plant, rather than a specific case. Experimentation reveals that a sampling rate of 10 units (minutes) yields a good compromise between simplicity, in the form of a relatively low order state vector, while ensuring the ability to maintain tight control at all times. In order to obtain suitable models for control system design, three open-loop experiments are carried out on the continuous-time simulation, the responses of which are illustrated in Figure 7.11.

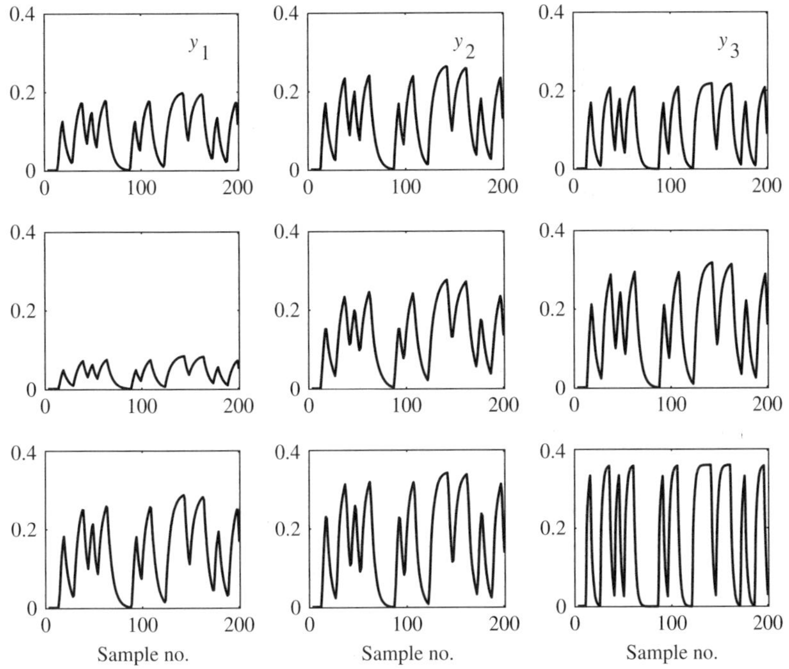

Figure 7.11 Open-loop response of the Shell Heavy Oil Fractionator simulation model in Example 7.5

Each row in Figure 7.11 represents an open-loop simulation experiment, in which one of the control input variables is a *Pseudo Random Binary Signal* (PRBS), while the other two are held at zero and there are no disturbances. As usual, each column represents one output variable. Figure 7.11 clearly illustrates the high degree of coupling between the variables, confirmed by the similar steady-state gains and time constants for many of the input–output pathways. Although PIP control will, of course, achieve steady-state decoupling of the outputs, Figure 7.11 suggests that good transient decoupling will be a difficult objective to achieve.

Statistical analysis of simulation data for each input–output pathway, utilising the SRIV algorithm developed in Chapter 8, yields discrete-time TF models that follow the simulated response almost exactly, as would be expected since the simulation is a linear model with no noise. Any time delays of the continuous-time model that are a non-integral number of sampling intervals are approximated, where necessary, by the estimation of two numerator parameters. These models are converted into the following MFD form based on equation (7.11):

$$\boldsymbol{A}_1 = \begin{bmatrix} -1.6637 & 0 & 0 \\ 0 & -2.4418 & 0 \\ 0 & 0 & -1.9578 \end{bmatrix}; \quad \boldsymbol{A}_2 = \begin{bmatrix} 0.6918 & 0 & 0 \\ 0 & 1.9864 & 0 \\ 0 & 0 & 1.2373 \end{bmatrix}$$

$$\boldsymbol{A}_3 = \begin{bmatrix} 0 & 0 & 0 \\ 0 & -0.5383 & 0 \\ 0 & 0 & -0.2487 \end{bmatrix}$$

$$\boldsymbol{B}_1 = \begin{bmatrix} 0 & 0 & 0 \\ 0 & 0.4173 & 0.5851 \\ 0 & 0 & 3.9022 \end{bmatrix}; \quad \boldsymbol{B}_2 = \begin{bmatrix} 0.1232 & 0 & 0.1789 \\ 0.9853 & -0.2036 & -0.0301 \\ 0.9993 & 0.8873 & -5.9895 \end{bmatrix} \tag{7.66}$$

$$\boldsymbol{B}_3 = \begin{bmatrix} 0.5079 & 0.2737 & 0.7375 \\ -1.6010 & -0.4722 & -1.1641 \\ -1.0715 & -0.6107 & 2.2950 \end{bmatrix}; \quad \boldsymbol{B}_4 = \begin{bmatrix} -0.5173 & -0.2241 & -0.7511 \\ 0.6493 & 0.2940 & 0.6520 \\ 0.1574 & -0.2103 & 0 \end{bmatrix}$$

$$\boldsymbol{B}_5 = \begin{bmatrix} 0 & 0 & 0 \\ 0 & 0 & 0 \\ 0.0495 & 0.1311 & 0 \end{bmatrix}$$

where the input and output vectors are:

$$\boldsymbol{y}(k) = [y_1(k) \quad y_2(k) \quad y_3(k)]^T; \quad \boldsymbol{u}(k) = [u_1(k) \quad u_2(k) \quad u_3(k)]^T \tag{7.67}$$

Here, the various input and output signals, which represent pressures, temperatures and so on, are defined by Sandoz *et al.* (2000). The NMSS model (7.19) and PIP-LQ controller (7.25) is obtained in the usual manner, to determine a 3×24 matrix of feedback gains $\boldsymbol{K}$. The response of the PIP-LQ controller is illustrated in Figure 7.12, in which the thin traces are based on identity matrices for both $\boldsymbol{Q}$ and $\boldsymbol{R}$, while the thick traces show the optimised response. In

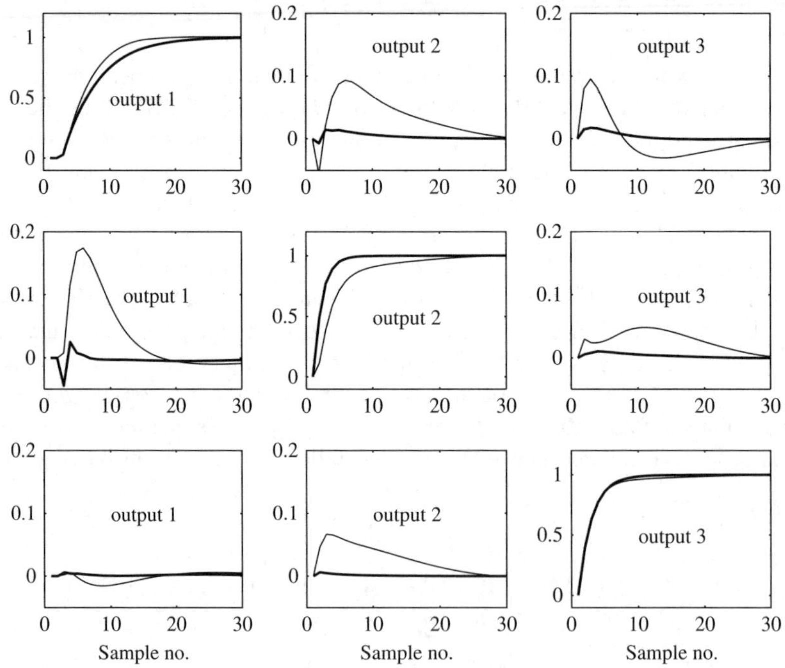

Figure 7.12 Closed-loop response of the multivariable system in Example 7.5, comparing design using diagonal unity weights (thin traces) and optimised weights (thick traces)

a similar manner to the earlier simulation examples, the subplots along the diagonal of the figure show the response to a unit step in the command input, with the off-diagonal subplots illustrating the various cross-coupled responses.

In order to obtain the optimised traces in Figure 7.12, the objective functions $\boldsymbol{F}(\boldsymbol{f})$ in (7.65) are based on the six cross-coupled responses: specifically, the sum of the absolute errors between the command input and the output that occur in two of the variables, when there is a step change in the third. The off-diagonal subplots in Figure 7.12 illustrate these cross-coupled responses. In other words, to obtain the values of the goals at each iterative step, the command response is determined for each output variable in turn, while the other two variables are controlled to zero. The associated (desired) goals in $\boldsymbol{F}^*$ are all set to zero. However, the desired speed of response is also included in $\boldsymbol{F}(\boldsymbol{f})$, in the form of the rise time for each of the three outputs. Without this balance, the decoupling objective could be achieved by simply ensuring an unacceptably slow command response. The rise times are obtained from the diagonal subplots in Figure 7.12. For the present example, the rise time goals in $\boldsymbol{F}^*$ are specified to be the same as those obtained with the initial PIP-LQ design based on identity weighting matrices. Hence, three of nine optimisation goals are to maintain these rise times, while the remainder are associated with the decoupling.

In this example, it is not necessary to employ the entire $\boldsymbol{Q}$ and $\boldsymbol{R}$ matrices in order to obtain reasonable transient decoupling. In fact, the optimisation is based mainly on the diagonal elements, together with the off-diagonals associated with the $z(k)$ states. For completeness, the numerical values of the optimised weights are quoted below. The diagonal elements of $\boldsymbol{R}$ are:

$$\text{diag}(\boldsymbol{R}) = [0.1187 \quad 0.0137 \quad 0.0471] \tag{7.68}$$

while

$$\begin{aligned}\text{diag}(\boldsymbol{Q}) = [&2.1039, 2.8282, 1.7701, 2.1050, 2.8279, 1.7714, 2.1031, 2.8281,\\ &1.7714, 0.1187, 0.0137, 0.0471, 0.1039, 0.0133, 0.0461, 0.1185,\\ &0.0137, 0.0458, 0.1190, 0.0138, 0.0463, 0.2711, 5.2028, 1.3661]\end{aligned} \tag{7.69}$$

and the 3×3 submatrix of $\boldsymbol{Q}$ associated with the $z(k)$ states is:

$$\boldsymbol{Q}(22:24,\ 22:24) = \begin{bmatrix} 0.2711 & 0.1102 & 0.0236 \\ 0.1102 & 5.2028 & 0.0606 \\ 0.0236 & 0.0606 & 1.3661 \end{bmatrix} \tag{7.70}$$

All the other elements of $\boldsymbol{Q}$ and $\boldsymbol{R}$ are set to zero (they are not optimised). It is clear from Figure 7.12 that the improved decoupling performance has been achieved without any significant reduction in the speed of response. Note the different scales for the axis of the off-diagonal cross-coupling subplots in this figure. Note also that Figure 7.12 is based on the discrete-time model (7.66) but similar results are obtained using the original continuous-time system.

7.5 Proportional-Integral-Plus Decoupling Control by Algebraic Pole Assignment

As we have seen, an important advantage of model-based multivariable controllers over multiple-loop SISO controllers is their ability to dynamically decouple the various control channels, exploiting information about the interactions contained in the control model. Although the examples in the previous subsection show that the standard PIP-LQ controller can often provide good closed-loop performance, with relatively low cross-coupling terms, the present section considers how PIP control systems can be used for *full* dynamic decoupling.

Since Morgan (1964) initiated research on the design and synthesis of non-interacting control systems for multivariable systems, various techniques for decoupling a multivariable system by SVF (e.g. Falb and Wolovich 1967; Gilbert 1969; Morse and Wonham 1970) and output feedback (e.g. Wolovich 1975; Bayoumi and Duffield 1977; Argoun and van de Vegte 1980) have appeared in the literature. Similarly, numerous approaches have been presented for analytical decoupling of PI and PID type controllers. Examples are given by Plummer and Vaughan (1997), Wang *et al.* (2000, 2002), Gilbert *et al.* (2003), Liu and Zhang (2007) and Liu *et al.* (2007).

In the remainder of this chapter, we formulate two different methods for non-minimal SVF decoupling design. The first method uses a combined algebraic pole assignment and decoupling algorithm to achieve complete decoupling, with the closed-loop response shaped by the desired pole positions. The second method algebraically decouples the open-loop system, so that multiple loop univariate PIP control systems can be employed to obtain the required dynamics for each input–output channel.

From equation (7.19) and equation (7.25), the closed-loop system becomes:

$$\boldsymbol{x}(k) = (\boldsymbol{F} - \boldsymbol{G}\boldsymbol{K})\boldsymbol{x}(k-1) + \boldsymbol{D}\boldsymbol{y}_d(k) \tag{7.71}$$

and the characteristic polynomial associated with this equation is given by:

$$\det(\lambda \boldsymbol{I} - \boldsymbol{F} + \boldsymbol{G}\boldsymbol{K}) \tag{7.72}$$

The poles of the closed-loop system can be arbitrarily assigned by SVF provided that the system is completely controllable, i.e. the pair $[\boldsymbol{F}, \boldsymbol{G}]$ is controllable (Theorem 7.1). Unfortunately, assignment of the closed-loop poles of a multivariable system does not, in itself, uniquely specify the feedback gain matrix $\boldsymbol{K}$. Consequently, a number of algorithms have been proposed for obviating this difficulty, e.g. the use of dyadic feedback (Gopinath 1971; Young and Willems 1972) or the Luenberger canonical form (Luenberger 1967; Munro 1979; Kailath 1980). The details of the Luenberger canonical form for the multivariable NMSS discrete-time system are developed by Wang (1988) and Chotai *et al.* (1991). However, in this book, we consider instead a special PIP pole assignment algorithm that takes advantage of the extra design freedom in the multivariable case to achieve complete decoupling of the output variables.

7.5.1 Decoupling Algorithm I

In order to develop this joint decoupling and pole assignment problem, the nominal PIP control algorithm (7.25) is modified as follows:

$$\boldsymbol{u}(k) = -\boldsymbol{K}\boldsymbol{x}(k) - \boldsymbol{M}_0\boldsymbol{u}(k) \tag{7.73}$$

where $\boldsymbol{M}_0$ is an additional control gain matrix. Hence, using equation (7.29) and equation (7.73), the modified control law is:

$$\boldsymbol{u}(k) = -\boldsymbol{L}(z^{-1})\boldsymbol{y}(k) - \boldsymbol{M}_o\boldsymbol{u}(k) - \boldsymbol{M}(z^{-1})\boldsymbol{u}(k) + \boldsymbol{K}_I\boldsymbol{z}(k) \tag{7.74}$$

Alternatively,

$$\boldsymbol{u}(k) = -\boldsymbol{L}(z^{-1})\boldsymbol{y}(k) - \boldsymbol{M}^*(z^{-1})\boldsymbol{u}(k) + \boldsymbol{K}_I\boldsymbol{z}(k) \tag{7.75}$$

where $\boldsymbol{L}(z^{-1})$ is defined by (7.28),

$$\boldsymbol{M}^*(z^{-1}) = \boldsymbol{M}_0 + \boldsymbol{M}_1 z^{-1} + \boldsymbol{M}_2 z^{-2} + \cdots + \boldsymbol{M}_{m-1} z^{-m+1} \tag{7.76}$$

and $\boldsymbol{K}_I$ is the integral gain matrix, as usual. Using (7.12), (7.15) and (7.74), we can determine the following relationship between $\boldsymbol{y}(k)$ and $\boldsymbol{y}_d(k)$:

$$\bar{\boldsymbol{A}}(z^{-1})\boldsymbol{y}(k) = \bar{\boldsymbol{B}}(z^{-1})\boldsymbol{y}_d(k) \tag{7.77}$$

where

$$\begin{aligned}\bar{\boldsymbol{A}}(z^{-1}) &= \Delta\left\{\boldsymbol{A}(z^{-1}) + \boldsymbol{B}(z^{-1})\left[\boldsymbol{I} + \boldsymbol{M}^*(z^{-1})\right]^{-1}\boldsymbol{L}(z^{-1})\right\} + \boldsymbol{B}(z^{-1})\left[\boldsymbol{I} + \boldsymbol{M}^*(z^{-1})\right]^{-1}\boldsymbol{K}_I \\ \bar{\boldsymbol{B}}(z^{-1}) &= \boldsymbol{B}(z^{-1})\left[\boldsymbol{I} + \boldsymbol{M}^*(z^{-1})\right]^{-1}\boldsymbol{K}_I\end{aligned} \tag{7.78}$$

with $\Delta = 1 - z^{-1}$. Consequently, the combined decoupling and pole assignment algorithm can be obtained if the matrices $\boldsymbol{L}(z^{-1})$, $\boldsymbol{M}^*(z^{-1})$ and $\boldsymbol{K}_I$ are chosen such that:

(i) $\bar{\boldsymbol{B}}(z^{-1})$ is diagonal and non-singular; and
(ii) $\bar{\boldsymbol{A}}(z^{-1})$ is diagonal and its zeros (i.e. the closed-loop poles) are located at desired positions in the complex z-plane.

If $\boldsymbol{B}^{-1}(z^{-1})$ exists and $\boldsymbol{B}_1$ is not a null matrix, then we can choose $\boldsymbol{M}^*(z^{-1})$ such that:

$$\boldsymbol{B}(z^{-1})\left(\boldsymbol{I} + \boldsymbol{M}^*(z^{-1})\right)^{-1} = \boldsymbol{I}z^{-1} \tag{7.79}$$

where $\boldsymbol{I}$ is the identity matrix. Furthermore, $\bar{\boldsymbol{A}}(z^{-1})$ in equations (7.78) can be written as:

$$\bar{\boldsymbol{A}}(z^{-1}) = \Delta\left(\boldsymbol{A}(z^{-1}) + z^{-1}\boldsymbol{L}(z^{-1})\right) + z^{-1}\boldsymbol{K}_I \tag{7.80}$$

where

$$\bar{A}(z^{-1}) = I + \bar{A}_1 z^{-1} + \cdots + \bar{A}_{n+1} z^{-(n+1)} \tag{7.81}$$

is the desired closed-loop polynomial matrix, in which $\bar{A}_i (i = 1, 2, \ldots, n+1)$ are diagonal matrices. In this case, $L(z^{-1})$ and K_I can be selected such that (7.80) is satisfied, to yield the following combined decoupling and pole assignment algorithm:

$$\begin{aligned} M_0 &= B_1 - I \\ M_i &= B_{i+1} && i = 1, 2, \ldots, m-1 \\ L_i &= -\sum_{j=i+2}^{n+1} \bar{A}_j - A_{i+1} && i = 0, 1, \ldots, n-1 \\ K_I &= I + \sum_{j=1}^{n+1} \bar{A}_j \end{aligned} \tag{7.82}$$

It is important to note, however, that this approach has limitations: in particular, the transmission poles and zeros of the system are cancelled by the controller, so that it is not applicable to systems with unstable poles or non-minimum phase zeros.

7.5.2 Implementation Form

In structural terms, the decoupling algorithm is particularly straightforward to implement by incorporating the new M_0 component into the nominal PIP gain matrix K. Here, K is defined by equation (7.26), with the various control gain matrices obtained using (7.82). In this case, rearranging from the modified control law (7.73) yields:

$$u(k) = -(I + M_0)^{-1} K x(k) = -\bar{K} x(k) \tag{7.83}$$

where $\bar{K}$ is the combined decoupling and pole assignment control gain matrix, now in a form that can be implemented using standard PIP methods. In other words, equation (7.83) takes the same form as the PIP state variable feedback algorithm (7.25), and hence can also be implemented in using any of the other control structures derived from (7.25), including the incremental form (7.30), Figure 7.2 or Figure 7.3.

Example 7.6 Pole Assignment Decoupling of a Two-Input, Two-Output System Consider again the multivariable system from Example 7.1, represented by the TFM (7.2), which can be factorised into the MFD form shown by equation (7.5) and equation (7.6). It is clear from $G(z^{-1})$ and the open-loop unit step responses illustrated in Figure 7.1, that the natural cross-coupling effects are large. Since $B^{-1}(z^{-1})$ exists and there are no unstable poles or non-minimum phase zeros, we can use the algorithm above for full decoupling and pole assignment. For example, define the desired closed-loop diagonal polynomial matrix (7.81):

$$\bar{A}(z^{-1}) = \begin{bmatrix} 1 & 0 \\ 0 & 1 \end{bmatrix} + \begin{bmatrix} -0.5 & 0 \\ 0 & -0.5 \end{bmatrix} z^{-1} \tag{7.84}$$

Here, we require both channels to have the same, first order response. In fact, (7.84) yields a PIP pole assignment control system with a similar speed of response as the PIP-LQ design discussed in Example 7.2, as illustrated by the diagonal subplots of Figure 7.4.

The decoupling and pole assignment algorithm (7.82) yields:

$$\begin{aligned} \boldsymbol{M}_0 &= \boldsymbol{B}_1 - \boldsymbol{I} = \begin{bmatrix} 0.9 & 2 \\ 1.1 & 1 \end{bmatrix} - \begin{bmatrix} 1 & 0 \\ 0 & 1 \end{bmatrix} = \begin{bmatrix} -0.1 & 2 \\ 1.1 & 0 \end{bmatrix} \\ \boldsymbol{L}_0 &= -\boldsymbol{A}_1 = \begin{bmatrix} 0.5 & 0 \\ 0 & 0.8 \end{bmatrix}; \quad \boldsymbol{K}_I = \boldsymbol{I} + \bar{\boldsymbol{A}}_1 = \begin{bmatrix} 0.5 & 0 \\ 0 & 0.5 \end{bmatrix} \end{aligned} \tag{7.85}$$

Hence, the control law (7.83) becomes:

$$\boldsymbol{u}(k) = -(\boldsymbol{I} + \boldsymbol{M}_0)^{-1}[\boldsymbol{L}_0 - \boldsymbol{K}_I]\boldsymbol{x}(k) = -\begin{bmatrix} 0.9 & 2 \\ 1.1 & 1 \end{bmatrix}^{-1} \begin{bmatrix} 0.5 & 0 & -0.5 & 0 \\ 0 & 0.8 & 0 & -0.5 \end{bmatrix} \boldsymbol{x}(k) \tag{7.86}$$

Resolving (7.86) yields:

$$\boldsymbol{u}(k) = -\begin{bmatrix} -0.3846 & 1.2308 & 0.3846 & -0.7692 \\ 0.4231 & -0.5538 & -0.4231 & 0.3462 \end{bmatrix} \boldsymbol{x}(k) \tag{7.87}$$

The earlier Figure 7.4 compares the response of the combined decoupling and pole assignment algorithm (7.87) with the PIP-LQ design (7.42) discussed previously. As noted above, both approaches yield a similar damped first order response to a unit step in the command level, but only (7.87) achieves complete decoupling control (shown by the thick traces).

7.5.3 Decoupling Algorithm II

In this second approach to analytic decoupling design, we assume that $\boldsymbol{A}(z^{-1})$ is a diagonal polynomial matrix (see section 3.3.3) and that there exists an open-loop, feed-forward compensation polynomial matrix $\boldsymbol{K}_c(z^{-1})$, such that $\boldsymbol{P}(z^{-1}) = \boldsymbol{B}(z^{-1})\boldsymbol{K}_c(z^{-1})$ is also diagonal. The necessary and sufficient conditions for the minimal order of $\boldsymbol{K}_c(z^{-1})$ are given by Teng and Ledwich (1992). Since $\boldsymbol{A}(z^{-1})$ is diagonal, if $\boldsymbol{K}_c(z^{-1})$ can be found such that $\boldsymbol{P}(z^{-1})$ is also diagonal, then the plant outputs of each channel are affected only by the input associated with the same channel, i.e. the open-loop system is decoupled. For closed-loop pole assignment design, we can then utilise the standard SISO algorithm of Chapter 5, to obtain the required dynamics for each channel, using either pole assignment or optimal LQ control as usual. In contrast to Algorithm I, this approach can be used even when $\det(\boldsymbol{B}(z^{-1}))$ is zero; and the design method can be applied to stable, unstable or non-minimum phase systems.

7.6 Concluding Remarks

This chapter has generalised the PIP control system to address multivariable systems. We have seen how, as in the univariate case, the NMSS setting of the control problem allows for the use of full SVF, involving only the measured input and output variables and their past values, so avoiding the need for an explicit state reconstruction filter or observer. The resulting multivariable PIP control system is designed on the basis of either pole assignment or LQ optimisation.

In the pole assignment case, the PIP control law can be defined so as to exactly dynamically decouple the various control channels. Of course, this result assumes an ideal model with no mismatch. We have also introduced a multi-objective optimisation procedure for optimal PIP-LQ design, in which the weighting matrices in the quadratic cost function are tuned by numerical goal attainment. This latter approach to dynamic decoupling has the advantage that it is more generally applicable and less sensitive to both modelling errors and stochastic disturbances than the analytical pole assignment solution. Although utilised here for multivariable decoupling, the approach is generic in nature and enables the designer to simultaneously meet other objectives, such as the combination of rapid response, smooth input activation and robustness. Furthermore, the approach can be applied to both multivariable (as here) and univariate (see e.g. Taylor *et al.* 2001) systems.

References

Albertos, P. and Sala, A. (2004) *Multivariable Control Systems: An Engineering Approach*, Springer-Verlag, London.

Argoun, M.B. and van de Vegte, J. (1980) Output feedback decoupling in the frequency domain, *International Journal of Control*, **31**, pp. 665–675.

Bayoumi, M. and Duffield, T. (1977) Output feedback decoupling and pole placement in linear time-invariant systems, *IEEE Transactions on Automatic Control*, **22**, pp. 142–143.

Chotai, A., Young, P.C. and Behzadi, M.A. (1991) Self-adaptive design of a non-linear temperature control system, *IEE Proceedings: Control Theory and Applications*, **38**, pp. 41–49.

Chotai, A., Young, P., McKenna, P. and Tych, W. (1998) Proportional-Integral-Plus (PIP) design for delta operator systems. Part 2: MIMO systems, *International Journal of Control*, **70**, pp. 149–168.

Dixon, R. (1996) *Design and Application of Proportional-Integral-Plus (PIP) Control Systems*, PhD thesis, Environmental Science Department, Lancaster University.

Dixon, R., Chotai, A., Young, P.C and Scott, J.N. (1997) The automation of piling rig positioning utilising multivariable proportional-integral-plus (PIP) control, *12th International Conference on Systems Engineering*, 9–11 September, Coventry University, UK, pp. 211–216.

Dixon, R. and Lees, M. (1994) Implementation of multivariable proportional-integral-plus (PIP) control design for a coupled drive system, *10th International Conference on Systems Engineering*, 6–8 September, Coventry University, UK, pp. 286–293.

Dixon, R. and Pike, A.W. (2006) Alstom Benchmark Challenge II on Gasifier Control, *IEE Proceedings: Control Theory Applications*, **153**, pp. 254–261.

Exadaktylos, V. and Taylor, C.J. (2010) Multi-objective performance optimisation for model predictive control by goal attainment, *International Journal of Control*, **83**, pp. 1374–1386.

Falb, P. and Wolovich, W. (1967) Decoupling in the design and synthesis of multivariable control systems, *IEEE Transactions on Automatic Control*, **12**, pp. 651–659.

Fleming, P. and Pashkevich, A. (1986) Application of multi-objective optimisation to compensator design for SISO control systems, *Electronics Letters*, **22**, pp. 258–259.

Gevers, M. and Wertz, V. (1984) Uniquely identifiable state-space and ARMA parametrizations for multivariable linear systems, *Automatica*, **20**, pp. 333–347.

Gilbert, A., Yousef, A., Natarajan, K. and Deighton, S. (2003) Tuning of PI Controllers with one-way decoupling in 2×2 MIMO systems based on finite frequency response data, *Journal of Process Control*, **13**, pp. 553–567.

Gilbert, E.G. (1969) The decoupling of multivariable systems by state feedback, *SIAM Journal of Control*, **7**, 50–63.

Gopinath, B. (1971) On the control of linear multiple input–output systems, *Bell System Technical Journal*, **50**, 1063–1081.

Green, M. and Limebeer, D.J.N. (1995) *Linear Robust Control*, Prentice Hall, Englewood Cliffs, NJ.

Kailath, T. (1980) *Linear Systems*, Prentice Hall, Englewood Cliffs, NJ.

Kuo, B.C. (1997) *Digital Control Systems*, Second Edition, The Oxford Series in Electrical and Computer Engineering, Oxford University Press, New York.

Lees, M.J., Young, P.C., Chotai, A. and Tych, W. (1994) Multivariable true digital control of glasshouse climate. In R. Whalley (Ed.), *Application of Multivariable System Techniques*, Mechanical Engineering Publications, London, pp. 255–267.

Liu, Y. and Zhang, W. (2007) Analytical Design of Two degree–of–freedom decoupling control scheme for two–by–two systems with integrator(s), *IET Proceedings: Control Theory and Applications*, **1**, pp. 1380–1389.

Liu, T., Zhang, W. and Gao, F. (2007) Analytical decoupling control strategy using a unity feedback control structure for MIMO processes with time delays, *Journal of Process Control*, **17**, pp. 173–186.

Luenberger, D.G. (1967) Canonical forms for linear multivariable systems, *IEEE Transactions on Automatic Control*, **12**, pp. 290–293.

Luenberger, D.G. (1971) An introduction to observers, *IEEE Transactions on Automatic Control*, **16**, pp. 596–603.

Morgan, B.S. (1964) The synthesis of linear multivariable systems by state-variable feedback, *IEEE Transactions on Automatic Control*, **9**, pp. 405–411.

Morse, A.S. and Wonham, W.M. (1970) Decoupling and pole assignment by dynamic compensation, *SIAM Journal of Control*, **8**, pp. 317–337.

Munro, N. (1979) Pole assignment, *IEE Proceedings: Control Theory and Applications*, **126**, pp. 549–554.

Plummer, A. and Vaughan, N. (1997) Decoupling pole-placement control with application to a multi-channel electrohydraulic servosystem, *Control Engineering Practice*, **5**, pp. 313–323.

Rossiter, J.A. (2003) *Model Based Predictive Control: A Practical Approach*, CRC Press, Boca Raton, FL.

Sandoz, D., Desforges, M., Lennox, B. and Goulding, P. (2000) Algorithms for industrial model predictive control, *Computing and Control Engineering Journal*, **11**, pp. 125–134.

Simm, A. and Liu, G.P. (2006) Improving the performance of the ALSTOM baseline controller using multiobjective optimisation, *IEE Proceedings: Control Theory Applications*, **153**, pp. 286–292.

Skogestad, S. and Postlethwaite, I. (2005) *Multivariable Feedback Control: Analysis and Design*, Wiley-Blackwell, Chichester.

Taylor, C.J., Chotai, A. and Young, P.C. (2001) Design and application of PIP controllers: robust control of the IFAC93 benchmark, *Transactions of the Institute of Measurement and Control*, **23**, pp. 183–200.

Taylor, C.J., McCabe, A.P., Young, P.C. and Chotai, A. (2000) Proportional-Integral-Plus (PIP) control of the ALSTOM gasifier problem, *IMECHE Proceedings: Systems and Control Engineering*, **214**, pp. 469–480.

Taylor, C.J. and Shaban, E.M. (2006) Multivariable Proportional-Integral-Plus (PIP) control of the ALSTOM nonlinear gasifier simulation, *IEE Proceedings: Control Theory and Applications*, **153**, pp. 277–285.

Taylor, C.J., Young, P.C., Chotai, A. and Whittaker, J. (1998) Non-minimal state space approach to multivariable ramp metering control of motorway bottlenecks, *IEE Proceedings: Control Theory and Applications*, **145**, pp. 568–574.

Teng, F.C. and Ledwich, G.F. (1992) Adaptive decouplers for multivariable systems: input dynamics compensation and output feedback approaches, *International Journal of Control*, **55**, pp. 373–391.

Tych, W. and Taylor, C.J. (1996) True Digital Control CACSD package: decoupling control of the Shell heavy oil fractionator simulation, *Institution of Electrical Engineers Colloquium on Advances in Computer-Aided Control System Design*, 14 March, University of Sheffield.

Wang, C.L. (1988) *New Methods of the Direct Digital Control of Discrete-Time Systems*, PhD thesis, Environmental Science Department, Lancaster University.

Wang, Q.G., Huang, B. and Guo, X. (2000) Auto-tuning of TITO decoupling controllers from step tests, *ISA Transactions*, **39**, pp. 407–418.

Wang, Q.G., Zhang, Y. and Chiu, M.S. (2002) Decoupling internal model control for multivariable systems with multiple time delays, *Chemical Engineering Science*, **57**, pp. 115–124.

Whittle, P. (1981) Risk sensitive Linear/Quadratic/Gaussian control, *Advances in Applied Probability*, **13**, pp. 764–777.

Wojsznis, W., Mehta, A., Wojsznis, P., Thiele, D. and Blevins, T. (2007) Multi-objective optimisation for Model Predictive Control, *ISA Transactions*, **46**, pp. 351–361.

Wolovich, W. (1975) Output feedback decoupling, *IEEE Transactions on Automatic Control*, **20**, pp. 148–149.

Wolovich, W.A. (1974) *Linear Multivariable Systems*, Springer-Verlag, Berlin.

Young, P.C., Chotai, A., McKenna, P.G. and Tych, W. (1998) Proportional-Integral-Plus (PIP) design for delta operator systems: Part 1, SISO systems, *International Journal of Control*, **70**, 149–168.

Young, P.C., Lees, M., Chotai, A., Tych, W. and Chalabi, Z.S. (1994) Modelling and PIP control of a glasshouse micro-climate, *Control Engineering Practice*, **2**, pp. 591–604.

Young, P.C. and Willems, J.C. (1972) An approach to the linear multivariable servomechanism problem, *International Journal of Control*, **15**, pp. 961–979.

8

Data-Based Identification and Estimation of Transfer Function Models

So far in this book, it has been assumed that the model of the system is available to the control systems designer. Sometimes such a model may be in the form of a simulation model that has been based on a mechanistic analysis of the system and has been 'calibrated' in some manner. However, in the context of *True Digital Control* (TDC) design, as formulated in this book, it seems more appropriate if the model has been obtained on the basis of experimental or monitored, digitally sampled data obtained directly from the system. In this chapter, therefore, we consider such data-based modelling methods and illustrate, mainly by means of simulation examples, how they are able to provide stochastic models that are well suited to the control system design methods described in previous chapters. They also indicate how these stochastic models allow for the use of *Monte Carlo Simulation* (MCS) analysis based on the estimated uncertainty in the model parameters, as quantified by the model estimation procedures. Previous chapters have demonstrated how MCS provides a useful means for assessing the robustness of the TDC designs to such uncertainty.

In particular, the chapter provides an introduction to the *en bloc* and recursive algorithms that are used in the previous chapters of the book for estimating parameters in *Transfer Function* (TF) models of stochastic dynamic systems. Although there are numerous approaches in the literature (a brief overview is given in section 8.2), the present book initially focuses on three algorithms for the estimation of *Single-Input, Single-Output* (SISO) discrete-time TF models, namely the *Recursive Least Squares* (RLS) (**I**), standard instrumental variable (**II**) and optimal *Refined Instrumental Variable* (RIV; **III**) algorithms, as considered in section 8.1, section 8.2 and section 8.3 respectively[1].

[1] The *en bloc* versions of each algorithm are denoted **Ie**, **IIe** and **IIIe** and a special 'symmetric' version of the latter is called **IIIs** (see later).

True Digital Control: Statistical Modelling and Non-Minimal State Space Design, First Edition.
C. James Taylor, Peter C. Young and Arun Chotai.

The present authors have utilised RIV estimation for most of the practical *Proportional-Integral-Plus* (PIP) control applications cited in earlier chapters, hence important variations of this algorithm, in the context of TDC design, are discussed in section 8.3. A general procedure for model structure identification is suggested and illustrated by several case studies (section 8.4). The algorithms are extended to the multivariable (section 8.5), continuous-time (section 8.6) and closed-loop (section 8.7) situation. All these algorithms are described in a tutorial style that avoids unnecessary theoretical rigour and is intended to provide the reader with a basic understanding of their development and application.

Finally, the main identification and estimation algorithms considered in the chapter are available as computational routines in either the authors' CAPTAIN Toolbox (Appendix G; Taylor *et al.* 2007a; Young 2011a) or the MATLAB® System Identification Toolbox, both available for use in the MATLAB®[2] software environment.

8.1 Linear Least Squares, ARX and Finite Impulse Response Models

Let us consider the following, slightly modified version of the deterministic, discrete-time model (2.11) introduced in Chapter 2. In the more general modelling context considered here, the model may contain a pure time-delay effect of $\delta \geq 0$ samples[3]. Depending upon the convention that is preferred by the analyst, the TF form of the model is:

$$y(k) = \frac{b_\delta z^{-\delta} + b_{\delta+1} z^{-(\delta+1)} + b_{\delta+2} z^{-(\delta+2)} + \cdots + b_{\delta+m} z^{-(\delta+m)}}{1 + a_1 z^{-1} + a_2 z^{-2} + \cdots + a_n z^{-n}} u(k) = \frac{B(z^{-1})}{A(z^{-1})} u(k) \quad (8.1a)$$

or

$$y(k) = \frac{b_0 + b_1 z^{-1} + b_2 z^{-2} + \cdots + b_m z^{-m}}{1 + a_1 z^{-1} + a_2 z^{-2} + \cdots + a_n z^{-n}} u(k-\delta) = \frac{B(z^{-1})}{A(z^{-1})} u(k-\delta) \quad (8.1b)$$

Here, the numerator polynomial $B(z^{-1})$ has been modified from that shown in (2.13) so that it includes the explicit presence of a pure time delay of δ samples. Also, when $\delta = 0$, this definition allows the input to affect the output instantaneously through the addition of the b_0 coefficient. Note that the structure of the model (8.1) can be characterised by the triad $[n\ m\ \delta]$, which defines the orders of the polynomials and the size of the time delay. However, since the number of parameters in the numerator is $m+1$, the modified triad $[n\quad (m+1)\quad \delta\,]$ is used in the CAPTAIN Toolbox and the examples below.

It is easy to see that the discrete-time equation associated with (8.1b) takes the form:

$$\begin{aligned} y(k) &= -a_1 y(k-1) - a_2 y(k-2) - \cdots - a_n y(k-n) \\ &\quad + b_0 u(k-\delta) + \cdots + b_m u(k-\delta-m) \end{aligned} \quad (8.2)$$

[2] MATLAB®, The MathWorks Inc., Natick, MA, USA.

[3] Earlier chapters have denoted the sampled time delay by τ. In this chapter τ will be reserved for the time delay of *continuous-time* models, such as equation (8.100); for discrete-time systems, as here, we will instead use δ (as also used by many of the publications on system identification cited in this chapter).

which can be written concisely in the following vector inner product terms:

$$y(k) = \boldsymbol{\phi}^T(k)\,\boldsymbol{\rho} \tag{8.3}$$

where

$$\begin{aligned}\boldsymbol{\phi}^T(k) &= [-y(k-1), -y(k-2), \ldots, -y(k-n), u(k-\delta), u(k-\delta-1), \ldots,\\ &\qquad u(k-\delta-m)]\\ \boldsymbol{\rho} &= [a_1\; a_2 \ldots a_n\; b_0\; b_1 \ldots b_m]^T\end{aligned} \tag{8.4}$$

Alternatively, for the model (8.1a), $\boldsymbol{\rho} = [a_1\; a_2 \ldots a_n\; b_\delta\; b_{\delta+1} \ldots b_{\delta+m}]^T$.

In statistical identification and estimation, it is necessary to assume that the data $\{y(k),\ u(k)\}$, $k = 1, 2, \ldots, N$, are corrupted in some manner by errors or noise. The simplest such assumption in this case is to add a noise variable $\xi(k)$ to equation (8.3), i.e.

$$y(k) = \boldsymbol{\phi}^T(k)\,\boldsymbol{\rho} + \xi(k) \tag{8.5a}$$

where $\xi(k)$ represents all those components in the measured output $y(k)$ that are not caused by the input excitation $u(k)$: for example, the effects of measurement noise, unmeasured stochastic inputs and modelling errors. In TF terms, the reader can easily verify that this model is:

$$y(k) = \frac{B(z^{-1})}{A(z^{-1})}u(k-\delta) + \frac{1}{A(z^{-1})}\xi(k) \tag{8.5b}$$

which exposes its rather special form, with the noise $\xi(k)$ being filtered by the TF defined by $1/A(z^{-1})$. It is not the same, for example, as the more obvious (and, as we shall see, more practically useful) *general* TF model:

$$y(k) = \frac{B(z^{-1})}{A(z^{-1})}u(k-\delta) + \xi(k) \tag{8.5c}$$

We might expect $\xi(k)$ in the above models to have quite general stochastic properties that require a similarly general stochastic representation and this will be considered later. To begin with, however, let us make a simplifying assumption and consider the model (8.5a) where $\xi(k) = e(k)$, in which $e(k)$ is a simple 'white noise' process:

$$E\{e(k)\} = 0; \quad E\{e(j)e(k)\} = \sigma^2\delta_{jk} \tag{8.6}$$

where E is the expectation operator and δ_{jk} is the Kronecker delta function:

$$\delta_{jk} = \begin{cases} 1 \text{ if } j = k \\ 0 \text{ if } j \neq k \end{cases} \tag{8.7}$$

In other words, $e(k)$ is a sequence of uncorrelated random variables with zero mean value and variance σ^2. In this special case, the model (8.5a) takes the special form:

$$y(k) = \boldsymbol{\phi}^T(k)\,\boldsymbol{\rho} + e(k) \tag{8.8a}$$

Note, for later reference, that this equation could be written in the operator form:

$$A(z^{-1})y(k) = B(z^{-1})u(k-\delta) + e(k) \tag{8.8b}$$

This is normally referred to as the *Auto Regressive eXogenous variable* (ARX) model because it is in the form of a linear regression relationship, in which the output $y(k)$ depends upon past values of itself (autoregressive terms), as well as present and past values of the input or 'exogenous' variable $u(k)$.

8.1.1 En bloc *LLS Estimation*

The assumption that the stochastic noise $\xi(k)$ in (8.5a) is the simple white noise process $e(k)$ in (8.6) is important, albeit rather restrictive, because it considerably simplifies the problem of estimating the parameter vector $\boldsymbol{\rho}$ that characterises the model (8.8a). In particular, the *Linear Least Squares* (LLS) estimate $\hat{\boldsymbol{\rho}}(N)$ of the vector $\boldsymbol{\rho}$, based on the N data samples, is very easy to compute by minimisation of the least squares cost function in the 'prediction error' $e(k)$, i.e.

$$\hat{\boldsymbol{\rho}}(N) = \arg\min_{\rho} \; J_2(\boldsymbol{\rho}) \qquad J_2(\boldsymbol{\rho}) = \sum_{k=1}^{N} [e(k)]^2 = \sum_{k=1}^{N} [y(k) - \boldsymbol{\phi}^T(k)\boldsymbol{\rho}]^2 \tag{8.9}$$

The minimum is obtained in the usual manner by partially differentiating with respect to each element of the estimated parameter vector $\boldsymbol{\rho}$, and then setting these derivatives to zero. This analysis, using matrix algebra (see Appendix A), is as follows:

$$\nabla_\rho(J_2) = \frac{\partial}{\partial \boldsymbol{\rho}}\left\{\sum_{k=1}^{N}\left[y(k) - \boldsymbol{\phi}^T(k)\boldsymbol{\rho}\right]^2\right\} = -2\sum_{k=1}^{N}\boldsymbol{\phi}(k)\left[y(k) - \boldsymbol{\phi}^T(k)\boldsymbol{\rho}\right] \tag{8.10}$$

so that

$$\frac{1}{2}\nabla_\rho(J_2) = -\sum_{k=1}^{N}\boldsymbol{\phi}(k)y(k) + \left[\sum_{k=1}^{N}\boldsymbol{\phi}(k)\boldsymbol{\phi}^T(k)\right]\boldsymbol{\rho} = 0 \tag{8.11a}$$

or

$$\left[\sum_{k=1}^{N}\boldsymbol{\phi}(k)\boldsymbol{\phi}^T(k)\right]\boldsymbol{\rho} = \sum_{k=1}^{N}\boldsymbol{\phi}(k)y(k) \tag{8.11b}$$

where $\nabla_\rho(J_2)$ denotes the gradient of J_2 with respect to all the elements of $\boldsymbol{\rho}$.

The vector-matrix equation (8.11b) represents $n+m+1$ equations in the $n+m+1$ elements of the parameter estimate vector $\boldsymbol{\rho}$. These equations are usually referred to as the 'normal equations' of LLS analysis. Consequently, provided that the matrix $\sum \boldsymbol{\phi}(k)\boldsymbol{\phi}^T(k)$ is non-singular and can be inverted, the *en bloc* LLS estimate $\hat{\boldsymbol{\rho}}(N)$ based on the N data samples is given by the solution of these equations, which can be written concisely as:

$$\hat{\boldsymbol{\rho}}(N) = \boldsymbol{P}(N)\boldsymbol{b}(N) \qquad \text{(i) } \mathbf{Ie}$$

where the $(n+m+1) \times (n+m+1)$ matrix $\boldsymbol{P}(N)$ and the $(n+m+1) \times 1$ vector $\boldsymbol{b}(N)$ are defined as follows:

$$\boldsymbol{P}(N) = \left[\sum_{i=1}^{N} \boldsymbol{\phi}(i)\boldsymbol{\phi}^T(i)\right]^{-1} \; ; \quad \boldsymbol{b}(N) = \sum_{i=1}^{N} \boldsymbol{\phi}(i)y(i) \qquad \text{(ii) } \mathbf{Ie}$$

At any intermediate sampling instant k, the estimate $\hat{\boldsymbol{\rho}}(k)$ is given by:

$$\hat{\boldsymbol{\rho}}(k) = \boldsymbol{P}(k)\boldsymbol{b}(k) \tag{8.12}$$

This is referred to in the next section, which considers the *recursive or sequentially updated* version of LLS estimation.

8.1.2 Recursive LLS Estimation

In model-based automatic control system design of the kind discussed in this book, it is sometimes useful to be able to 'tune' or 'adapt' the design online and in real-time. One way of doing this is to estimate the parameters on the basis of data being received online from sensors, and then use these updated parameters to update the control system gains based on the PIP design algorithm. This can be achieved by developing a recursive version of the *en bloc* solution (**Ie**). Here, the estimate $\hat{\boldsymbol{\rho}}(k)$ at the kth sampling instant is updated on the basis of (i) the previous estimate $\hat{\boldsymbol{\rho}}(k-1)$ at the previous $(k-1)$th sampling instant and (ii) the error between the predicted model output $\boldsymbol{\phi}^T(k)\hat{\boldsymbol{\rho}}(k-1)$ and the measured output $y(k)$, i.e. the 'recursive residual' or 'model prediction error':

$$\varepsilon(k) = y(k) - \boldsymbol{\phi}^T(k)\hat{\boldsymbol{\rho}}(k-1)$$

This is often called the 'innovation' error because it provides the new (latest) information on the quality of the estimates. In order to develop this *Recursive Least Squares* (RLS) estimation algorithm, note from equations (**Ie**) that, after $k < N$ sampling intervals:

$$\boldsymbol{P}^{-1}(k) = \boldsymbol{P}^{-1}(k-1) + \boldsymbol{\phi}(k)\boldsymbol{\phi}^T(k) \tag{8.13}$$

and

$$\boldsymbol{b}(k) = \boldsymbol{b}(k-1) + \boldsymbol{\phi}(k)y(k) \tag{8.14}$$

Now pre-multiply equation (8.13) by $\boldsymbol{P}(k)$ and post-multiply by $\boldsymbol{P}(k-1)$ to give:

$$\boldsymbol{P}(k-1) = \boldsymbol{P}(k) + \boldsymbol{P}(k)\boldsymbol{\phi}(k)\boldsymbol{\phi}^T(k)\boldsymbol{P}(k-1) \tag{8.15}$$

Post multiplying by $\boldsymbol{\phi}(k)$ then yields:

$$\begin{aligned}\boldsymbol{P}(k-1)\boldsymbol{\phi}(k) &= \boldsymbol{P}(k)\boldsymbol{\phi}(k) + \boldsymbol{P}(k)\boldsymbol{\phi}(k)\boldsymbol{\phi}^T(k)\boldsymbol{P}(k-1)\boldsymbol{\phi}(k) \\ &= \boldsymbol{P}(k)\boldsymbol{\phi}(k)[1+\boldsymbol{\phi}^T(k)\boldsymbol{P}(k-1)\boldsymbol{\phi}(k)]\end{aligned} \tag{8.16}$$

Post-multiplying by $[1 + \boldsymbol{\phi}^T(k)\boldsymbol{P}(k-1)\boldsymbol{\phi}(k)]^{-1}\boldsymbol{\phi}^T(k)\boldsymbol{P}(k-1)$:

$$\boldsymbol{P}(k-1)\boldsymbol{\phi}(k)[1 + \boldsymbol{\phi}^T(k)\boldsymbol{P}(k-1)\boldsymbol{\phi}(k)]^{-1}\boldsymbol{\phi}^T(k)\boldsymbol{P}(k-1) = \boldsymbol{P}(k)\boldsymbol{\phi}(k)\boldsymbol{\phi}^T(k)\boldsymbol{P}(k-1) \tag{8.17}$$

Hence, we obtain finally:

$$\boldsymbol{P}(k) = \boldsymbol{P}(k-1) - \boldsymbol{P}(k-1)\boldsymbol{\phi}(k)[1+\boldsymbol{\phi}^T(k)\boldsymbol{P}(k-1)\boldsymbol{\phi}(k)]^{-1}\boldsymbol{\phi}^T(k)\boldsymbol{P}(k-1) \tag{8.18}$$

which is termed the 'matrix inversion lemma' (Bodewig 1956; Ho 1962) since, at each sampling instant, it provides the inverse of the accumulated 'cross-product' matrix, i.e.

$$\boldsymbol{P}(k) = \left[\sum_{i=1}^{k} \boldsymbol{\phi}(i)\boldsymbol{\phi}^T(i)\right]^{-1} \tag{8.19}$$

By now substituting in equation (8.12) from (8.18) and (8.14), it is a simple matter to obtain the equivalent recursive equation for $\hat{\boldsymbol{\rho}}(k)$, i.e.

$$\begin{aligned}\hat{\boldsymbol{\rho}}(k) &= \{\boldsymbol{P}(k-1) - \boldsymbol{P}(k-1)\boldsymbol{\phi}(k)[1 + \boldsymbol{\phi}^T(k)\boldsymbol{P}(k-1)\boldsymbol{\phi}(k)]^{-1}\boldsymbol{\phi}^T(k)\boldsymbol{P}(k-1)\} \\ &\quad \{\boldsymbol{b}(k-1) + \boldsymbol{\phi}(k)y(k)\} \\ &= \hat{\boldsymbol{\rho}}(k-1) - \boldsymbol{P}(k-1)\boldsymbol{\phi}(k)[1+\boldsymbol{\phi}^T(k)\boldsymbol{P}(k-1)\boldsymbol{\phi}(k)]^{-1}\boldsymbol{\phi}^T(k)\hat{\boldsymbol{\rho}}(\boldsymbol{k}-\mathbf{1}) \\ &\quad + \boldsymbol{P}(k-1)\boldsymbol{\phi}(k)y(k) - \boldsymbol{P}(k-1)\boldsymbol{\phi}(k)[1 + \boldsymbol{\phi}^T(k)\boldsymbol{P}(k-1)\boldsymbol{\phi}(k)]^{-1}\boldsymbol{\phi}^T(k) \\ &\quad \boldsymbol{P}(k-1)\boldsymbol{\phi}(k)y(k)\end{aligned}$$

So that finally:

$$\hat{\boldsymbol{\rho}}(k) = \hat{\boldsymbol{\rho}}(k-1) + \mathbf{g}(k)[y(k) - \boldsymbol{\phi}^T(k)\hat{\boldsymbol{\rho}}(k-1)] \tag{8.20a}$$

where

$$\boldsymbol{g}(k) = \boldsymbol{P}(k-1)\boldsymbol{\phi}(k)[1+\boldsymbol{\phi}^T(k)\boldsymbol{P}(k-1)\boldsymbol{\phi}(k)]^{-1}$$

An alternative expression for $\boldsymbol{g}(k)$ can be obtained straightforwardly by manipulation, i.e.

$$\begin{aligned}\boldsymbol{g}(k) &= [\boldsymbol{P}(k)\boldsymbol{P}^{-1}(k)]\boldsymbol{P}(k-1)\boldsymbol{\phi}(k)[1+\boldsymbol{\phi}^T(k)\boldsymbol{P}(k-1)\boldsymbol{\phi}(k)]^{-1} \\ &= \boldsymbol{P}(k)[\boldsymbol{P}^{-1}(k-1) + \boldsymbol{\phi}(k)\boldsymbol{\phi}^T(k)]\boldsymbol{P}(k-1)\boldsymbol{\phi}(k)[1 + \boldsymbol{\phi}^T(k)\boldsymbol{P}(k-1)\boldsymbol{\phi}(k)]^{-1} \\ &= \boldsymbol{P}(k)[\boldsymbol{\phi}(k) + \boldsymbol{\phi}(k)\boldsymbol{\phi}^T(k)\boldsymbol{P}(k-1)\boldsymbol{\phi}(k)][1+\boldsymbol{\phi}^T(k)\boldsymbol{P}(k-1)\boldsymbol{\phi}(k)]^{-1} \\ &= \boldsymbol{P}(k)\boldsymbol{\phi}(k)[1+\boldsymbol{\phi}^T(k)\boldsymbol{P}(k-1)\boldsymbol{\phi}(k)][1 + \boldsymbol{\phi}^T(k)\boldsymbol{P}(k-1)\boldsymbol{\phi}(k)]^{-1} = \boldsymbol{P}(k)\boldsymbol{\phi}(k)\end{aligned}$$

Thus an alternative form of the recursion (8.20a) is the following:

$$\hat{\rho}(k) = \hat{\rho}(k-1) + \boldsymbol{P}(k)[\boldsymbol{\phi}(k)y(k) - \boldsymbol{\phi}(k)\boldsymbol{\phi}^T(k)\hat{\rho}(k-1)] \tag{8.20b}$$

This second form exposes nicely that the RLS algorithm is a 'gradient' algorithm, since the expression in the square brackets will be recognised as being proportional to the gradient of the instantaneous squared error $(y(k) - \boldsymbol{\phi}^T(k)\,\hat{\rho})^2$. However, equation (8.20a) is usually preferred in computational terms.

To summarise, the RLS algorithm consists of equation (8.20a) and equation (8.18), i.e.

$$\left.\begin{aligned}
\hat{\rho}(k) &= \hat{\rho}(k-1) + \boldsymbol{P}(k-1)\boldsymbol{\phi}(k)[1+\boldsymbol{\phi}^T(k)\boldsymbol{P}(k-1)\boldsymbol{\phi}(k)]^{-1} \\
&\qquad\qquad [y(k) - \boldsymbol{\phi}^T(k)\hat{\rho}(k-1)]^{-1} \quad \text{(i)} \\
\boldsymbol{P}(k) &= \boldsymbol{P}(k-1) - \boldsymbol{P}(k-1)\boldsymbol{\phi}(k)[1+\boldsymbol{\phi}^T(k)\boldsymbol{P}(k-1)\boldsymbol{\phi}(k)]^{-1}\boldsymbol{\phi}^T(k)\boldsymbol{P}(k-1) \quad \text{(ii)}
\end{aligned}\right\} \ \mathbf{I}$$

Since the algorithm **I** is recursive, it is necessary to specify starting values $\hat{\rho}(0)$ and $\boldsymbol{P}(0)$ for the vector $\hat{\rho}(k)$ and the matrix $\boldsymbol{P}(k)$, respectively. This presents no real problem, however, since it can be shown that the criterion function–parameter hypersurface is unimodal and that an arbitrary finite $\hat{\rho}(0)$ [say $\hat{\rho}(0) = \mathbf{0}$] coupled with a $\boldsymbol{P}(0)$ having large diagonal elements (say 10^6 in general) will yield convergence and performance commensurate with the stage-wise solution of the same problem (Lee 1964).

8.1.3 Statistical Properties of the RLS Algorithm

As pointed out by Ho (1962), $\boldsymbol{P}(k)$ is a strictly decreasing function of k and this results in the matrix having a smoothing effect on the innovation error $\varepsilon(k) = y(k) - \boldsymbol{\phi}^T(k)\hat{\rho}(k-1)$ in the algorithm **I**(i). At the beginning of the RLS estimation, when the $\boldsymbol{P}(k)$ matrix is still quite large, it takes a lot of notice of the innovation error, since this is most likely to be due to parameter estimation error. However, as k increases and $\boldsymbol{P}(k)$ becomes much smaller, it is more likely that the error is due to the noise $e(k)$. Note that $\varepsilon(k)$ is different from the final *a posteriori model residuals* or *least squares residuals*:

$$\hat{e}(k) = y(k) - \boldsymbol{\phi}^T(k)\hat{\rho}(N) \tag{8.21}$$

obtained in the same manner but with $\hat{\boldsymbol{\phi}}(k-1)$ replaced by the final *en bloc* estimate $\hat{\rho}(N)$.

It is also interesting to note that the instantaneous gradient of the squared innovation error can be written in the form:

$$\frac{1}{2}\nabla_{\hat{\rho}}[\hat{e}^2(k)] = \boldsymbol{\phi}(k)\{y(k) - \boldsymbol{\phi}^T(k)\hat{\rho}(k-1)\} = \nabla_{\hat{\rho}}[\hat{e}(k)]\hat{e}(k) \tag{8.22}$$

revealing that this gradient measure can also be considered as the product of the derivative $\nabla_{\hat{\rho}}[\hat{e}(k)]$ and the innovation error $\hat{e}(k)$. As a result, and noting equation (8.20b), the recursive estimation equation **I**(i) can be written in the form:

$$\hat{\rho}(k) = \hat{\rho}(k-1) + \boldsymbol{P}(k)\nabla_{\hat{\rho}}[\hat{e}(k)]\hat{e}(k) \tag{8.23}$$

This is a more general form of the recursive algorithm, one that has relevance to situations where the error variable $\hat{e}(k)$ is not, as here, a simple linear function of the unknown parameters.

The algorithm **I** provides some considerable advantage over the stage-wise solution of equation (8.12). In addition to the now convenient recursive form, which provides for a minimum of computer storage, note that the term $[1 + \boldsymbol{\phi}^T(k)\boldsymbol{P}(k-1)\boldsymbol{\phi}(k)]$ is simply a *scalar* quantity. As a result, there is no requirement for direct matrix inversion, even though the repeated, stage-wise solution of the equivalent classical *en bloc* solution (8.12) entails inverting an $(m+n+1)\times(m+n+1)$ matrix for each solution update.

Finally, it is beneficial to introduce the additional statistical assumption that the white noise $e(k)$ has a Gaussian normal probability distribution function. This has the advantage that it is completely defined by the first two moments: the mean value (here zero) and the variance σ^2. It can then be shown that the estimate $\hat{\boldsymbol{\rho}}(k)$ of $\boldsymbol{\rho}(k)$, whether obtained by *en bloc* or recursive estimation, is asymptotically unbiased, consistent and statistically efficient (minimum variance). This means that, in statistical terms, the asymptotically unbiased nature of the estimate can be written as:

$$\underset{k\to N}{\mathrm{p.\,lim}}\, \hat{\boldsymbol{\rho}}(k) = \boldsymbol{\rho} \quad \text{or} \quad \underset{k\to N}{\mathrm{p.\,lim}}\, \tilde{\boldsymbol{\rho}}(k) = 0 \tag{8.24}$$

where $\tilde{\boldsymbol{\rho}} = \hat{\boldsymbol{\rho}} - \boldsymbol{\rho}$ is the estimation error and p. lim denotes the 'probability-in-the-limit', in this case as the sample size $N \to \infty$. Young (2011a) provides the mathematical background and offers a fuller discussion of these results. Put simply, however, they show that the recursive estimate $\hat{\boldsymbol{\rho}}(k)$ becomes more accurate (in the sense that the probability of it being close to the true value $\boldsymbol{\rho}$ increases) as the sample size k increases; and that for large k, it has minimum variance. Moreover, because the sole source of the stochasticity in the model, $e(k)$, has a normal probability density function, it is straightforward to show that the LLS estimate can be considered as a *multivariate normal* probability density function. In this case its first moment, the *vector mean*, is defined by the estimated parameter vector $\hat{\boldsymbol{\rho}}(k)$ and its second moment, the *covariance matrix*, is conveniently given by $\hat{\sigma}^2\boldsymbol{P}(k)$. Here $\boldsymbol{P}(k)$ is the matrix computed in the above estimation algorithms and $\hat{\sigma}^2$ is an estimate of the noise variance σ^2. These results will be considered in more detail later but we will first make use of them in Example 8.1 and Example 8.2.

Example 8.1 Estimation of a Simple ARX Model Consider the following simple ARX model:

$$y(k) = -a_1 y(k-1) + b_1 u(k-1) + e(k) \quad k = 1,\ 2, \ldots, 1000 \tag{8.25}$$

with $a_1 = -0.5$ and $b_1 = 0.5$. Noting the triad $[n \ \ (m+1) \ \ \delta]$ defined above, this will be referred to as the ARX [1 1 1] model (i.e. $n = \delta = 1$ and $m = 0$). For the purpose of this example, the input $u(k)$ is a *Pseudo Random Binary Signal* (PRBS) with switching interval of 5 samples and an amplitude 2.0; and $e(k)$ is a zero mean white noise process with variance $\sigma^2 = 0.3$, giving a *Noise–Signal Ratio* (NSR) based on the *Standard Deviations* (SDs) of the $e(k)$ and $y(k)$ of 0.52 (52% noise by SD[4]), i.e. $\mathrm{NSR}_y = \mathrm{std}\{e(k)\}/\mathrm{std}\{y(k)\} = 0.52$. A typical section of the input–output data is shown in Figure 8.1 where the dashed line is the

[4] Quoting the noise level in decibels is often the tradition in the control literature but we feel that the NSR defined in terms of the SDs is a more transparent and physically obvious definition.

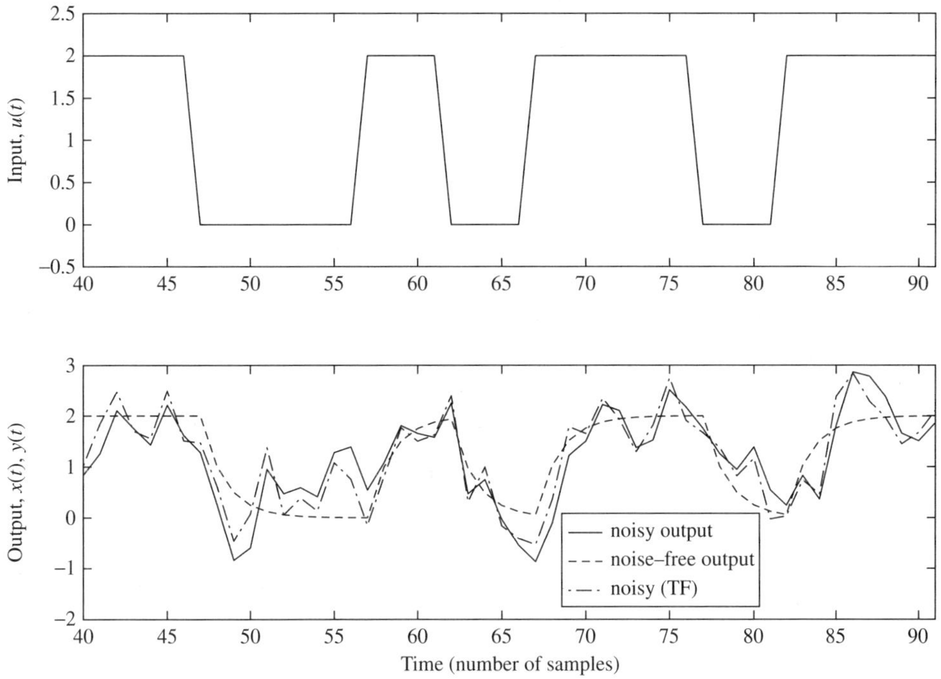

Figure 8.1 Representative section of the input–output data for the ARX model in Example 8.1

'noise-free' output obtained by generating $y(k)$ from equation (8.25) with $e(k)$ zero for all k. The $\text{NSR}_x = \text{std}\{e(k)\}/\text{std}\{x(k)\}$ based on this noise-free signal $x(k)$ is 0.61. The noisy output shown as a dash–dot line [referred to as 'noisy (TF)' in Figure 8.1] is discussed in Example 8.2. The ARX results obtained by LLS using arx in the MATLAB® System Identification Toolbox, or by using MATLAB® directly to compute the estimates, are as follows:

$$\hat{a}_1 = -0.513\ (0.019); \quad \hat{b}_1 = 0.484\ (0.019) \tag{8.26}$$

where the figures in parentheses are the estimated *Standard Errors* (SEs) computed from the square root of the diagonal elements of the estimated covariance matrix $\sigma^2 \boldsymbol{P}(k)$, as supplied by the algorithm.

This model has a coefficient of determination $R_T^2 = 0.63$. As discussed more fully in section 8.4, R_T^2 is a statistical measure of how well the simulated model explains the data: if the variance of the simulated model residuals is low compared with the variance of the data, then R_T^2 tends towards unity. In this case, the result $R_T^2 = 0.63$ is based on the error between the measured output $y(k)$ and the ARX model output $\hat{x}(k)$, where

$$\hat{x}(k) = -\hat{a}_1\hat{x}(k-1) + \hat{b}_1 u(k-1) \quad \text{or} \quad \hat{x}(k) = \frac{\hat{b}_1}{1 + \hat{a}_1 z^{-1}} u(k-1) \tag{8.27}$$

However, when evaluated in relation to the noise-free output $x(k)$, $R_T^2 = 0.999$; i.e. 99.9% of the noise-free output is explained by the ARX model. In other words, the ARX procedure has estimated the parameters very well and so it is able to obtain a very good estimate of the underlying dynamics and output of the system.

Example 8.2 Estimation of a Simple TF Model As already emphasised, the LLS estimation of the ARX model is quite restrictive in practice. In order to illustrate these limitations, let us see what LLS estimation yields if the model (8.25) is modified slightly to the following TF form:

$$y(k) = \frac{b_1}{1 + a_1 z^{-1}} u(k-1) + e(k); \qquad k = 1, 2, \ldots, 1000 \tag{8.28a}$$

still with $a_1 = -0.5$ and $b_1 = 0.5$. Note that this model can be written in the alternative, decomposed form:

$$\begin{aligned} &\text{System equation: } x(k) = \frac{b_1}{1 + a_1 z^{-1}} u(k-1) \\ &\text{Output equation: } y(k) = x(k) + e(k) \end{aligned} \tag{8.28b}$$

and it results in the following discrete-time difference equation form:

$$y(k) = -a_1 y(k-1) + b_1 u(k-1) + e(k) + a_1 e(k-1) \tag{8.28c}$$

which reveals that the main change in the data generation process is the inclusion of an additional noise term $a_1 e(k-1)$ *that is dependent on the model parameter* a_1. As we shall see, although relatively small at first sight, this change has a profound and deleterious effect on the ARX estimates.

The output of the model (8.28) with the same noise as used in Example 8.1, is shown as the dash-dot line in Figure 8.1. Despite the apparently very small change in the data generation process and the input–output data, the ARX estimates are now found to be very poor, with:

$$\hat{a}_1 = -0.307\ (0.022)\,; \quad \hat{b}_1 = 0.655\ (0.022) \tag{8.29}$$

Not only are the estimates badly biased away from their true values but the estimated SEs are much too optimistic: in other words, the modeller seeing these results would be very misled about the nature of the system. The reason for this bias is discussed more fully in section 8.2 when the estimation of more general TF models is considered. The poor quality of these results is also illustrated in Figure 8.2. Here, Figure 8.2a compares the output of the first order ARX estimated model [ARX(1): solid line] with the noise-free output $x(k)$ (dashed trace), as obtained from (8.28b), over a small, step response portion of the data; while Figure 8.2b shows the associated model error [for the model (8.29), these errors are shown as the solid line: the other lines are explained later].

It is possible to obtain better results with the above example using ARX estimation, but only at the cost of estimating a higher order model, i.e. by using LLS methods to estimate an ARX model that is higher order than the [1 1 1] model used to generate the input–output data! Åström and Eykhoff (1971) refer to this approach as 'repeated least squares' but a better name is 'high order ARX modelling'. Here, ARX models of increasing order are estimated against the data, each time checking on the significance of the result in some manner. Åström and Eykhoff suggested evaluating the significance of the decrease in the sum of the squares using a

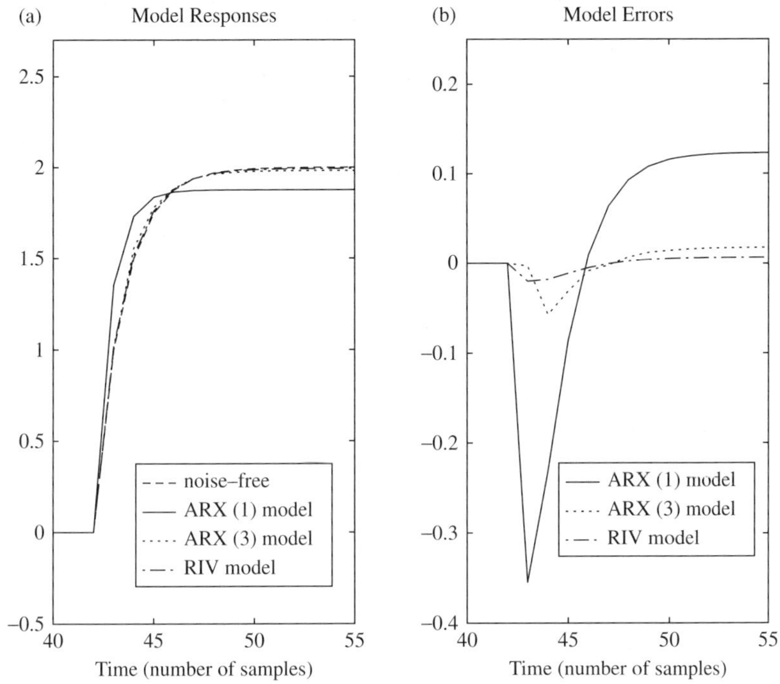

Figure 8.2 Comparison of model outputs with the noise-free output from samples 40 to 55, showing the limitations of the ARX estimated model in Example 8.2

Student *t*-test (see e.g. Kendall and Stuart 1961). An alternative approach, exploited below, is to utilise *AutoCorrelation Function* (ACF) and *Partial AutoCorrelation Function* (PACF) tests (e.g. Box and Jenkins 1970), as well as the *Akaike Information Criterion* (AIC; Akaike 1974).

Given the ARX model, it is clear that the model residuals $\hat{e}(k)$ provide an estimate of the white noise input $e(k)$ and so should possess similar statistical characteristics, as specified in equation (8.6) and equation (8.7). These require that $\hat{e}(k)$ should have no significant autocorrelation at any lag. The ACF and PACF of $\hat{e}(k)$ obtained from the above analysis, using the model structure ARX [1 1 1] show that the series represents 'coloured', rather than white noise. In particular, there is significant lag 1 autocorrelation; and significant partial autocorrelation up to lag 3, the latter suggesting an autoregressive model of at least order 3, i.e. AR(3) is required to model $\hat{e}(k)$ (Box and Jenkins 1970). This is confirmed by the AIC (see section 8.4), which also clearly identifies an AR(3) model for $\hat{e}(k)$. These results suggest that estimation of an ARX model [3 3 1] may yield white residuals and this is indeed the case, with the ACF of the residuals $\hat{e}(k)$ showing that the significant autocorrelations have been removed.

The ARX [3 3 1] model parameters are estimated as:

$$\begin{aligned}&\hat{a}_1 = -0.0455\ (0.032);\ \hat{a}_2 = -0.0440\ (0.030);\ \hat{a}_3 = -0.0606\ (0.0230)\\&\hat{b}_1 = 0.5094\ (0.029);\ \hat{b}_2 = 0.2037\ (0.043);\ \hat{b}_3 = 0.1260\ (0.040)\end{aligned} \tag{8.30}$$

where the $\hat{a}_i, i = 1, 2$ are not significantly different from their SEs, suggesting that the model is not very well identified. Nevertheless, it explains the data quite well, as shown by the dotted lines in Figure 8.2. Here, we see that the model response $\hat{x}(k)$ in Figure 8.2a (dotted line) matches the actual noise-free output $x(k)$ very well; and in Figure 8.2b, that the model error $x(k) - \hat{x}(k)$ (again the dotted line) is quite small. Indeed this error is only a little larger than that obtained using an optimal, RIV estimation algorithm (dash-dot line) to estimate the [1 1 1] model, as discussed in section 8.3 (Example 8.5). Of course, the difference is that, whereas the two optimal RIV parameter estimates are, as we shall see, statistically very well defined because the model is parametrically efficient (or 'parsimonious'), the six ARX [3 3 1] parameter estimates (8.30) are poorly defined, suggesting that the model is considerably over-parameterised (as is obviously the case for this simulation example, which is based on data generated using an ARX [1 1 1] model). We will consider why this over-parameterised [3 3 1] model is able to explain the data well in Example 8.4.

8.1.4 *The FIR Model*

The only other modelling approach that can directly exploit simple LLS is the estimation of the *Finite Impulse Response* (FIR) model, which takes the form:

$$\begin{aligned} y(k) &= \frac{B(z^{-1})}{A(z^{-1})}u(k-\delta) + e(k) \approx G(z^{-1})\,u(k-\delta) + e(k) \\ G(z^{-1}) &= g_0 + g_1 z^{-1} + g_2 z^{-2} + \cdots + g_p z^{-p} \end{aligned} \tag{8.31}$$

Here, the coefficients $g_i,\ i = 0, 1, 2, \ldots, p$ provide a finite dimensional approximation to the infinite dimensional, discrete-time impulse response, as obtained when $B(z^{-1})$ is divided by $A(z^{-1})$. In other words, (8.31) is the discrete-time approximation of the well-known continuous-time convolution integral equation model, where $y(k)$ is simply a weighted linear sum of a finite number of past inputs $u(k-\delta), u(k-\delta-1), \ldots, u(k-\delta-p)$, with p chosen to provide a reasonable description of the impulse response and the system dynamics.

The FIR model can be formulated in a similar vector inner product manner to the ARX model (8.8), i.e.

$$y(k) = \boldsymbol{\phi}^T(k)\boldsymbol{\rho} + e(k) \tag{8.32a}$$

but now

$$\begin{aligned} \boldsymbol{\phi}^T(k) &= [u(k-\delta)\ u(k-\delta-1)\ u(k-\delta-2)\ldots u(k-\delta-p)] \\ \boldsymbol{\rho}^T &= [g_\delta\ g_{\delta+1}\ g_{\delta+2}\cdots g_{\delta+p}] \end{aligned} \tag{8.32b}$$

As a result, the FIR parameters can be estimated by LLS using either the *en bloc* or recursive solution. However, there are two main problems in this case. First, the model is not in a TF form and so is not particularly useful in relation to the PIP design methods discussed in previous chapters of this book. Secondly, it is not a parsimonious description, requiring a sufficient number of parameters to reasonably describe the impulse response *at the sampling interval of the data*. This will often mean that the model is severely over-parameterised, particularly when the system is characterised by long time constants. In such cases, the parameter estimates will not be well estimated, as shown in Example 8.3.

Example 8.3 Estimation of a Simple FIR Model Using the same data set as in Example 8.1 (i.e. Figure 8.1 with the TF model output shown as a dash-dot line), there is little advantage in terms of the coefficient of determination for FIR lengths $p > 5$ samples. The resulting LLS estimates and SEs for this FIR(5) model are:

$$\begin{aligned} \hat{g}_1 &= 0.513\ (0.029);\ \hat{g}_2 = 0.225\ (0.039);\ \hat{g}_3 = 0.108\ (0.039)\ ; \\ \hat{g}_4 &= 0.099\ (0.040);\ \hat{g}_5 = 0.044\ (0.029) \end{aligned} \tag{8.33}$$

Although this model does not explain the data as well as the ARX [3 3 1] and RIV [1 1 1] models, the result is reasonably acceptable. However, if the first order model parameters in equation (8.25) are changed to $a_1 = -0.95$ and $b_1 = 0.05$, the FIR model is now highly over-parameterised and performs poorly. For example, let us consider the results obtained with this model using a PRBS input $u(k)$ with a switching interval of 65 samples and amplitude 2.0. The noise $e(k)$ is once again a zero mean white noise process with variance $\sigma^2 = 0.3$, giving a NSR commensurate with that used in Example 8.1. Here, in order to obtain a reasonable explanation of the data, the FIR model has to have 94 parameters which, not surprisingly, are estimated with high variance. As a result, the model does not estimate the noise-free output $x(k)$ very well with a quite high error variance $\text{var}\{x(k) - \hat{x}(k)\} = 0.046$ and $R_T^2 = 0.928$. By contrast, the RIV estimated [1 1 1] model (see section 8.3, Example 8.5) has an error variance of 0.0007 and $R_T^2 = 0.999$; and its two parameters are estimated very well indeed. This poor FIR estimation performance arises because the high variance associated with the FIR parameter estimates means that the estimated impulse response is very noisy indeed.

8.2 General TF Models

How can we explain the results presented in Example 8.2, in which the LLS estimates are biased away from their true values? In order to answer this question, it is first necessary to consider the form of the general ARX model in TF terms. In this case, equation (8.8b) takes the form:

$$y(k) = \frac{B(z^{-1})}{A(z^{-1})} u(k - \delta) + \frac{1}{A(z^{-1})} e(k) \tag{8.34}$$

revealing that the additive noise, in this important TF context, is not white noise but coloured noise that is strongly influenced by the nature of the TF denominator polynomial $A(z^{-1})$. The only exception is the case where $A(z^{-1}) = 1.0$, which then forms the special FIR model discussed above.

Example 8.4 Poles and Zeros of the Estimated ARX [3 3 1] Model If the ARX [3 3 1] model from Example 8.2, with the estimated parameters (8.30), is now considered in the form of equation (8.34), then:

$$\begin{aligned} y(k) &= \frac{\hat{B}(z^{-1})}{\hat{A}(z^{-1})} u(k-1) + \frac{1}{\hat{A}(z^{-1})} e(k) \\ &= \frac{\hat{b}_1 + \hat{b}_2 z^{-1} + \hat{b}_3 z^{-2}}{1 + \hat{a}_1 z^{-1} + \hat{a}_2 z^{-2} + \hat{a}_3 z^{-3}} u(k-1) + \frac{1}{1 + \hat{a}_1 z^{-1} + \hat{a}_2 z^{-2} + \hat{a}_3 z^{-3}} e(k) \end{aligned} \tag{8.35}$$

A clue to what is happening is obtained if the poles and zeros of the TF are computed based on the estimated parameters in (8.30). This yields:

$$\text{Roots of } \hat{A}(z^{-1})\text{: } -0.2008+/-0.3086j; 0.4471$$

$$\text{Roots of } \hat{B}(z^{-1})\text{: } -0.2000+/-0.4553j$$

There is a complex conjugate pair of poles that are similar to a complex conjugate pair of zeros. These will tend to cancel each other out, resulting in a model that behaves approximately as a first order, rather than a third order, dynamic system. Moreover, if the MATLAB® function deconv is used to divide $\hat{A}(z^{-1})$ by $\hat{B}(z^{-1})$, we obtain:

$$\frac{\hat{A}(z^{-1})}{\hat{B}(z^{-1})} = 1.9631\ (1 - 0.8743z^{-1}) + \text{remainder} \tag{8.36}$$

Consequently, the model can be written in the approximate form:

$$y(k) \approx \frac{0.509}{1 - 0.445z^{-1}} u(k-1) + e(k) \tag{8.37}$$

This reduced order model (8.37) can be compared with the original simulated TF (8.28a), with $a_1 = -0.5$ and $b_1 = 0.5$, and we see now why the ARX [3 3 1] model is able to achieve the good response matching shown in Figure 8.2. In fact, if more data are used, then the approximation becomes closer: e.g. for a very large N=100 000, the two parameter estimates are –0.504 and 0.496.

As a general approach to TF modelling, the problem with high order ARX estimation is two-fold. First, it is a rather 'messy' ad hoc procedure. Secondly, and more important, it yields a model with rather poorly defined estimates. For example, if we use MCS analysis to carry out the above model reduction, based on the covariance matrix of the ARX [3 3 1] model estimates and 5000 realisations, then the mean estimate of the a_1 parameter is –0.450, with a high SE of 0.1. As we shall see later, this is ten times worse than the RIV algorithm MCS estimate of −0.4997 (0.0182). So, although LLS estimation of an ARX model seems, at first sight, to be a reasonable approach to TF model estimation, it has proven necessary to develop other, less restrictive and more generally applicable algorithms.

8.2.1 *The Box–Jenkins and ARMAX Models*

The most common stochastic TF models are obtained from the model form (8.5c) by assuming that the additive noise $\xi(k)$ has rational spectral density, i.e. it is coloured noise obtained here by passing white noise through a linear TF model consisting of a ratio of rational polynomials. These include, most importantly, the *Box–Jenkins* (BJ) model:

$$y(k) = \frac{B(z^{-1})}{A(z^{-1})} u(k-\delta) + \frac{D(z^{-1})}{C(z^{-1})} e(k); \quad e(k) = \mathcal{N}(0, \sigma^2); \quad k = 1, 2, \ldots, N \tag{8.38}$$

and the *Auto-Regressive Moving Average eXogenous variables* (ARMAX) model:

$$A(z^{-1})y(k) = B(z^{-1})u(k-\delta) + D(z^{-1})e(k);\ e(k) = \mathcal{N}(0,\sigma^2);\ k = 1,2,\ldots,N \quad (8.39a)$$

or [cf. equation (6.43)]

$$y(k) = \frac{B(z^{-1})}{A(z^{-1})}u(k-\delta) + \frac{D(z^{-1})}{A(z^{-1})}e(k);\ e(k) = \mathcal{N}(0,\sigma^2);\ k = 1,2,\ldots,N \quad (8.39b)$$

where N is the available sample size and $e(k) = \mathcal{N}(0,\sigma^2)$ denotes a zero mean, independent and identically distributed, white noise process with Gaussian normal amplitude distribution.

In the above models, the noise TF or *Auto-Regressive Moving Average* (ARMA) model polynomials $C(z^{-1})$ and $D(z^{-1})$ are defined as follows:

$$\begin{aligned} C(z^{-1}) &= 1 + c_1 z^{-1} + c_2 z^{-2} + \cdots + c_p z^{-p} \\ D(z^{-1}) &= 1 + d_1 z^{-1} + d_2 z^{-2} + \cdots + d_q z^{-q} \end{aligned} \quad (8.40)$$

and the associated noise model parameter vector is denoted by:

$$\eta = [c_1\ c_2 \cdots c_p\ d_1\ d_2 \cdots d_q]^T \quad (8.41)$$

With the additional noise model polynomials in (8.40) introduced into the stochastic model, the model structure is now defined by the pentad $[n\ m\ \delta\ p\ q]$[5]. Note that both BJ and the ARMAX models can be extended straightforwardly to include the effects of more than one input, as we will see in section 8.5 which deals with such *Multi-Input, Single-Output* (MISO) and full *Multi-Input, Multi-Output* (MIMO) models.

8.2.2 A Brief Review of TF Estimation Algorithms

The estimation problem associated with time series models such as (8.38) and (8.39) is well known and various procedures have been suggested over the past 50 years. A good survey of these techniques was first given by Åström and Eykhoff (1971) and Eykhoff (1974), although there have been many books on the topic published since then (e.g. Goodwin and Payne 1977; Ljung and Söderström 1983; Young 1984; Norton 1986; Söderström and Stoica 1989; Wellstead and Zarrop 1991; Young 2011a). Probably the best known method of estimation is the *Maximum Likelihood* (ML) method suggested by Box and Jenkins in the 1960s (Box and Jenkins 1970); and by Åström and Bohlin (1966) for ARMAX models. Here, the problem of bias is obviated by *simultaneously* estimating the parameters of all the polynomials A, B, C and D in the BJ model; or A, B and D in the ARMAX model. This is clearly a nonlinear estimation problem that, in general, requires either numerical hill-climbing (gradient) procedures to determine the parameters that maximise the likelihood function, or some alternative iterative approach of the kind considered in section 8.3 and subsequently in the present chapter. Here,

[5] As noted earlier, the number of parameters in the polynomial $B(z^{-1})$ is given by $m+1$, hence the modified pentad $[n\ (m+1)\ \delta\ p\ q]$ is used in the CAPTAIN Toolbox and Example 8.8 and Example 8.9.

the data are normally analysed in block form so that online utilisation and time variable parameter estimation are not nearly so straightforward as in the LLS case.

If a recursive solution with online and time variable parameter estimation potential is required, then there are a number of alternatives. One approach is to reformulate the problem in state space terms and consider the *Extended Kalman Filter* (EKF) estimation methods of analysis (Jazwinski 1970) or some of the more recent methods based on discrete-time state space models (see Chapter 10 in Young 2011a). Another is to use the recursive form of *Generalised Least Squares* (GLS) analysis as suggested by Hastings-James and Sage (1969), which applies to the so-called 'Dynamic Adjustment' model with autoregressive residuals. The iterative GLS approach was discussed originally by Johnston (1963) and applied to time series analysis by Clarke (1967).

Since 1970, a number of other recursive approaches have been suggested (see Chapter 7 in Eykhoff 1974), such as the Extended Matrix method (Talmon and Van den Boom 1973, which is related to the *Approximate Maximum Likelihood* (AML) procedure (Young 1968; Panuska 1969) and is also referred to as the *Recursive Maximum Likelihood 1* (RML1) by Söderström *et al.* (1974); the *Recursive Maximum Likelihood 2* (RML2), which is applied to the ARMAX model (8.39); the *Prediction Error Recursion* (PER) approach of Ljung (1979; see also Ljung and Söderström 1983 and Ljung 1987, 1999), which is similar to the RML2 for the ARMAX model but can be applied to general time series model forms; the various recursive algorithms suggested by Landau (1976) which are closely related to some of the other procedures mentioned above, but are derived from an alternative theoretical standpoint based on hyperstability theory; and, finally, the real-time recursive version of the RIV algorithm discussed in section 8.3.

In the present context, it would serve little useful purpose to describe all of these alternative methods. It will suffice to note that all of the above procedures require *simultaneous* estimation of the system $[a_i; b_i]$ and noise $[c_i; d_i]$ parameters and are, therefore, relatively complex to implement. There is one procedure, however, that does not require this and is able to obtain consistent, asymptotically unbiased and *relatively* efficient (i.e. low, but not in general minimum variance) *en bloc* or recursive estimates of the $[a_i; b_i]$ parameters *without* simultaneous noise model parameter estimation. This is the *Instrumental Variable* (IV) method proposed by the second author of the present book (Young 1965, 1969, 1970) in connection with continuous-time TF model estimation and by Wong and Polak (1967) and again by the second author (Young 1974) in relation to discrete-time TF models (see also Mayne 1963). Other publications that deal in detail with such IV estimation are by Durbin (1954), Kendall and Stuart (1961), Young (1984, 2011a) and Söderström and Stoica (1989).

If the noise can be assumed to have rational spectral density then the IV method can be extended into an optimal or RIV form mentioned above (Young 1976, 1984, 2011a; Young and Jakeman 1979) that allows for simultaneous noise model estimation and yields asymptotically efficient, ML parameter estimates. As we shall see in section 8.3, one attraction of the RIV approach is that the associated algorithms retain some of the appearance and characteristics of LLS estimation, so the estimation solution can always be implemented naturally in a recursive manner. In contrast to this, the *Recursive Prediction Error Method* (RPEM) approach represents an approximation to *en bloc* PEM optimisation and so does not produce parameter estimates that are exactly the same as those produced by the stage-wise PEM solution.

8.2.3 Standard IV Estimation

The general TF model (8.5c) can be written in the alternative form:

$$\begin{aligned} x(k) &= \frac{B(z^{-1})}{A(z^{-1})}u(k) && \text{(i)} \\ y(k) &= x(k) + \xi(k) && \text{(ii)} \end{aligned} \tag{8.42a}$$

where $x(k)$ can be considered as underlying 'noise-free' output of the system. Now, the vector $\boldsymbol{\phi}(k)$ in (8.8a) can be decomposed in the following manner, where $\boldsymbol{\phi}(k)$ is the 'noise-free' vector containing $x(k-i)$ in place of $y(k-i)$, for $i = 1,2,\ldots n$, i.e.

$$\boldsymbol{\phi}(k) = \overset{\circ}{\boldsymbol{\phi}}(k) + \boldsymbol{\xi}(k) \tag{8.42b}$$

in which

$$\begin{aligned} \overset{\circ}{\boldsymbol{\phi}}(k) &= [-x(k-1) - x(k-2) - \cdots - x(k-n)\; u(k-\delta)\; u(k-\delta-1)\cdots u(k-\delta-m)]^T \\ \boldsymbol{\xi}^T(k) &= [-\xi(k-1)\; -\xi(k-2)\cdots -\xi(k-n)\; 0\; 0\cdots 0] \end{aligned}$$

Equations (8.42) reveal that the estimation problem is one that has been termed *errors-in-variables*[6] and that the model (8.5a) is a 'structural' rather than a 'regression' model. The structural nature of the model (8.5a) has serious consequences that become apparent if (8.42b) is used to modify the LLS normal equations (8.11b) to the following form:

$$\left[\sum_{k=1}^{N} [\overset{\circ}{\boldsymbol{\phi}}(k) + \boldsymbol{\xi}(k)][\overset{\circ}{\boldsymbol{\phi}}(k) + \boldsymbol{\xi}(k)]^T(k)\right]\boldsymbol{\rho} - \sum_{k=1}^{N} [\overset{\circ}{\boldsymbol{\phi}}(k) + \boldsymbol{\xi}(k)](k)y(k) = 0 \tag{8.43}$$

In general, because the noise matrix $\sum_{k=1}^{N} \boldsymbol{\xi}(k)\boldsymbol{\xi}^T(k)$ arising from the expansion of the left-hand side of this equation has noise-squared elements on its first n diagonal elements, the following result applies:

$$\underset{N \to \infty}{p\lim} \frac{1}{N} \sum_{k=1}^{N} \boldsymbol{\xi}(k)\boldsymbol{\xi}^T(k) \neq \mathbf{0} \tag{8.44}$$

even if the noise $\boldsymbol{\xi}(k)$ is zero mean, white noise. We have seen in Example 8.2 that, in the case of general TFs, the estimates obtained from the solution of these equations can be badly biased [see the estimates in equation (8.29)]. It can be shown (e.g. Young 1984, 2011a; Söderström and Stoica 1989) that this bias is induced in direct consequence of the result (8.44). And this is the reason why, in general, estimation methods other than simple LLS are required for TF model estimation.

---

[6] The situation where there is noise also on the input variable $u(k)$ is not considered here since it is assumed that the control input will normally be measured without noise.

Although there are a number of such methods available (see those mentioned in section 8.2.2), the only solution to the above structural model problem that retains much of the simplicity and robustness of LLS estimation is the IV method. This is an extremely simple technique that *does not require detailed a priori information on the noise statistics in order to yield consistent, asymptotically unbiased estimates.* It entails modifying the LLS normal equations (8.11b), for the purposes of estimation, to the following IV form:

$$\left[\sum_{k=1}^{N} \hat{\boldsymbol{\phi}}(k)\boldsymbol{\phi}^T(k)\right]\boldsymbol{\rho} - \sum_{k=1}^{N} \hat{\boldsymbol{\phi}}(k)y(k) = 0 \tag{8.45}$$

so that the IV estimate is obtained as:

$$\hat{\boldsymbol{\rho}} = \left[\sum_{k=1}^{N} \hat{\boldsymbol{\phi}}(k)\boldsymbol{\phi}^T(k)\right]^{-1} \sum_{k=1}^{N} \hat{\boldsymbol{\phi}}(k)y(k) \tag{8.46a}$$

or

$$\hat{\boldsymbol{\rho}} = [\hat{\Phi}^T(k)\Phi(k)]^{-1}\hat{\Phi}^T(k)\mathbf{y} \tag{8.46b}$$

where $\hat{\Phi}(k)$ is a $k \times N$ matrix with rows $\hat{\boldsymbol{\phi}}^T(k)$; $\Phi(k)$ is a similarly dimensioned matrix with rows $\boldsymbol{\phi}^T(k)$, $k = 1, 2, \ldots, N$; and $[\hat{\Phi}^T(k)\Phi(k)]$ is called the *Instrumental Product Matrix* (IPM). In equation (8.46a), $\hat{\boldsymbol{\phi}}(k)$ is a vector of 'instrumental variables' (variables that are 'instrumental' in solving the estimation problem) which are chosen to be as highly correlated as possible with the equivalent variables in the noise-free vector $\mathring{\boldsymbol{\phi}}(k)$ in equation (8.42b); but *totally* statistically independent of the noise, i.e.

$$E\{\hat{\boldsymbol{\phi}}(i)\boldsymbol{\phi}(j)\} >> \mathbf{0} \quad \text{and} \quad E\{\hat{\boldsymbol{\phi}}(i)e(j)\} = \mathbf{0} \qquad \forall i, j \tag{8.47}$$

Of course, the highest correlation occurs when $\hat{\boldsymbol{\phi}}(k) = \mathring{\boldsymbol{\phi}}(k),\ \forall k$. It is not surprising, therefore, that this notion is important in developing good IV algorithms which attempt to generate and exploit, in various ways, a variable that is as good an approximation as possible to the noise-free output.

The *en bloc* solution (8.46) can be written in the form:

$$\hat{\boldsymbol{\rho}}(k) = \hat{\mathbf{P}}(k)\hat{\mathbf{b}}(k) \tag{IIe}$$

where

$$\hat{\mathbf{P}}(k) = \hat{\mathbf{C}}(k)^{-1} = \left[\sum_{i=1}^{k} \hat{\boldsymbol{\phi}}(i)\boldsymbol{\phi}^T(i)\right]^{-1} \tag{8.48}$$

$$\hat{\mathbf{b}}(k) = \sum_{i=1}^{k} \hat{\boldsymbol{\phi}}(i)y(i) \tag{8.49}$$

As a result, it is a simple matter to obtain a recursive IV algorithm using an approach similar to that employed in the RLS algorithm discussed earlier. We will leave the reader to derive these equations, which take the following form (cf. algorithm **I**):

$$\hat{\rho}(k) = \hat{\rho}(k-1) + \hat{\boldsymbol{P}}(k-1)\hat{\boldsymbol{\phi}}(k)[1 + \boldsymbol{\phi}^T(k)\hat{\boldsymbol{P}}(k-1)\hat{\boldsymbol{\phi}}(k)]^{-1}\{y(k) - \boldsymbol{\phi}^T(k)\hat{\rho}(k-1)\}$$
$$\hat{\boldsymbol{P}}(k) = \hat{\boldsymbol{P}}(k-1) - \hat{\boldsymbol{P}}(k-1)\hat{\boldsymbol{\phi}}(k)[1 + \boldsymbol{\phi}^T(k)\hat{\boldsymbol{P}}(k-1)\hat{\boldsymbol{\phi}}(k)]^{-1}\boldsymbol{\phi}^T(k)\hat{\boldsymbol{P}}(k-1) \qquad \text{(i) } \mathbf{II}$$

The major problem with the general application of algorithm **II** is that of choosing the IVs themselves: how can we obtain or generate variables with such specific statistical properties? The difficulty of answering this question in any general manner has acted as a strong deterrent to the widespread use of the IV approach in statistics (Kendall and Stuart 1961). But the question is surprisingly easy to answer in the TF modelling context considered here. The IV $\hat{x}(k)$, at the jth iteration of an iterative (or 'relaxation') estimation algorithm, is generated as the output of an iteratively updated 'auxiliary model':

$$\hat{x}(k) = \frac{\hat{B}(z^{-1}, \hat{\boldsymbol{\rho}}^{j-1})}{\hat{A}(z^{-1}, \hat{\boldsymbol{\rho}}^{j-1})} u(k-\delta) \qquad \text{(ii) } \mathbf{II}$$

where $\hat{A}(z^{-1}, \hat{\boldsymbol{\rho}}^{j-1})$ and $\hat{B}(z^{-1}, \hat{\boldsymbol{\rho}}^{j-1})$ are the TF polynomials based on the IV estimate $\hat{\boldsymbol{\rho}}_{j-1}$ of the TF model parameter vector $\boldsymbol{\rho}$ at the $(j-1)$ th iteration of the algorithm. Thus, if the algorithm converges, the IV $\hat{x}(k)$ will converge on the noise-free output $x(k)$ (recall our earlier comments) and the IV estimates will have good, albeit not optimal, statistical properties. This standard IV algorithm is summarised as follows:

Overview of the Standard IV Algorithm

Step 1: Use LLS or RLS to estimate the parameters in the ARX model.

Step 2: Iterative or recursive-iterative IV estimation:

for j = 1 : convergence

1. If the estimated TF model is unstable, reflect the unstable eigenvalues of the estimated $\hat{A}_j$ polynomial into the stable region of the complex z-plane (using, for example, the polystab routine in MATLAB®). This is not essential to the functioning of the algorithm: it allows for rarely occurring situations, normally with very poor data, when the initially estimated ARX model is found to be unstable. Generate the IV series $\hat{x}(k)$ from the system auxiliary model **II**(ii), with the polynomials based on the estimated parameter vector $\hat{\boldsymbol{\rho}}^{j-1}$ obtained at the previous iteration of the algorithm; for $j = 1$, $\hat{\boldsymbol{\rho}}^0$ is the estimate obtained at step 1.
2. Compute the estimate $\hat{\boldsymbol{\rho}}^j$ of the TF system model parameter vector using the *en bloc* algorithm **IIe**, or the recursive equivalent of this **II**(i).

end

The iteration is continued for $j = 1, 2, \ldots, i_t$, where i_t is defined so that the IV parameter estimate $\hat{\rho}_j$ converges in some sense (e.g. an insignificant change between iterations). Normally, only 3–4 iterations are required for a reasonable sample length N.

8.3 Optimal RIV Estimation

The standard recursive (or *en bloc*) IV algorithm is simply a modification of the RLS (or *en bloc* LLS) algorithm applied to the TF model (8.5c), without making, or exploiting, any statistical assumptions about the nature of the noise $\xi(k)$. In fact, the standard IV algorithm is no longer used very much because the more sophisicated alternatives that are based on such assumptions perform much better in practice. However, its successful iterative implementation served as a primary stimulus to the development of its successor, the more advanced RIV algorithm for the estimation of parameters in the full BJ model (8.38) which include, of course, the parameters of the ARMA noise model defined in (8.40). The reason for considering this BJ model rather than the ARMAX model, which is probably better known in the control literature, is that it has some advantages in statistical estimation terms; advantages that are exploited in RIV estimation. These advantages are exposed by an important theorem due to Pierce (1972: see Appendix H) concerning the asymptotic statistical properities of the ML estimates for the BJ model parameters. In particular, it reveals that the system and noise model parameter estimates $\hat{\rho}$ and $\hat{\eta}$ are asymptotically independent, so that the covariance matrix of the parametric estimation errors is block diagonal, with zero off-diagonal blocks. As we shall see, this motivates and justifies an important structural aspect of the iterative RIV algorithm.

8.3.1 Initial Motivation for RIV Estimation

Before considering the development of the RIV algorithm in more detail, it is worthwhile considering the motivation for the algorithm in the simplest terms. Bearing in mind equations (8.9) and considering the BJ model (8.38), minimisation of a least squares criterion function of the form:

$$J_2 = \sum_{k=1}^{k=N} [e(k)]^2 \tag{8.50}$$

provides the basis for stochastic estimation and is the criterion function used by Box and Jenkins (1970). From equation (8.39b):

$$e(k) = \frac{C}{D}\left[y(k) - \frac{B}{A}u(k-\delta)\right] = \frac{C}{DA}\left[Ay(k) - Bu(k-\delta)\right] \qquad k = 1, 2, \ldots, N$$

where A, B, C, D denote the model polynomials with the z^{-1} argument removed for simplicity. However, since the polynomial operators commute in this linear case, the filter C/DA can be taken inside the square brackets:

$$e(k) = Ay_f(k) - Bu_f(k-\delta) \tag{8.51}$$

or [cf. equation (8.2)]

$$e(k) = y_f(k) + a_1 y_f(k-1) \cdots + a_n y_f(k-n) - b_0 u_f(k-\delta) \cdots - b_m u_f(k-\delta-m) \tag{8.52}$$

where $y_f(k)$ and $u_f(k)$ are 'prefiltered' output and input signals, respectively, i.e.

$$y_f(k) = \frac{C}{AD} y(k) \quad \text{and} \quad u_f(k) = \frac{C}{AD} u(k) \tag{8.53}$$

and the associated estimation equation at the kth sampling instant can be written as:

$$y(k) = \boldsymbol{\phi}^T(k)\boldsymbol{\rho} + e(k) \tag{8.54}$$

where

$$\begin{aligned} \boldsymbol{\phi}^T(k) &= [-y_f(k) - y_f(k-1) \; - \cdots - \; y_f(k-n) \; u_f(k-\delta) \cdots u_f(k-\delta-m)] \\ \boldsymbol{\rho} &= [a_1 \quad a_2 \ldots a_n \quad b_0 \quad \ldots \quad b_m]^T \end{aligned} \tag{8.55}$$

This is of the same form as the equivalent simple regression model (8.8a) and (8.8b) but here all the variables are prefiltered. Thus, provided we assume that A, $C$and D are known *a priori*, the estimation model forms a basis for the definition of a likelihood function and ML estimation.

The most interesting feature of the estimation model (8.54) is the introduction of the prefilters. In this regard, it is important to note that, although the prefiltering operations arise naturally from the above analysis, they make sense in a more heuristic manner and confirm the veracity of earlier heuristic IV algorithms that exploited prefilters for continuous-time TF estimation (e.g. see section 8.6 and Young 1981). In particular, the prefilter C/AD will pass signals with components within the frequency pass-band of the system, as defined by the transfer function $1/A$, while attenuating components at higher frequencies that are likely to constitute noise on the data. At the same time, the 'inverse' noise filter C/D will 'prewhiten' those noise components within the pass-band of the system that have not been attenuated by the $1/A$ part of the prefilter. Such a prewhitening procedure is either explicit or implicit in many optimal time series estimation algorithms because it yields serially uncorrelated residuals and so improves the statistical efficiency (reduces the variance) of the resulting estimates. It achieves the same objective in the present context.

The major problem with this formulation of the estimation problem in terms of the estimation equation (8.54) is, of course, that the polynomials A, C and D are *not* known *a priori*: indeed it is the estimation of their parameters, together with those of B, that is the whole purpose of the estimation procedure! Fortunately, as in the case of the standard IV algorithm, it is possible to construct an iterative algorithm where these polynomials are updated at each iteration, based on the estimation results obtained at the previous iteration. The formulation of this algorithm, within the context of ML, is discussed in the following subsections.

8.3.2 The RIV Algorithm in the Context of ML

The standard IV algorithm is simply a modification of the LLS algorithm and so, unlike the LLS algorithm, it is not defined in optimal statistical terms. In order to consider the RIV algorithm in optimal terms, therefore, it is necessary to consider the ML approach to the estimation of the BJ model (Box and Jenkins 1970), where the log-likelihood function for the observations takes the form (Young 1984, 2011a):

$$L(\boldsymbol{\rho}, \boldsymbol{\psi}, \sigma^2, \boldsymbol{y}, \boldsymbol{u}) = -\frac{T}{2}\log_e(2\pi) - \frac{T}{2}\log_e \sigma^2 - \frac{1}{2\sigma^2}\left[\frac{C}{D}\boldsymbol{y} - \frac{BC}{AD}\boldsymbol{u}\right]^T \left[\frac{C}{D}\boldsymbol{y} - \frac{BC}{AD}\boldsymbol{u}\right] \tag{8.56}$$

Maximisation of this log-likelihood function clearly requires the minimisation of the final term on the right-hand side of (8.56), i.e.

$$J_2 = \sum_{k=1}^{k=N}[e(k)]^2 = \boldsymbol{e}^T\boldsymbol{e} = \left[\frac{C}{D}\boldsymbol{y} - \frac{BC}{AD}\boldsymbol{u}\right]^T \left[\frac{C}{D}\boldsymbol{y} - \frac{BC}{AD}\boldsymbol{u}\right] \tag{8.57}$$

In this cost function, $\boldsymbol{e}$ is the vector of white noise inputs, i.e.

$$\boldsymbol{e} = \mathcal{N}(\boldsymbol{0}, \sigma^2\boldsymbol{I}); \quad \boldsymbol{e} = [e_1\, e_2 \ldots e_N]^T$$

while $\boldsymbol{y}$ and $\boldsymbol{u}$ are defined as follows:

$$\boldsymbol{y} = [y_1 y_2 \ldots y_N]^T; \; \boldsymbol{u} = [u_1 u_2 \ldots u_N]^T$$

As shown, the inner product $\boldsymbol{e}^T\boldsymbol{e}$ is just another representation of the sum of the squares of the noise $e(k)$. Thus we need to find those estimates $\hat{A}$, $\hat{B}$, $\hat{C}$ and $\hat{D}$ of the polynomials A, B, C and D which minimise the sum of the squares of $e(k)$ – a classical *nonlinear* least squares problem that makes obvious sense.

If, for the moment, we assume that C and D are known, then the conditions for the minimisation of J_2 in (8.57) are obtained in the usual manner by partially differentiating with respect to $\hat{a}_i$, $i = 1, 2, \ldots, n$, $\hat{b}_i$, $i = 0, 1, \ldots, m$ and $\hat{\sigma}^2$, respectively; and then setting these derivatives to zero. This yields the following three equations:

$$\frac{\partial L}{\partial \hat{a}_i} = \frac{1}{\sigma^2}\sum_{k=2n+1}^{T}\left[\frac{C}{D}y(k) - \frac{BC}{AD}u(k)\right]\frac{BC}{A^2D}z^{-i}u(k) = 0 \tag{8.58a}$$

$$\frac{\partial L}{\partial \hat{b}_i} = \frac{1}{\sigma^2}\sum_{k=2n+1}^{T}\left[\frac{C}{D}y(k) - \frac{BC}{AD}u(k)\right]\frac{C}{AD}z^{-i}u(k) = 0 \tag{8.58b}$$

$$\frac{\partial L}{\partial \hat{\sigma}^2} = -\frac{T}{\sigma^2} + \frac{1}{\sigma^4}\sum_{k=2n+1}^{T}\left[\frac{C}{D}y(k) - \frac{BC}{AD}u(k)\right]^2 = 0 \tag{8.58c}$$

These equations are highly nonlinear in the a_i and b_i parameters but are linear in the b_i for a given a_i. Moreover, if estimates of the a_i and b_i are available, an estimate of $\hat{\sigma}^2$ can be obtained easily from condition (8.58c).

The best known method for obtaining estimates of the a_i and b_i that satisfy conditions (8.58) is to search the whole parameter space of the BJ model in some manner; for example, using a numerical optimisation scheme. However, bearing in mind the prefiltering operations in (8.53), together with an equivalent prefiltered auxiliary model **II**(ii), i.e.

$$y_f(k) = \frac{C}{AD}y(k); \quad u_f(k) = \frac{C}{AD}u(k); \quad \hat{x}_f(k) = \frac{B}{A}u_f(k) \tag{8.59}$$

equation (8.58a) and equation (8.58b) can be rewritten as follows in terms of these prefiltered variables, where for simplicity m is assumed equal to n:

$$\sum [Ay_f(k) - Bu_f(k)]\hat{x}_f(k-i) = 0; \quad \text{for } i = 1, 2, \ldots, n \tag{8.60a}$$

$$\sum [Ay_f(k) - Bu_f(k)]u_f(k-i) = 0; \quad \text{for } i = 0, 1, 2, \ldots, n \tag{8.60b}$$

which are now *linear* in the a_i and b_i parameters, provided we assume knowledge of the prefiltered or transformed variables $y_f(k)$, $u_f(k)$ and $\hat{x}_f(k)$; in which case, they can be solved to yield ML estimates of these parameters.

Perhaps the most interesting aspect of the equations (8.60) is their remarkable similarity to the equivalent IV equations for the same problem. This becomes clear if we examine the IV normal equations (8.45) again, which represent a set of $2n+1$ linear simultaneous equations. Upon expansion, the first n equations can be written in the form:

$$\sum y(k) + a_1 y(k-1) + \cdots + a_n y(k-n) - b_0 u(k) - b_n u(k-n)]\hat{x}(k-i) = 0, \quad i = 0, 1, \ldots, n \tag{8.61}$$

while the subsequent $n+1$ equations become:

$$\sum y(k) + a_1 y(k-1) + \cdots + a_n y(k-n) - b_0 u(k) - b_n u(k-n)]u(k-i) = 0, \quad i = 0, 1, \ldots, n \tag{8.62}$$

which, on introduction of the operator notation, and the A and B polynomials, becomes:

$$\sum [Ay(k) - Bu(k)]\,\hat{x}(k-i) = 0 \quad i = 0, 1, \ldots, n \tag{8.63a}$$

$$\sum [Ay(k) - Bu(k)]\,u(k-i) = 0 \quad i = 0, 1, \ldots, n \tag{8.63b}$$

Comparison of equation (8.60) and equation (8.63) shows immediately that the former can be interpreted as the IV normal equations for the system, with the input $u(k)$, the output $y(k)$ and the auxiliary model output $\hat{x}(k)$ replaced by their prefiltered equivalents $u_f(k)$, $y_f(k)$ and $\hat{x}_f(k)$, respectively.

As a result of the similarities between equation (8.60) and equation (8.63), it is clear that the iterative and recursive methods of solution used in the standard IV case, can also be utilised in the present context. Moreover, upon convergence of the iterations, this solution represents a solution of the ML optimisation equations (8.57) and the estimate provides a ML estimate of the parameters in the BJ model (8.38). However, the implementation is obviously more complex in this case. In particular, it requires the introduction of the additional prefiltering operations; and the parameters of these prefilters need to be updated, either iteratively or recursively, using similar adaptive mechanisms to those used to update the auxiliary model parameters in the basic IV case. Furthermore, in order to do this, it is necessary to introduce an algorithm to estimate the noise model parameters C and D that are required to implement the prefilters, as discussed in the following subsection.

8.3.3 Simple AR Noise Model Estimation

The estimation of the noise model parameters can be aided by two simplifications. First, we note, again from the Pierce Theorem (Appendix H), that the ML estimate of $\boldsymbol{\eta}$ (8.41) is asymptotically independent of the system parameter vector $\boldsymbol{\rho}$ (8.55), suggesting that the estimation of $\boldsymbol{\eta}$ can be carried out by a separate algorithm that follows the RIV estimation of $\boldsymbol{\rho}$. Secondly, we can approximate the ARMA noise model in the BJ model by a simpler *Auto-Regressive* (AR) model, whose parameters can be estimated, as we shall see, by simple LLS or RLS algorithms. This latter modification can also be justified because any ARMA process, as defined by a transfer function say *G/F*, can be represented by an AR model *1/H*, where *H* is obtained from the polynomial division *F/G* [i.e. *G/F=1/(F/G)*]. Although this AR model will usually be of infinite order, arising from the polynomial division, a finite order approximation will normally suffice, particularly in the present context, where the main objective of the noise model estimation is to define the nature of the prewhitening filter.

With this AR modification, the BJ model takes the following simpler form:

$$y(k) = \frac{B(z^{-1})}{A(z^{-1})}u(k-\delta) + \frac{1}{C(z^{-1})}e(k) \qquad e(k) = \mathcal{N}(0,\sigma^2) \tag{8.64}$$

This model structure can be defined by the tetrad $[n\ m\ \delta\ p]$. Following the above reasoning, and assuming that we have consistent estimates $\hat{A}(z^{-1})$ and $\hat{B}(z^{-1})$ of the TF model polynomials, an estimate $\hat{\xi}(k)$ of the noise series $\xi(k)$ can be obtained from:

$$\hat{\xi}(k) = y(k) - \frac{\hat{B}(z^{-1})}{\hat{A}(z^{-1})}u(k-\delta) \tag{8.65}$$

An AR model for $\hat{\xi}(k)$ can then be obtained from either the LLS *en-bloc* algorithm **Ie** or the RLS algorithm **I**, with the data vector $\boldsymbol{\phi}(k)$ and parameter vector $\boldsymbol{\rho}$ replaced by $\boldsymbol{\psi}(k)$ and $\boldsymbol{\eta}$, respectively, where:

$$\begin{aligned} \boldsymbol{\psi}^T(k) &= [-\hat{\xi}(k-1) - \hat{\xi}(k-2) \cdots - \hat{\xi}(k-p)] \\ \boldsymbol{\eta} &= [c_1\ c_2 \cdots c_p]^T \end{aligned} \tag{8.66a}$$

These define the AR noise model, which is written in the vector form:

$$\hat{\xi}(k) = \boldsymbol{\psi}^T(k)\boldsymbol{\eta} + e(k) \tag{8.66b}$$

Since this is a linear regression model in which all of the elements in the vector are available, the parameter vector $\boldsymbol{\eta}$ can be estimated using the LLS or RLS algorithms. The *Refined Instrumental Variable* algorithm with *Auto-Regressive* noise (RIVAR) model is now simply the conjunction of the IV algorithm **II**, incorporating the appropriate prefilters on the input, output and IV variables, and this AR estimation algorithm, linked through the noise estimation defined in equation (8.65).

8.3.4 RIVAR Estimation: RIV with Simple AR Noise Model Estimation

Based on the above arguments, the separate *en bloc* algorithms for the system and noise models take the following form:

$$\text{System model estimation: } \hat{\boldsymbol{\rho}}(N) = \hat{\boldsymbol{P}}(N)\hat{\boldsymbol{b}}(N) \qquad \text{(i) } \mathbf{IIIe}$$

$$\text{Noise model estimation: } \hat{\boldsymbol{\eta}}(N) = \boldsymbol{P}_\eta(N)\boldsymbol{b}_\eta(N) \qquad \text{(ii) } \mathbf{IIIe}$$

where

$$\begin{aligned}
\hat{\boldsymbol{P}}(N) &= \hat{\boldsymbol{C}}(N)^{-1} = \left[\sum_{k=1}^{N} \hat{\boldsymbol{\phi}}(k)\boldsymbol{\phi}^T(k)\right]^{-1} \quad \text{and} \quad \hat{\boldsymbol{b}}(N) = \sum_{k=1}^{N} \hat{\boldsymbol{\phi}}(k)y_f(k) \\
\boldsymbol{P}_\eta(N) &= \boldsymbol{C}_\eta(N)^{-1} = \left[\sum_{k=1}^{N} \boldsymbol{\psi}(k)\boldsymbol{\psi}^T(k)\right]^{-1} \quad \text{and} \quad \boldsymbol{b}_\eta(N) = \sum_{k=1}^{N} \boldsymbol{\psi}(k)\xi(k)
\end{aligned} \tag{8.67}$$

The two equivalent recursive algorithms are given below.
System model estimation:

$$\begin{aligned}
\hat{\boldsymbol{\rho}}(k) &= \hat{\boldsymbol{\rho}}(k-1) + \hat{\boldsymbol{P}}(k-1)\hat{\boldsymbol{\phi}}(k)[1 + \boldsymbol{\phi}^T(k)\hat{\boldsymbol{P}}(k-1)\hat{\boldsymbol{\phi}}(k)]^{-1} \\
&\quad [y_f(k) - \boldsymbol{\phi}^T(k)\hat{\boldsymbol{\rho}}(k-1)] && \text{(i)} \\
\hat{\boldsymbol{P}}(k) &= \hat{\boldsymbol{P}}(k-1) - \hat{\boldsymbol{P}}(k-1)\hat{\boldsymbol{\phi}}(k)[1 + \boldsymbol{\phi}^T(k)\hat{\boldsymbol{P}}(k-1)\hat{\boldsymbol{\phi}}(k)]^{-1}\boldsymbol{\phi}^T(k)\hat{\boldsymbol{P}}(k-1) && \text{(ii)}
\end{aligned} \quad \mathbf{III}$$

Noise model estimation:

$$\begin{aligned}
\hat{\boldsymbol{\eta}}(k) &= \hat{\boldsymbol{\eta}}(k-1) + \boldsymbol{P}_\eta(k-1)\boldsymbol{\psi}(k)[1 + \boldsymbol{\psi}^T(k)\boldsymbol{P}_\eta(k-1)\boldsymbol{\psi}(k)]^{-1} \\
&\quad [\xi(k) - \boldsymbol{\psi}^T(k)\hat{\boldsymbol{\eta}}(k-1)]^{-1} && \text{(iii)} \\
\boldsymbol{P}_\eta(k) &= \boldsymbol{P}_\eta(k-1) - \boldsymbol{P}_\eta(k-1)\boldsymbol{\psi}(k)[1 + \boldsymbol{\psi}^T(k)\boldsymbol{P}_\eta(k-1) \\
&\quad \boldsymbol{\psi}(k)]^{-1}\boldsymbol{\psi}^T(k)\boldsymbol{P}_n(k-1) && \text{(iv)}
\end{aligned} \quad \mathbf{III}$$

Here, the prefiltered $\hat{\boldsymbol{\phi}}(k)$ and $\boldsymbol{\phi}(k)$ vectors are defined:

$$\begin{aligned}\hat{\boldsymbol{\phi}}(k) &= [-\hat{x}_f(k-1)\ -\hat{x}_f(k-2)\cdots-\hat{x}_f(k-n)\ u_f(k-\delta)\ u_f(k-\delta-1)\cdots\\ &\quad u_f(k-\delta-m)]^T\\ \boldsymbol{\phi}(k) &= [-y_f(k-1)\ -y_f(k-2)\cdots-y_f(k-n)\ u_f(k-\delta)\ u_f(k-\delta-1)\cdots\\ &\quad u_f(k-\delta-m)]^T\end{aligned} \tag{8.68}$$

in which the prefiltered IV $\hat{x}(k)$, $u(k)$ and $y(k)$ are given by:

$$\hat{x}_f(k) = \frac{\hat{C}(z^{-1})}{\hat{A}(z^{-1})}\hat{x}(k); \quad u_f(k) = \frac{\hat{C}(z^{-1})}{\hat{A}(z^{-1})}u(k); \quad y_f(k) = \frac{\hat{C}(z^{-1})}{\hat{A}(z^{-1})}y(k) \tag{8.69}$$

The RIVAR algorithm uses the above component algorithms in an iterative manner similar to the standard IV algorithm but modified in a manner motivated by the Pierce Theorem (Appendix H). This shows that the system and noise model parameter estimates obtained by ML optimisation are asymptotically independent, so that for large sample size, they can be estimated sequentially, linked by the noise estimation equation (8.65). As far as the authors are aware, the small sample performance of the ML method has not been analysed theoretically so this independence cannot necessarily be guaranteed for small sample size. However, extensive MCS analysis and experience over many years has revealed no discernible difference between the results obtained using the RIVAR algorithm and the standard, non-recursive ML and PEM algorithms that utilise alternative gradient methods of optimisation.

Based on these considerations, the main steps in the iterative RIVAR algorithm are outlined below. As in the standard IV algorithm, the iteration is continued for $j = 1, 2, \ldots, i_t$, where i_t is defined so that the IV parameter estimate $\hat{\boldsymbol{\rho}}_j$ converges in some sense (e.g. an insignificant change between iterations). And again, normally only 3–4 iterations are required for a reasonable sample length N. The RIVAR algorithm is summarised as follows:

Overview of the RIVAR Algorithm

Step 1: Use LLS or RLS to estimate the parameters in the ARX model.

Step 2: Iterative or recursive-iterative IV estimation with prefilters:

for j = 1: convergence

1. If the estimated TF model is unstable, reflect the unstable eigenvalues of the estimated $\hat{A}_j$ polynomial into the stable region of the complex z-plane. This is not essential to the functioning of the algorithm: it allows for rarely occurring situations, normally with very poor data, when the initially estimated ARX model is found to be unstable. Generate the IV series $\hat{x}(k)$ from the system auxiliary model:

$$\hat{x}(k) = \frac{\hat{B}(z^{-1}, \hat{\boldsymbol{\rho}}^{j-1})}{\hat{A}(z^{-1}, \hat{\boldsymbol{\rho}}^{j-1})}u(k-\delta) \tag{8.70}$$

with the polynomials based on the estimated parameter vector $\hat{\boldsymbol{\rho}}^{j-1}$ obtained at the previous iteration of the algorithm; for $j = 1$, $\hat{\boldsymbol{\rho}}^{0}$ is the estimate obtained at step 1.

2. Use standard LLS estimation to obtain the latest estimate $\hat{\boldsymbol{\eta}}^j$ of the AR noise model parameter vector based on the estimated noise sequence $\hat{\xi}(k)$ from the equation $\hat{\xi}(k) = y(k) - \hat{x}(k)$ and the regression model (8.66b).
3. Prefilter the input $u(k)$, output $y(k)$ and instrumental variable $\hat{x}(k)$ signals by the filter:

$$f_1(z^{-1}, \hat{\boldsymbol{\rho}}^{j-1}, \hat{\boldsymbol{\eta}}^j) = \frac{\hat{C}(z^{-1}, \hat{\boldsymbol{\eta}}^j)}{\hat{A}(z^{-1}, \hat{\rho}^{j-1})} \tag{8.71}$$

with the polynomials based on the estimated parameter vector $\hat{\boldsymbol{\rho}}^{j-1}$ obtained at the previous iteration of the algorithm and $\hat{\boldsymbol{\eta}}^j$ obtained in (2); for $j = 1$, $\hat{\boldsymbol{\rho}}^{0}$ is the estimate obtained at step 1.

4. Based on these prefiltered data, compute the estimate $\hat{\boldsymbol{\rho}}^j$ of the TF system model parameter vector using the *en bloc* algorithm **III**e(i), or the recursive equivalent of this, **III**(i) and (ii).

end

Step 3: Subsequent to convergence of the estimates, carry out a final iteration using the following 'symmetric' version (Young 1970, 1984) of the algorithm [cf. **III**(i) and (ii)]:

$$\begin{aligned}
\hat{\boldsymbol{\rho}}(k) &= \hat{\boldsymbol{\rho}}(k-1) + \hat{\boldsymbol{P}}(k-1)\hat{\boldsymbol{\phi}}(k)[1 + \hat{\boldsymbol{\phi}}^T(k)\hat{\boldsymbol{P}}(k-1)\hat{\boldsymbol{\phi}}(k)]^{-1} \\
&\quad [y^f(k) - \boldsymbol{\phi}^T(k)\hat{\boldsymbol{\rho}}(k-1)] && \text{(i)} \\
\hat{\boldsymbol{P}}(k) &= \hat{\boldsymbol{P}}(k-1) - \hat{\boldsymbol{P}}(k-1)\hat{\boldsymbol{\phi}}(k)[1 + \hat{\boldsymbol{\phi}}^T(k)\hat{\boldsymbol{P}}(k-1)\hat{\boldsymbol{\phi}}(k)]^{-1}\hat{\boldsymbol{\phi}}^T(k)\hat{\boldsymbol{P}}(k-1) && \text{(ii)}
\end{aligned} \qquad \textbf{III}\text{s}$$

This generates superior estimates of the covariance matrix $\boldsymbol{P}^*(N)$ in equation (H.1) of the Pierce Theorem (Appendix H), i.e. at $k = N$:

$$\boldsymbol{P}^*(N) = \frac{\sigma^2}{N}\left[p\lim \frac{1}{N}\sum_{k=1}^{N} \hat{\boldsymbol{\phi}}(k)\hat{\boldsymbol{\phi}}^T\right]^{-1} \approx \hat{\sigma}^2\hat{\boldsymbol{P}}(N) \tag{8.72}$$

so providing an estimate of the statistical properties of $\hat{\boldsymbol{\rho}}(N)$, with the SEs on the individual parameter estimates obtained as the square root of the diagonal elements of $\boldsymbol{P}^*(N)$. The covariance matrix $\boldsymbol{P}^*_\eta(N)$ associated with the AR(p) noise model parameters is obtained in the usual manner for AR estimation, i.e.

$$\boldsymbol{P}^*_\eta(N) = \hat{\sigma}^2\left[\sum_{i=1}^{N} \boldsymbol{\psi}(k)\boldsymbol{\psi}^T(k)\right]^{-1} \tag{8.73}$$

Here, the estimate $\hat{\sigma}^2$ of the residual noise variance σ^2 is obtained from the estimated AR model residuals $\hat{e}(k) = \hat{\xi}(k) - \boldsymbol{\psi}^T(k)\,\hat{\boldsymbol{\eta}}(N)$, i.e.

$$\hat{\sigma}^2 = \frac{1}{N}\sum_{k=1}^{N} \hat{e}(k)^2$$

In the case of recursive estimation, the covariance matrices at all recursive steps are obtained in the same manner as (8.72) and (8.73) but with $N = k, k = 1, 2, \ldots, N$.

8.3.5 Additional RIV Algorithms

Some simplifications and additions to the RIVAR algorithm are presented below, and linked to their implementation in the CAPTAIN Toolbox (Appendix G).

1. First, if it is assumed that the noise is zero mean white noise, then AR noise model estimation is not included and the algorithm is called the *Simplified Refined Instrumental Variable* (SRIV) algorithm. This is an extremely useful algorithm in practical terms, being robust to assumptions about the noise and computationally very efficient: as a result, it is useful during the initial stages of model structure identification (section 8.4); for initiation of the RIVAR algorithm rather than LLS/RLS estimation of the ARX model; and for the estimation of deterministic models for control system design. In the CAPTAIN Toolbox, SRIV estimation is available as an option using either the riv or rivbj routines (see item 2).
2. A similar but more complex version of the RIVAR algorithm can be obtained by replacing the AR noise model estimation by full ARMA noise model estimation. In the CAPTAIN Toolbox, the relevant algorithm is called RIVBJ (and the associated function similarly rivbj) to differentiate it from RIVAR (called riv in CAPTAIN) as described above. In RIVBJ, the ARMA estimation can be carried out using the *Instrumental Variable ARMA* (IVARMA) algorithm (Young 2008, 2011a), or an alternative ARMA estimation algorithm, such as the ARMA option of the armax routine from the MATLAB® System Identification Toolbox, if the latter is available to the user. Of course, this RIVBJ algorithm is computationally more expensive than the RIVAR algorithm if used in its full form with the ARMA noise model included. However, it reverts to RIVAR if AR noise is selected by the user; and SRIV if the white noise option is used. Consequently, it is recommended that rivbj is used in preference to riv in CAPTAIN because it has more options.

Example 8.5 SRIV Estimation of a Simple TF model The model (8.28) is an example where the SRIV algorithm produces optimal results because the additive noise is white. Using the same noise as in Example 8.1, the SRIV estimation results are as follows:

$$\begin{aligned}\hat{a}_1 &= -0.4883\ (0.0176)\\ \hat{b}_1 &= 0.5100\ (0.0168)\end{aligned};\quad \text{cov}(\tilde{\mathbf{a}}) = 10^{-3}\begin{bmatrix}0.2507 & 0.2224\\ 0.2224 & 0.2342\end{bmatrix} \tag{8.74}$$

MCS analysis provides a more discerning evaluation of estimation methods than a single run, such as this. Below are the results obtained using such analysis based on 5000 realisations, i.e.

5000 separate SRIV estimation runs, each using the same $u(k)$ and $x(k)$ sequences but with an independent noise sequence $\xi(k) = e(k), \quad k = 1, 2, \ldots, 1000$ added for each realisation.

$$\text{Mean: } \begin{matrix} \hat{a}_1 = -0.4997\ (0.0167) \\ \hat{b}_1 = 0.5003\ (0.0160) \end{matrix}; \text{ Covariance: } \text{cov}\,(\tilde{\boldsymbol{a}}) = 10^{-3} \begin{bmatrix} 0.2790 & 0.2466 \\ 0.2466 & 0.2554 \end{bmatrix} \tag{8.75}$$

As might be expected, since they are based on an ML approach to estimation, these SRIV results are virtually the same as those obtained using the PEM algorithm in the MATLAB® System Identification Toolbox, which is also an ML-based estimation algorithm.

8.3.6 RIVAR and IV4 Estimation Algorithms

The RIVAR algorithm is sometimes confused with the IV4 algorithm in the MATLAB® System Identification Toolbox (Söderström and Stoica 1989, p. 276) when, in fact, they are quite different. As we shall see, the IV4 algorithm is a four-step (*not* an iterative) algorithm and it is not nearly as robust as the RIVAR algorithm. It is based on the following modified ARMAX-type model but with AR rather than moving average noise:

$$A(z^{-1})y(k) = B(z^{-1})u(k-\delta) + \frac{1}{C(z^{-1})}e(k) \tag{8.76a}$$

or

$$y(k) = \frac{B(z^{-1})}{A(z^{-1})}u(k-\delta) + \frac{1}{A(z^{-1})C(z^{-1})}e(k) \tag{8.76b}$$

where we see that the model of the noise $\xi(k)$ is effectively constrained to include the TF denominator polynomial $A(z^{-1})$. Following Söderström and Stoica (1989, p. 276), IV4 is the following four-step algorithm:

1. Initialisation: use LLS or RLS to estimate the parameters in the ARX model.
2. Obtain first stage, ordinary IV estimates $\hat{A}(z^{-1})$ and $\hat{B}(z^{-1})$ of the $A(z^{-1})$ and $B(z^{-1})$ polynomials, with the auxiliary model parameters set to the ARX estimates obtained in (1).
3. Compute the residuals of the IV model obtained in (2), and estimate an AR model of order $n + m$ for these. Prefilter the input and output signals by a moving average filter based on this estimated AR(n+m) polynomial $\hat{C}(z^{-1})$.
4. Obtain the IV estimates based on these prefiltered data.

Note that the prefilter of IV4 is of a high pass, pure moving average type based on $\hat{C}(z^{-1})$, rather than the prefilter used in RIVAR, which is of the ARMA type with the additional AR part based on $\hat{A}(z^{-1})$. Thus the IV4 prefilter will tend to amplify high frequency noise, while the low-pass element in the RIVAR pre-filter will compensate for this and attenuate the high frequency noise components. Such a smoothing effect, based adaptively on the pass-band of the system, is obviously an advantage in estimation terms, as we see in Example 8.6. Finally, note that the RIVAR algorithm is based on the TF model (8.64), where the noise model is only constrained to AR form and so is able to exploit the special properties revealed by the Pierce

Theorem (Appendix H). Moreover, full RIVBJ estimation of the BJ model with ARMA noise is completely unconstrained.

Example 8.6 A Full RIVBJ Example Here, the RIV algorithm is applied to N=1700 samples of simulated data generated by the following BJ TF model:

$$\begin{aligned} y(k) &= \frac{0.016 + 0.026z^{-1} - 0.0375z^{-2}}{1 - 1.6252z^{-1} + 0.642z^{-2}}u(k) + \frac{1 + 0.5z^{-1}}{1 - 0.85z^{-1}}e(k) \\ u(k) &= N(0, 8.8) \qquad e(k) = N(0, 0.0009) \end{aligned} \tag{8.77}$$

This is a 'stiff' dynamic system, with widely spaced eigenvalues, and it has a reasonable noise level (NSR of 0.33 by SD). The TF can be decomposed into a parallel connection of three TFs: one a simple gain, 0.015964; and two with first order TFs having time constants (Appendix B) of 2.6 and 18.7 samples. This is a common model form in the environmental and natural sciences but it appears to pose some problems for the PEM algorithm.

Table 8.1 compares the results of single run and MCS analysis, based on 100 realisations, for the RIVBJ, SRIV, PEM and IV4 algorithms. It also presents the results obtained with the RIV algorithm when the ARMA model is approximated by an AIC identified AR(5) model (i.e.

Table 8.1 Estimation results for the model (8.77) [SR refers to a single run]

Method	Value	$\hat{a}_1$	$\hat{a}_2$	$\hat{b}_0$	$\hat{b}_1$	$\hat{b}_2$	$\hat{c}_1$	$\hat{d}_1$	Fails
	True	−1.6252	0.642	0.016	0.026	−0.0375	−0.85	0.5	
RIVBJ	Estimated	−1.611	0.633	0.0162	0.026	−0.0375	0.845	0.494	
(SR)	(SE)	0.045	0.033	0.0003	0.0007	0.0018	0.010	0.021	
RIVBJ	Estimated	−1.626	0.642	0.0160	0.0260	−0.0375	−0.847	0.501	
(MCS)	(SD)	0.0254	0.0200	0.0002	0.0004	0.0010	0.011	0.021	
SRIV	Estimated	−1.631	0.649	0.0165	0.0256	−0.0379	−0.848	0.485	
(SR)	(SE)	0.0199	0.0159	0.0007	0.0014	0.0009	0.011	0.020	
SRIV	Estimated	−1.616	0.635	0.0159	0.0262	−0.0372	—	—	
(MCS)	(SD)	0.0535	0.0428	0.0005	0.0013	0.0014	—	—	
PEM	Estimated	−1.594	0.620	0.0161	0.0266	−0.0367	−0.849	0.502	
(SR)	(SE)	0.0650	0.0480	0.0003	0.0010	0.0026	0.010	0.021	
PEM	Estimated	−1.617	0.635	0.0160	0.0261	−0.0371	−0.849	0.502	9
(MCS: 91/100)	(SD)	0.0400	0.0300	0.0003	0.0007	0.00164	0.011	0.021	
RIVAR	Estimated	−1.611	0.6328	0.0162	0.0263	−0.0375	—[a]	—	
(SR)	(SE)	0.0440	0.0328	0.0003	0.0007	0.0017			
RIVAR	Estimated	−1.619	0.6371	0.0161	0.0261	−0.0373	—[a]	—	
(MCS)	(SD)	0.0356	0.0270	0.0002	0.0006	0.0014			
IV4	Estimated	−1.568	0.6019	0.0162	0.0270	−0.0357	—	—	
(SR)	(SE)	0.0991	0.0716	0.0003	0.0016	0.0040			
IV4	Estimated	−1.609	0.6300	0.0160	0.0262	−0.0369	—	—	
(MCS)	(SD)	0.0513	0.0384	0.0002	0.0008	0.0021			

[a]High order so estimates not cited here.

RIVAR estimation). In Table 8.1 and later, the estimated SEs from the single runs and the MCS mean SDs are stated below the parameter estimates. As would be expected because of their common basis in ML estimation, both of the RIVBJ and PEM algorithms perform similarly when convergence occurs, with the single run predicted SEs on the parameter estimates matching reasonably the SDs computed from the MCS analysis. However, in this example, the PEM algorithm fails to converge in 9 of the 100 MCS realisations (these realisations were removed in computing the statistics shown in Table 8.1), while the RIVBJ algorithm does not fail at all.

The SRIV algorithm also performs well: in fact, in terms of the single run and MCS estimated mean parameter estimates, it performs better than PEM in the sense that it has no failures. However, we see that its estimated SEs are too optimistic, as might be expected. Also, note that an estimate of an ARMA model for the noise can be obtained by applying the IVARMA algorithm separately to the residual noise estimate obtained from the SRIV estimation results. For illustration, this is shown in Table 8.1 only for the single run case, i.e the row labelled SRIV (SR). Of course, it had no influence on the TF system model parameter estimates and it was not computed at all in the MCS results presented in the following row, SRIV (MCS). Note that this SRIV algorithm is computationally a quite efficient algorithm (a little faster than PEM in this case and only marginally slower than IV4) and so, as pointed out previously, it provides the best algorithm for initial model structure/order identification analysis.

The results in Table 8.1 are typical of the performance comparison in the case of examples such as that considered here. The somewhat poorer performance of the PEM algorithm appears to be due to the 'stiff' nature of the TF model in this example and a consequent failure to converge from the initial conditions specified for the parameter estimates in the PEM gradient optimisation algorithm. It is clear from the results that the RIVBJ algorithm does not suffer from this problem and is always providing statistically consistent and efficient estimates. It is, in other words, another approach to optimal estimation of discrete-time TF models of the BJ type (which includes the ARMAX model as a special constrained case, as pointed out previously). In more general 'non-stiff' examples, however, the results obtained by RIVBJ and PEM are very similar.

We see that the RIVAR results are comparable with those of RIVBJ. They demonstrate how this 'approximate' implementation of the RIVBJ algorithm, in the ARMA noise case, is appealing because it is computationally much more efficient than RIVBJ and yet performs similarly in most cases. For this reason, as pointed out previously, it is has been an implementation used in the CAPTAIN Toolbox for many years. In this case, the IV4 algorithm produces reasonable results but they are noticeably poorer than those of the RIVBJ/RIVAR/SRIV algorithms. However, like them, it has no failures amongst the MCS realisations. This is not always the case, however, as we see in Example 8.7.

Example 8.7 A More Difficult Example (Young 2008) This example is concerned with a simulation model based on a [2 2 4] TF model identified and estimated from the real effective rainfall–flow data shown in Figure 8.3. Figure 8.3b shows the hourly flow $y(k)$ measured in a river over most of a year (7500 h or 312.5 days); while Figure 8.3a shows the associated 'effective rainfall' $u(k)$ (Young 2008). The simulation data are generated by passing this effective rainfall input through the model, with its parameters set to those estimated from the real data. For the purposes of this example, the output is then contaminated by white noise with variance $\sigma^2 = 5$, giving a high NSR (by SD) of 0.62.

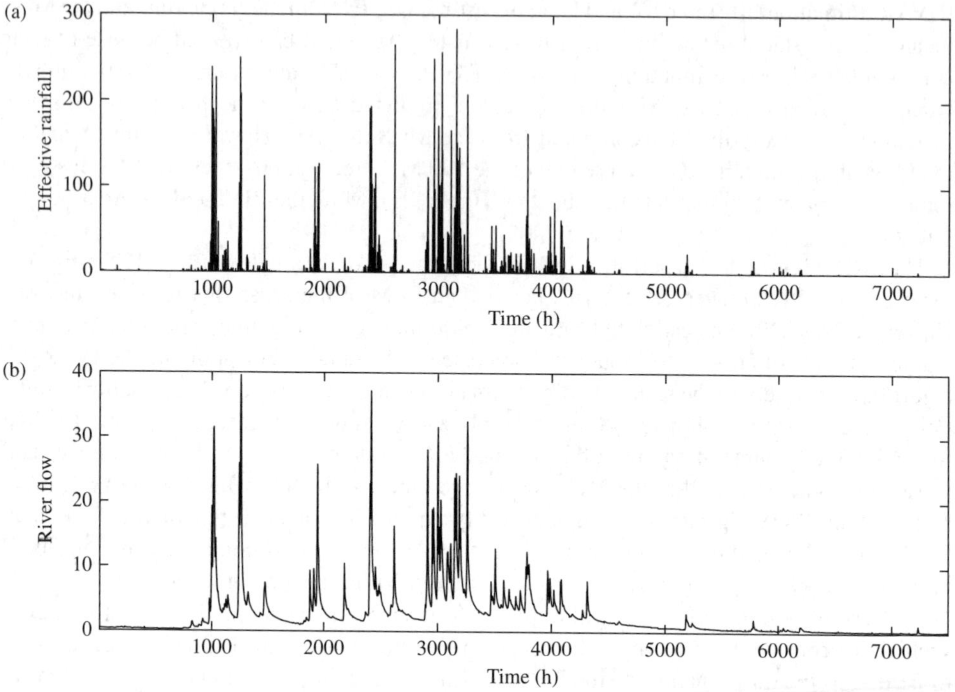

Figure 8.3 Hourly effective rainfall (a) and flow (b) from a UK catchment

Table 8.2 compares the results of the RIVBJ estimation (here effectively SRIV because the additive noise is white) with those obtained using the PEM and IV4 algorithms. Again, as would be expected, both of the RIVBJ and PEM algorithms perform similarly (*when convergence occurs*), with the single run predicted SEs on the parameter estimates matching the SDs computed from the MCS analysis. However, the PEM algorithm has a quite high

Table 8.2 Estimation results for the effective rainfall–flow simulated data in Example 8.7

Method	Value	$\hat{a}_1$	$\hat{a}_2$	$\hat{b}_0$	$\hat{b}_1$	Failures
	True	−1.8563	0.8565	0.0545	−0.0542	
RIVBJ (SR)	Estimated	−1.8575	0.8578	0.0543	−0.0541	—
	(SE)	(0.0028)	0.0027)	(0.0009)	(0.0009)	
RIVBJ (MCS)	Estimated	−1.8560	0.8563	0.0545	−0.0543	0
	(SD)	(0.0027)	(0.0026)	(0.0008)	(0.0008)	
PEM (SR)	Estimated	−1.8561	0.8563	0.0546	−0.0543	—
	(SE)	(0.0028)	(0.0028)	(0.0008)	(0.0008)	
PEM (MCS)	Estimated	−1.8585	0.8587	0.0541	−0.0539	24
	(SD)	(0.0027)	(0.0027)	(0.0008)	(0.0009)	
IV4 (MCS)	Estimated	0.0954	−0.8806	0.0585	0.0481	114
	(SD)	(13.0)	(11.6)	(0.0224)	(0.694)	

failure rate of 19% in the MCS analysis: it fails to converge satisfactorily in 24 of the 124 realisations (these realisations were removed in computing the statistics shown in Table 8.2) while, as in the previous example, the RIVBJ algorithm does not fail at all. The performance of IV4 is much worse: it fails to converge satisfactorily in 114 of the 124 realisations, an unacceptable failure rate of 92%.

This model is an even 'stiffer' dynamic system than the previous example, with time constants of 6.5 and 605 h[7] and PEM problems seem, once again, to be connected with this property. When the PEM algorithm fails to converge on the correct system, it most often converges on a false optimum with one root of the denominator polynomial $A(z^{-1})$ negative and very close to the unit circle; while the other is positive and just greater than 0.9 (a typical example is -0.99943, 0.91657). And, in all cases such as this, the explanation of the data is poor: e.g. the coefficient of determination R_T^2 based on the simulated noise-free output $\hat{x}(k)$, is only ca. 0.85, compared with values very close to unity when the correct convergence occurs (as in all the RIV estimated models).

8.4 Model Structure Identification and Statistical Diagnosis

Model structure identification is an important aspect of data-based model building. In the present linear modelling context, the model structure is defined by the orders of both the system (n, m) and noise (p, q) TF polynomials; and the size of any pure time delay, δ: i.e. the model structure pentad $[n\ m\ \delta\ p\ q]$; see e.g. equation (8.38). An essential adjunct to identification is the statistical diagnosis of the model, in particular the analysis of the model residuals and the evaluation of the model time and frequency response characteristics. Such statistical measures, in themselves, are rarely sufficient to completely and unambiguously identify the 'best' model structure: normally a number of models prove acceptable and it is necessary to look further at the physical nature of the system being modelled. In the context of the present book, the practical utility of the estimated model for model-based control system design is an additional consideration: in general, relatively low order, 'dominant mode' models are required.

An important aspect of model structure identification is that the model should, if at all possible, relate clearly to the physical nature of the system being modelled, where the term 'physical' is interpreted within the particular engineering, scientific or social scientific area under study. The *Data-Based Mechanistic* (DBM) approach to modelling (see Young and Lees 1993; Young 1998, 2011a, b, and references therein; Price *et al.* 1999) emphasises this importance by requiring that the model should only be considered credible and, therefore, fully acceptable, if it also is capable of interpretation in such a physically meaningful manner. This differs from 'grey-box' modelling because it is an inductive, rather than a hypothetico-deductive process (e.g. Young 2011b): in other words, the interpretation of the model in physical terms is carried out *subsequent* to inductive, data-based modelling based on a generic class of models (here linear TF models); in contrast to grey-box modelling, where the model interpretation and structure are assumed mainly prior to data-based modelling.

[7] Note that these time constants are sensitive to the estimated model parameter values and were computed from estimates with more decimal places than those shown in Table 8.2.

8.4.1 Identification Criteria

Model order identification can be a difficult task and can involve subjective judgement based on the experience of the user. With this caveat in mind, model order identification in DBM modelling is based around the *simulation coefficient of determination* R_T^2; the standard *coefficient of determination* R^2; and various model order identification criteria, such as the *Young Information Criterion* (YIC) and AIC, which are defined as follows, where $y(k)$ is the measured output of the system and N is the total number of samples in the data set:

(i)
$$R_T^2 = 1 - \frac{\hat{\sigma}_s^2}{\sigma_y^2} \tag{8.78}$$

where

$$\hat{\sigma}_s^2 = \frac{1}{N}\sum_{k=1}^{k=N}[y(k) - \hat{x}(k)]^2, \text{ in which } \hat{x}(k) \text{ is the simulated model output}$$

$$\sigma_y^2 = \frac{1}{N}\sum_{t=1}^{t=N}[y(k) - \bar{y}]^2; \quad \bar{y} = \frac{1}{N}\sum_{k=1}^{k-N} y(k)$$

(ii)
$$R^2 = 1 - \frac{\hat{\sigma}^2}{\sigma_y^2}$$

where

$$\hat{\sigma}^2 = \frac{1}{N}\sum_{k=1}^{k=N}\hat{e}(k)^2 \tag{8.79}$$

is the residual variance[8]

(iii)
$$\text{YIC} = \log_e \frac{\hat{\sigma}^2}{\sigma_y^2} + \log_e\{\text{NEVN}\}; \textit{ Normalised Error Variance Norm } (\text{NEVN})$$

$$= \frac{1}{np}\sum_{i=1}^{i=np}\frac{\hat{\sigma}^2 . p_{ii}^*}{\hat{a}_i^2} \tag{8.80}$$

(iv)
$$\text{AIC}(np) = N \log_e \hat{\sigma}^2 + 2.np \tag{8.81}$$

Here, in the case where the additive noise is white, or where only the system model parameters are being estimated, $np = n + m + 1$ is the number of parameters in the estimated system model parameter vector $\boldsymbol{\rho}(N)$; p_{ii}^* is the ith diagonal element of the $\boldsymbol{P}^*(N)$ covariance matrix obtained from the estimation analysis (so that $\hat{\sigma}^2 . \hat{p}_{ii}^*$ can be considered as an estimate of the variance of the estimated uncertainty on the ith parameter estimate); and $\hat{a}_i^2$ is the square of the ith parameter estimate in the $\hat{\boldsymbol{\rho}}(N)$ vector. In the case where the full BJ model is being estimated, $np = n + m + p + q + 1$ is the total number of parameters in the system and noise models.

[8] This also defines the variance of the *one step ahead prediction errors*, i.e. $\frac{1}{N}\sum_{k=1}^{k=N}[y(k) - \hat{y}(k|k-1)]^2$.

We see that R_T^2 is a statistical measure of how well the simulated model explains the data: if the variance of the simulated (deterministic) model residuals σ_s^2 is low compared with the variance of the data σ_y^2, then R_T^2 tends towards unity; while if $\hat{\sigma}^2$ is of similar magnitude to σ_y^2 then it tends towards zero (and can become negative). Note, however, that R_T^2 is based on the variance of the model errors $y(k) - \hat{x}(k)$, where

$$\hat{x}(k) = \frac{\hat{B}(z^{-1})}{\hat{A}(z^{-1})} u(k - \delta) \tag{8.82}$$

which is identical to the auxiliary model used in RIV/RIVBJ estimation. This needs to be differentiated from the more conventional coefficient of determination R^2, which is based on the variance of the noise model residual variance, which is identical to the one step ahead prediction errors. This is because R_T^2 is a more discerning measure than R^2 for TF model identification: R^2 can often be quite close to unity even if the model is relatively poor (i.e. it is easy to predict reasonably well only one sample ahead); while R_T^2 shows how much of the measured output is being explained by the *deterministic* output of the system model and will only be close to unity if the noise level on the measured output is low.

The YIC is more a more complex, heuristic criterion (Young 1989) based on the properties of the IPM (Wellstead 1978; Young *et al.* 1980), which is defined below equation (8.46b), and the statistical interpretation of its inverse $\hat{\boldsymbol{P}}(N)$ in (8.73). We see that the first term of YIC is simply a relative measure of how well the model explains the data: the smaller the model residuals the more negative the term becomes. The second term, on the other hand, provides a measure of the conditioning of the IPM, which needs to be inverted when the IV normal equations are solved: if the model is over-parameterised, then it can be shown that the IPM will tend to singularity and, because of its ill-conditioning, the elements of its inverse $\hat{\boldsymbol{P}}(N)$ will increase in value, often by several orders of magnitude. When this happens, the second term in the YIC tends to dominate the criterion function, indicating over-parameterisation. An alternative justification of the YIC can be obtained from statistical considerations (e.g. Young 1989).

Although heuristic, the YIC has proven very useful in practical identification terms over the past 20 years, notably for the estimation of control models. Indeed, the authors have utilised YIC for most of the practical PIP control applications cited in earlier chapters. However, it should not be used as a sole arbiter of model order since, in some cases, it tends to favour low order models that do not provide a sufficiently good explanation of the data compared with other models that have a low, albeit not the lowest, YIC value. For this reason, it should be used to select the best model only from those models that provide a satisfactory explanation of the data (see Example 8.8, Example 8.9 and Example 8.10).

The AIC is a well known identification criterion for AR processes (Akaike 1974) and is related to the *Final Prediction Error* (FPE). It is used to identify the order of AR or ARMA models for the noise process $\xi(k)$, based on the model residuals $\hat{e}(k)$ and the number of model parameters. Here, the first term is a measure of how well the model explains the data; while the second term is simply a penalty on the number of parameters. Thus, as in the YIC, the AIC seeks a compromise between the degree of model fit and the complexity of the model: the smaller the value of the AIC (that is the more negative it is since it is based on a logarithmic measure) the better identified is the model. As we shall see in later examples, it can also be used as an aid in identifying the whole BJ transfer function model.

Following from the publication of Akaike's seminal paper (Akaike 1974), the topic of model order identification criteria has received a lot of attention and several different modifications to the AIC have been suggested. All of these are in the general form (Shibata 1985):

$$T + \alpha.np \tag{8.83}$$

where T is a test statistic and the second term is the penalty term on the number of parameters in the model, with $\alpha > 1$. The test statistic T can take several forms, as discussed by Shibata, but the best known is the maximum log likelihood, where the dominant term is the one step prediction error variance $\hat{\sigma}^2$, as in the AIC defined above. The most controversial point is how to choose α: this is 2 in the case of the AIC, log(N) in the *Bayesian Information Criterion* (BIC) of Schwarz (1978) and c.loglog(N) for some c>2 in the criterion suggested by Hannan and Quinn (1979). We have found that the AIC and BIC both work well, although the AIC is more prone to over-fitting.

8.4.2 Model Structure Identification Procedure

Using the above statistical measures, a generally useful approach to model structure identification is as follows:

1. Use the SRIV estimation algorithm to estimate a range of different models for $\min(n) \leq n \leq \max(n)$ and $\min(m) \leq m \leq \max(m)$, where the maximum and minimum values of n and m are selected by the user, and sort these by the value of R_T^2, so exposing those model structures that best explain the data in a simulation sense. This can be accomplished easily using the rivbjid identification routine in CAPTAIN with the ARMA noise model orders p and q set to zero.
2. Amongst the best models in step 1, select one that has a *relatively* low YIC value: normally this will be a large negative value, since it is a logarithmic measure, but the choice is not critical *provided* the associated R_T^2 is relatively high compared with that of other models. But do not select a model that has a high R_T^2 and a relatively large YIC, since the YIC is then indicating over-parameterisation. Use this selected model to generate an estimate $\hat{\xi}(k)$ of the additive noise $\xi(k)$. Then estimate ARMA(p,q) models for this estimated noise over a range of p and q, $\min(p) \leq p \leq \max(p)$ and $\min(q) \leq q \leq \max(q)$, again with the maximum and minimum values selected by the user, and sort these in terms of the AIC or BIC. In CAPTAIN, this can be accomplished straightforwardly using the ivarmaid routine.
3. Re-estimate the full model, with the ARMA noise model included, over a smaller range of n, m, p and q, based on the results obtained in step 1 and step 2, using the full RIVBJ algorithm and sorting on the basis of the AIC or BIC. Select the best model from the results obtained in this manner.
4. In some cases, particularly if the system is a 'stiff' system characterised by widely spaced eigenvalues, it may be that the model estimated in step 3 provides a poor estimate of the deterministic system output compared with the simpler SRIV estimated output. In this case, it is better to use the SRIV estimate, together with the ARMA noise model estimated in step 2, to define the full stochastic model that is referred to as the SRIV-ARMA model by Young (2011a, p. 213 and the example on pp. 383–388). In this case, the covariance matrix provided by the SRIV algorithm will not necessarily provide a good estimate of the parametric uncertainty.

Typical examples of this model identification process are discussed in the following case studies, including one well-known laboratory example (Example 8.8) and selected control applications from earlier chapters (Example 8.9 and Example 8.10).

Example 8.8 Hair-Dryer Experimental Data For the first example, let us consider a well known demonstration in the MATLAB® System Identification Toolbox, where the data are shown in Figure 8.4. These data are derived from an experiment on a laboratory scale 'hair-dryer' system (see Ljung 1987, p. 440, who refers to this as a 'nice' example). A fan blows heated air through a tube and the air temperature is measured by a thermocouple at the outlet. The input $u(k)$ is the voltage over the heating device, which is just a mesh of resistor wires. The output $y(k)$ is the voltage from the thermocouple at the output. Finally, the sampling interval is $\Delta t = 0.08$ seconds. Model structure identification, based on the the first 500 samples of the data shown in Figure 8.4 and using the approach suggested above, yields the initial SRIV results shown in Table 8.3 (obtained here using the CAPTAIN Toolbox rivbjid routine). This suggests that, based on YIC considerations, the best model structure is [2 2 3 0 0]. However, a number of other models of second and third order perform well and the BIC identifies a [2 3 2 0 0] model, which has the same $R_T^2 = 0.9881$. In addition, the model structure with the highest $R_T^2 = 0.9886$ is [3 2 3 0 0]. So at the next stage in identification, we should consider a tighter range of models that include these as well as noise models.

When AIC and BIC identification is applied to the error series from both the [2 2 3 0 0] and [2 3 2 0 0] models they identify either a first order AR(1) or a second order AR(2) noise

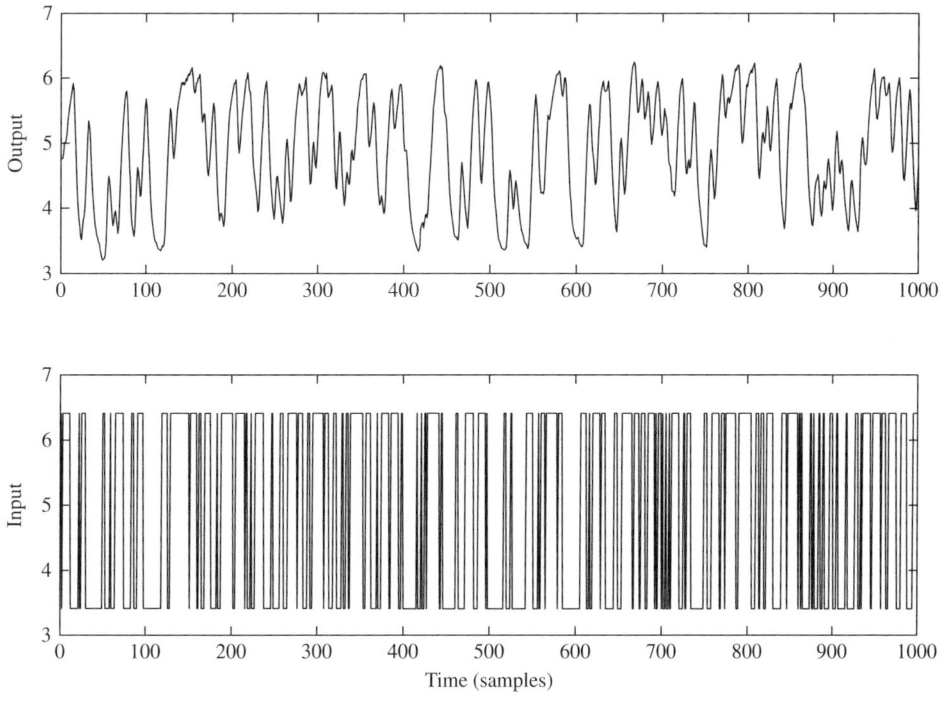

Figure 8.4 Hair-dryer experimental data for Example 8.8

Table 8.3 Initial SRIV structure identification for Example 8.8

Structure	YIC (TF)	R_T^2	$\hat{\sigma}_s^2$	NEVN
[2 2 3 0 0]	−7.806	0.9881	0.008280	−7.894
[2 2 2 0 0]	−6.815	0.9867	0.009273	−8.994
[3 3 3 0 0]	−5.561	0.9883	0.008181	−7.093
[2 3 3 0 0]	−5.056	0.9882	0.008229	−6.452
[3 3 2 0 0]	−4.302	0.9708	0.02042	−6.881
[3 2 3 0 0]	−3.890	0.9886	0.007952	−4.957
[2 3 2 0 0]	−0.783	0.9881	0.008283	−8.002
[3 2 2 0 0]	0.578	0.9884	0.008081	−6.290

model. So the final stage of model identification is applied to all models in the range [2 2 2 0 0] to [3 3 3 2 2] again using the rivbjid routine in CAPTAIN. The best five models identified in this manner by the BIC are shown in Table 8.4.

From Table 8.4, the best identified model structure is [2 3 2 1 0], which defines a general TF model of the form:

$$y(k) = \frac{b_0 + b_1 z^{-1} + b_2 z^{-2}}{1 + a_1 z^{-1} + a_2 z^{-2}} u(k-2) + \frac{1}{1 + c_1 z^{-1}} e(k) \tag{8.84}$$

The associated full RIVBJ estimated parameter estimates are given by:

$$\begin{aligned} \hat{a}_1 &= -1.3145\ (0.016); \quad \hat{a}_2 = 0.4325\ (0.015) \\ \hat{b}_0 &= 0.00173\ (0.0020); \quad \hat{b}_1 = 0.0655\ (0.0029); \quad \hat{b}_2 = 0.0406\ (0.0034) \\ \hat{c}_1 &= -0.9230\ (0.0229); \quad \hat{\sigma}^2 = 0.00142 \end{aligned} \tag{8.85}$$

As we see in Figure 8.5, this model explains the data very well, with a coefficient of determination $R_T^2 = 0.9873$ and the measured output (dots) is always within the 95% (twice the SD) confidence bounds. All the parameters of the system TF are well defined statistically, with SEs much lower than the estimated value.

However, it is necessary to evaluate the model in other regards before it is finally accepted. An important aspect of such evaluation is the investigation of the model errors and the final residuals. The model simulation errors, based on the difference between the measured output and the simulated model output have a near zero mean value, as required but they

Table 8.4 Final RIVBJ structure identification for Example 8.8

Structure	BIC	R_T^2	$\hat{\sigma}_s^2$	NEVN	R^2
[2 3 2 1 0]	−1955	0.9873	0.001421	−9.197	0.998
[3 3 2 1 0]	−1952	0.9879	0.001411	−5.656	0.998
[2 3 3 1 0]	−1951	0.9877	0.001416	−7.755	0.998
[2 3 2 2 0]	−1950	0.9873	0.001422	−9.214	0.998
[2 3 2 1 1]	−1949	0.9873	0.001423	−9.165	0.998

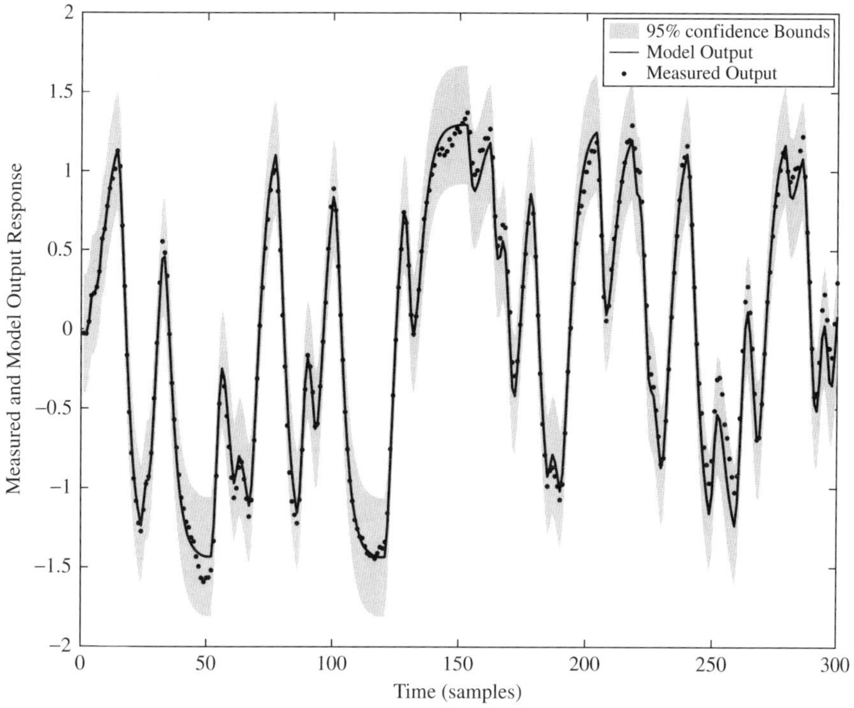

Figure 8.5 Estimation: model output (8.84) compared with the measured dryer data

are quite highly coloured. However, the ACF of the final model residuals $\hat{e}(k)$ (one step ahead prediction errors), resulting from the inclusion of the AR(1) noise model, have no significant lag correlation, as shown by Figure 8.6a. Furthermore, as shown in Figure 8.6b, the residuals are not significantly cross-correlated with the input series $u(k)$, confirming that the model is explaining the data very satisfactorily. The coefficient of determination based on the final residuals is $R_T^2 = 0.9980$ (i.e. 99.8% of the output data are explained by the full stochastic general TF model). The unexplained 0.02% remaining is accounted for by the final residuals, which have the properties of white noise and are, therefore, completely unpredictable.

Validation is important in data-based modelling: the model should not be accepted unless it continues to explain measured data that were not used in its identification. Figure 8.7 shows the validation results: here, the simulated deterministic output of the estimated model (8.84) is compared with the measured output from samples 500 to 900. The results are very good, fully confirming the efficacy of the model, with $R_T^2 = 0.9935$, which is better than that obtained over the estimation data; and the standard coefficient of determination based on the final residuals, which is also marginally better at $R^2 = 0.9983$.

The above results can be compared with those obtained by Ljung (1987) and those available as a demonstration example in the MATLAB® System Identification Toolbox. In both, an ARX model with structure [2 2 3 0 0] is selected, although with the caveat that a number of models provide a similar explanation of the data. This performs similarly to the RIVBJ

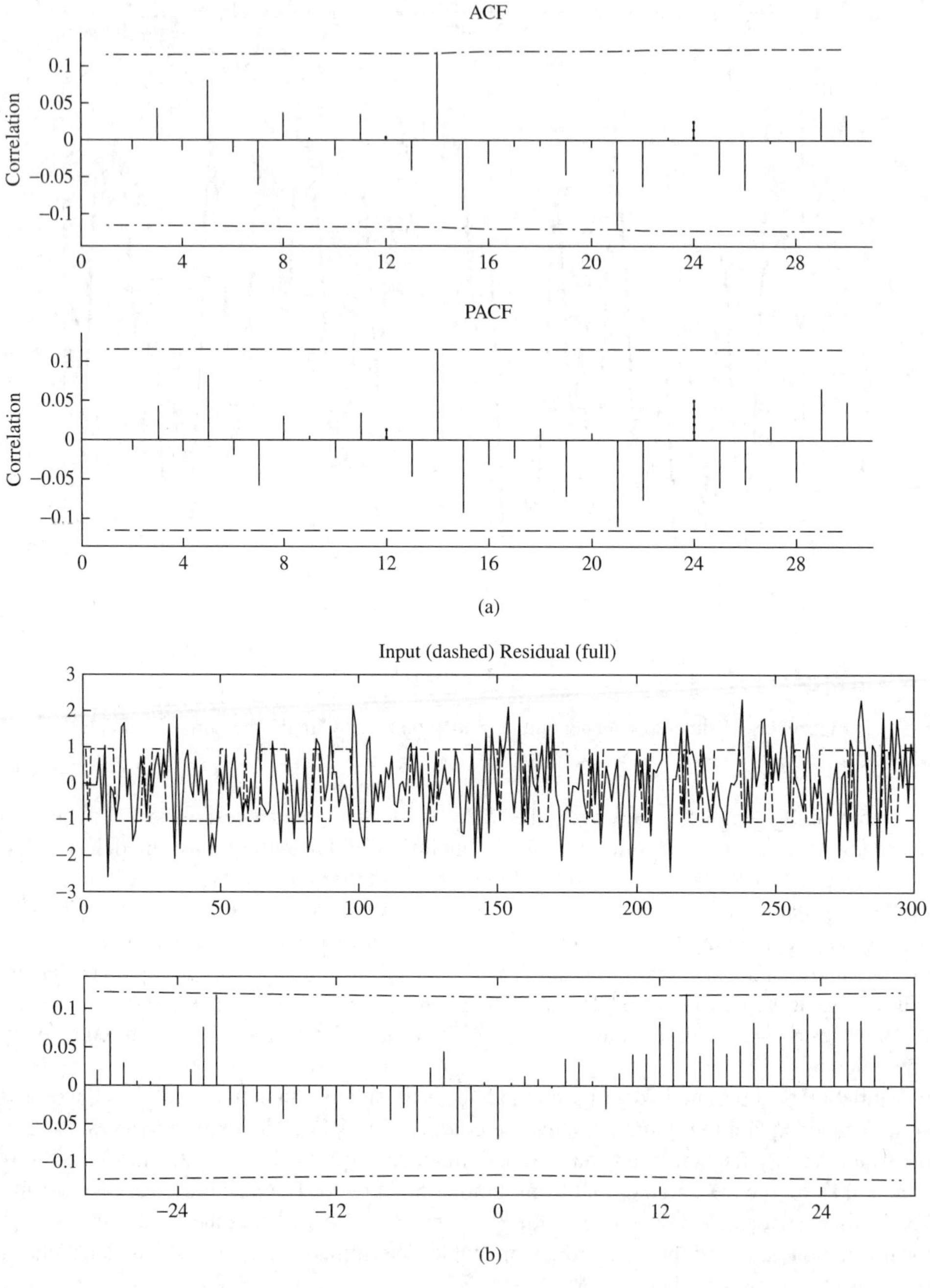

Figure 8.6 (a) Autocorrelation (ACF) and partial autocorrelation (PACF) functions; and (b) cross correlation function for the dryer model residuals in Example 8.8

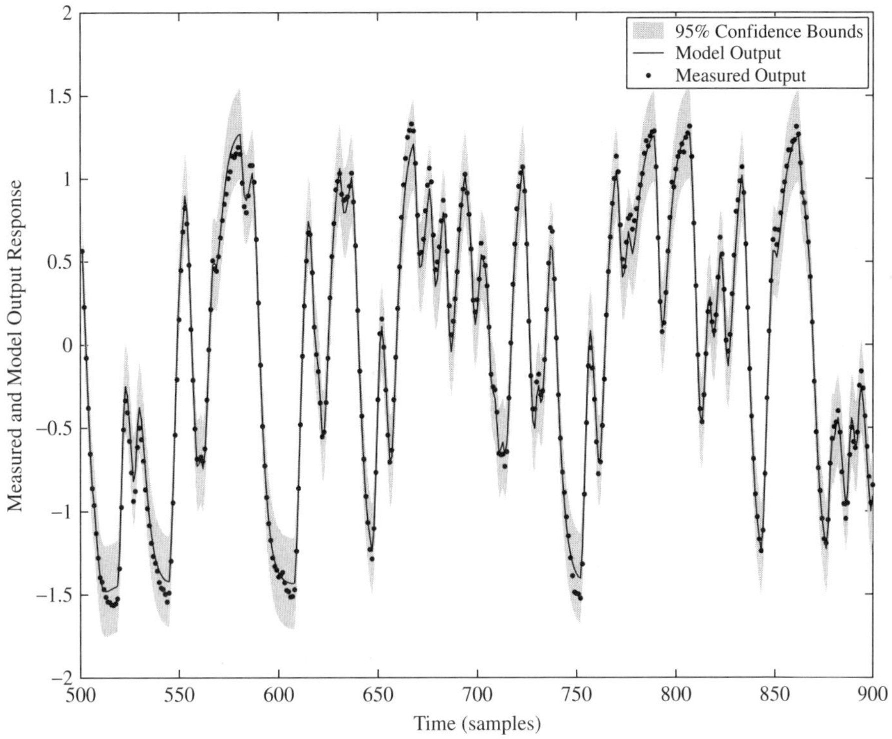

Figure 8.7 Validation: model output (8.84) compared with the validation data for Example 8.8

identified and estimated [2 3 2 1 0] model at the deterministic validation stage of the analysis but, not surprisingly, the ACF/PACF tests show that its residuals are significantly coloured, with the PACF showing a large lag one value of 0.9028, implying a first order AR model for the residuals. Also, the cross correlation test shows that the residual is significantly cross-correlated with the input. This suggests that the ARX model might well be acceptable for certain applications, such as deterministic control system design, but that for gaining understanding about the nature of the system, stochastic control system design or forecasting applications, the [2 3 2 1 0] model identified here would be a much better choice.

Perhaps the most important limitation of ARX models is the statistical inconsistency and noise-induced asymptotic bias on the parameter estimates, particularly when the noise is heavily coloured, as in this case. Here, the poles of the estimated ARX model are at 0.748 and 0.527, indicating associated time constants of 0.107 and 0.152 s. On the other hand, the RIVBJ estimate of the same structure system but with AR(1) noise, i.e. a [2 2 3 1 0] model, has complex poles of $0.657 \pm 0.028j$. Since the complex part is so small, this is effectively two real poles: in fact, the *ddamp* routine in MATLAB® reports two real roots at 0.658; i.e. two equal time constants of 0.122 s. This is confirmed by our preferred identified model (8.84), which has virtually the same estimated poles at $0.657 \pm 0.023j$. Further confirmation is obtained by continuous-time model estimation (see section 8.6, Example 8.13) and constrained RIVBJ estimation with the roots of the denominator polynomial constrained to be real, although not necessarily equal. The latter yields:

$$y(k) = \frac{0.001712 + 0.06534z^{-1} + 0.04120z^{-2}}{(1 - 0.6556z^{-1})^2} u(k-2) + \frac{1}{1 - 0.9215z^{-1}} e(k) \tag{8.86}$$

where $e(k)$ is a zero mean, serially uncorrelated, white noise input with variance $\hat{\sigma}^2 = 0.00142$, the same as that estimated for the unconstrained model (8.84).

Whether the differences in the models discussed above are important in practical terms will depend upon the use of the model. However, given the limitations of the ARX estimation algorithm, there is no doubt, in statistical terms, that we should have more confidence in the RIVBJ estimated model. As a result, the constrained model (8.86), which is more parsimonious than (8.84) with one less parameter, would seem to be the best model to use in initial PIP control system design, although subsequent practical implementation over a wider range of input perturbations might suggest re-estimation and evaluation.

Example 8.9 Axial Fan Ventilation Rate Chapter 2 (Example 2.8) considers the ventilation rate in a 1.0 m^2 by 2.0 m test chamber at Lancaster University (Taylor 2004). Illustrative open-loop experimental data based on a sampling rate $\Delta t = 2$ s are shown in Figure 8.8, where $y(k)$ is the air velocity (m s^{-1}) and $u(k)$ is the control input (the applied voltage to the fan, which is scaled to lie in the range zero to 100%, i.e. full power). It is clear that the noise levels

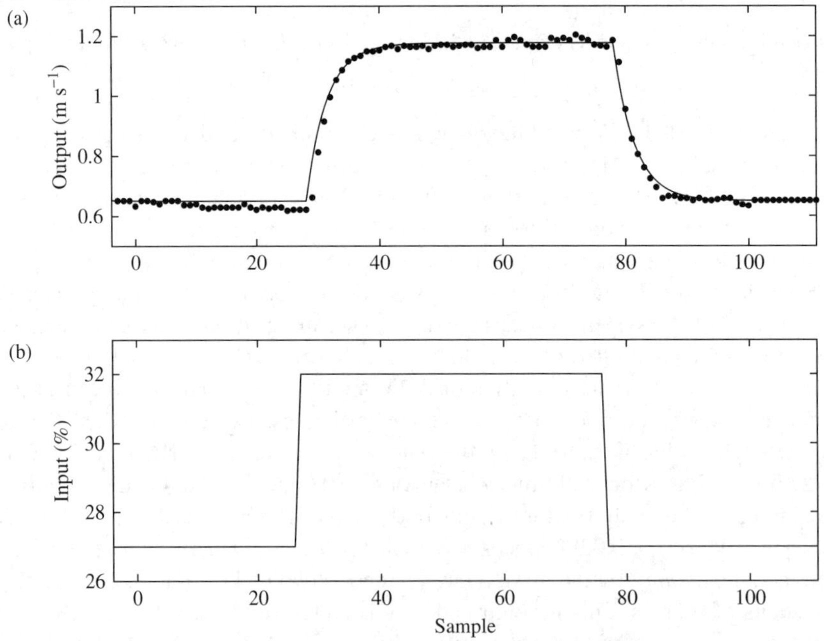

Figure 8.8 Axial fan experimental data in Example 8.9. (a) Ventilation rate (points) and model response (thin trace). (b) Control input. Sampling rate 2 s

are relatively low, especially when the small perturbation of the system for this particular experiment is noted (i.e. the input changes over only 5% of its full range).

In Example 2.8, the model was estimated using the CAPTAIN rivbj routine with the simpler SRIV settings and the model structure was intentionally restricted to [1 1 1] in order to simplify the control system design. Here, however, we will utilise the full rivbj analysis, which identifies a more complicated [1 3 1 4 0] model structure, based on the smallest BIC value:

$$y(k) = \frac{b_1 z^{-1} + b_2 z^{-2} + b_3 z^{-3}}{1 + a_1 z^{-1}} u(k) + \frac{1}{1 + c_1 z^{-1} + c_2 z^{-2} + c_3 z^{-3} + c_4 z^{-4}} e(k);$$

$$e(k) = \mathcal{N}(0, \sigma^2)$$

The estimated parameters and SEs are as follows:

$$\hat{a}_1 = -0.682(0.0063);\ \hat{b}_1 = -0.00177(0.0009);\ \hat{b}_2 = 0.0121(0.0015);\ \hat{b}_3 = 0.0230(0.0012)$$

$$\hat{c}_1 = -0.363(0.053);\ \hat{c}_2 = -0.137(0.055);\ \hat{c}_3 = -0.155(0.055);\ \hat{c}_4 = -0.116(0.053);$$

$$\hat{\sigma}^2 = 8.6 \times 10^{-5}$$

These estimates and the associated step response, suggests a more complicated initial response that includes a small non-minimum phase effect. The model explains the data very well, as can be seen from Figure 8.8, with a high $R_T^2 = 0.997$. The auto and cross correlation functions show that the estimates $\hat{e}(k)$ of the white noise $e(k)$ are serially uncorrelated and not correlated with the input signal $u(k)$, as required. By contrast and not surprisingly given the need for an AR(4) noise process in the above model, the SRIV estimated residuals of the first order [1 1 1] model are both serially correlated and significant correlated with $u(k)$, so the model is clearly not optimal in statistical terms. Nevertheless, it was clearly good enough for PIP control system design, showing how robust the PIP control system is to modelling errors. The above model is based on an illustrative operating level of ≈ 1 m s^{-1}. Similar models have been used for other operating conditions and have provided an excellent basis for successful PIP control system design (e.g. Taylor *et al.* 2004a,b).

Example 8.10 Laboratory Excavator Bucket Position The laboratory excavator bucket position discussed in Chapter 2 (Example 2.1) and Chapter 5 (Example 5.2) is a little unusual, since the system essentially behaves as a first order integrator. In fact, preliminary data-based analysis using the methods discussed above, suggests that an adequate description of the excavator bucket position is generally given by equation (8.1a) with $n = \delta = 1$, $m = 0$ and $a_1 \approx -1.0$. This is because the joint angle $y(k)$ changes at a relatively constant rate for a steady input voltage $u(k)$. Furthermore, the normalised voltage has been calibrated so that there is no movement of the manipulator when the input is zero. The deterministic control model is, therefore:

$$y(k) = \frac{b_1 z^{-1}}{1 - z^{-1}} u(k) \tag{8.87}$$

Assuming that the value of a_1 is known a priori to be very close to -1.0, it is possible to estimate b_1 in equation (8.87) using the recursive option of the rivbj routine in CAPTAIN.

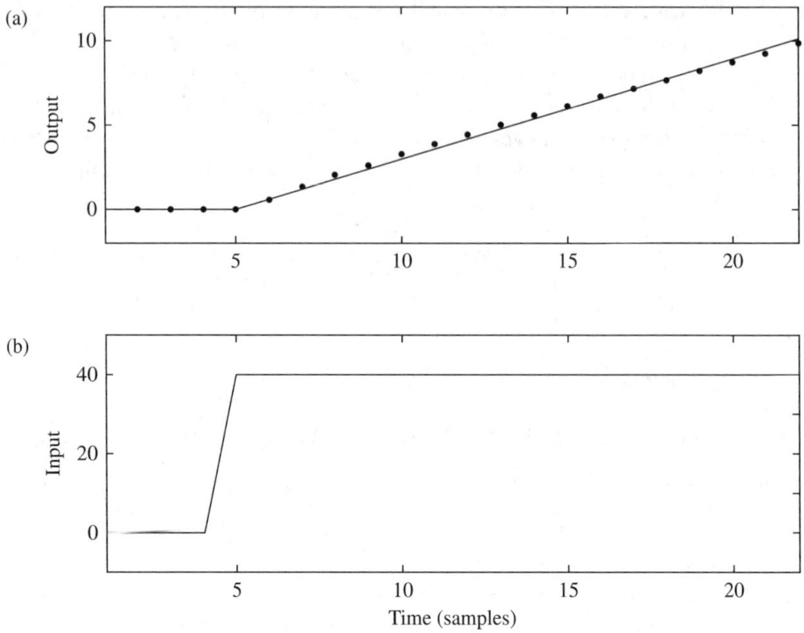

Figure 8.9 Open-loop experimental data collected from the laboratory excavator in Example 8.10. (a) Joint angle data (points) and model response (solid trace) in degrees. (b) Control input (scaled voltage signal). Sampling rate 0.1 s

This allows the user to specify the a priori conditions on the estimated parameter vector and its associated covariance matrix. In this case, these were set to $\hat{\boldsymbol{\rho}}(0) = [-0.99999 \;\; 0]$ and $\boldsymbol{P}(0) = \text{diag}[0 \;\; 10^6]$, so that $\hat{a}_1(k)$ does not change at all from its initially specified -0.99999 value because the associated variance $p_{11}(k)$ in $\boldsymbol{P}(k)$ is maintained at the its initially set zero value. However, the value of $p_{22} = 1 \times 10^6$ allows the algorithm to generate the recursive estimate $\hat{b}_1(k),\;\; k = 1, 2, \ldots, N$. Of course, equation (8.87) is an approximation of the actual system, as always, but similar models have been identified and successfully used for the control of this and other hydraulically actuated manipulators (Shaban *et al.* 2008). An illustrative experiment and model response are shown in Figure 8.9.

For the illustrated data set, the estimated b_1 parameter and its SE obtained using the SRIV settings in rivbj is $\hat{b}_1 = 0.01489\,(0.00012)$ and $R_T^2 = 0.997$, reflecting the good explanation of the data in Figure 8.9. However, further analysis suggests that the estimated value of b_1 can vary depending on the applied voltage used in the experiment. In fact, as discussed by Taylor *et al.* (2007b), b_1 is best represented as a graphical function of this voltage, an example of a *state-dependent* parameter. Further discussion of the concept is deferred until Chapter 9. At this juncture, it is sufficient to point out that, for the purposes of the linear PIP control experiments considered in Example 5.2, an optimised value for b_1 was used, based on several open-loop experiments at a range of different operating levels (i.e. voltages) and with PRBS inputs. The details of this latter analysis (and related state-dependent estimation) are somewhat peripheral to the present discussion and so are omitted (Shaban 2006).

8.5 Multivariable Models

The most general multivariable TF model is the stochastic version of the MIMO representation given previously by equation (7.9) with an additive noise vector, i.e.

$$\boldsymbol{y}(k) = \boldsymbol{G}(z^{-1})\boldsymbol{u}(k) + \boldsymbol{\xi}(k) \tag{8.88}$$

where

$$\begin{aligned} \boldsymbol{y}(k) &= [y_1(k)\; y_2(k) \cdots y_p(k)]^T \\ \boldsymbol{u}(k) &= [u_1(k)\; u_2(k) \cdots u_r(k)]^T \\ \boldsymbol{\xi}(k) &= [\xi_1(k)\; \xi_2(k) \cdots \; \xi_p(k)]^T \end{aligned} \tag{8.89}$$

Methods for estimating such models are available (Jakeman and Young 1979). From an estimation point of view, however, it is simpler to consider the MISO models representing each row of (8.88). In general, such a model can be written in the form (where time delays have been explicitly introduced for generality):

$$y_i(k) = \sum_{j=1}^{j=r} \frac{B_{ij}(z^{-1})}{A_{ij}(z^{-1})} u_j(k - \delta_j) + \xi_i(k) \tag{8.90}$$

where $\xi_i(k)$ is a coloured noise process and $i = 1, \ldots, p$. In the case where the noise has rational spectral density, this can be considered as an ARMA(p,q) process, i.e.

$$y_i(k) = \sum_{j=1}^{j=r} \frac{B_{ij}(z^{-1})}{A_{ij}(z^{-1})} u_j(k - \delta_j) + \frac{D(z^{-1})}{C(z^{-1})} e_i(k) \tag{8.91}$$

where $e_i(k)$ is a zero mean, serially uncorrelated, white noise process. This is the MISO version of the BJ model.

8.5.1 *The Common Denominator Polynomial MISO Model*

If the MIMO model (8.88) has been obtained from a state-space representation of the system, then all of the elemental TFs will have a common denominator polynomial (as is the case for the examples in Chapter 7), i.e.

$$\boldsymbol{y}(k) = \frac{1}{A(z^{-1})} \boldsymbol{G}_n(z^{-1})\boldsymbol{u}(k) + \boldsymbol{\xi}(k) \tag{8.92}$$

where $\boldsymbol{G}_n(z^{-1})$ is the associated matrix of numerator polynomials. In this case, the MISO model (8.90) becomes:

$$y_i(k) = \sum_{j=1}^{j=r} \frac{B_{ij}(z^{-1})}{A(z^{-1})} u_j(k - \delta_j) + \xi_i(k) \tag{8.93}$$

with a similar change to (8.91). One potential identifiability problem with this common denominator MISO model form is that, in general, it will involve some elemental TFs that have pole-zero cancellations. However, if this is not the case, it is the simplest model form to identify and estimate. In fact, in this form, the RIVBJ estimation algorithm is the same as in the SISO case: all that is required is to extend the relevant data vectors to include the additional input terms. For instance the ith vector becomes:

$$\begin{aligned}\boldsymbol{\phi}_i^T(k) = [&-y_i(k-1) - \cdots - y_i(k-n)\; u_{i1}(k-\delta_i)\cdots u_{i1}(k-\delta_i-m_i) + \cdots \\ &+ u_{ir}(k-\delta_i)\cdots u_{ir}(k-\delta_i-m_i)]\end{aligned} \tag{8.94}$$

with similar changes to all other data vectors, including the IV vector. And, of course, in this MISO environment, the IV auxiliary model will also be a common denominator MISO model.

Example 8.11 Multivariable System with a Common Denominator A simple example of the common denominator model identification and estimation is the following first order model:

$$\begin{bmatrix} y_1(k) \\ y_2(k) \end{bmatrix} = \frac{1}{1-0.5z^{-1}} \begin{bmatrix} z^{-1} & 2z^{-1} \\ z^{-1} & z^{-1} \end{bmatrix} \begin{bmatrix} u_1(k) \\ u_2(k) \end{bmatrix} + \begin{bmatrix} \xi_1(k) \\ \xi_2(k) \end{bmatrix} \tag{8.95}$$

or, in MISO form:

$$\begin{aligned} y_1(k) &= \frac{1}{1-0.5z^{-1}} u_1(k-1) + \frac{2}{1-0.5z^{-1}} u_2(k-1) + \xi_1(k) \\ y_2(k) &= \frac{1}{1-0.5z^{-1}} u_1(k-1) + \frac{1}{1-0.5z^{-1}} u_2(k-1) + \xi_2(k) \end{aligned} \tag{8.96}$$

where $u_1(k)$ and $u_2(k)$ are independent PRBS sequences with a switching period of 10; while $\xi_1(k)$ and $\xi_2(k)$ are additive measurement noise inputs, in the form of zero mean, white noise sequences with variances 0.55 and 0.20, respectively. For both MISO models, the resulting NSR by SD is 0.35 (35% noise) and a sample of the output data is shown in Figure 8.10.

MCS results for both MISO channels, based on 100 realisations, each with a sample size of 2000 samples, are compared with the results from an illustrative single estimation run in Table 8.5 and Table 8.6. It is clear that the RIVBJ estimates are well defined and that estimated SEs from the single runs are consistent with the MCS means and SDs. The single run results are good, with the simulation coefficient of determination, based on the error between the deterministic model output and the noise-free output data, very close to unity. For comparison, Table 8.5 and Table 8.6 also show the equivalent results obtained using the PEM algorithm in the MATLAB® System Identification Toolbox. To obtain the PEM results, each MISO model is best estimated in the following form[9]:

$$(1 - a_1 z^{-1}) y(k) = b_1 u_1(k) + b_2 u_2(k) + (1 - a_2 z^{-1}) \xi_i(k) \tag{8.97}$$

since the implementation of the PEM algorithm precludes direct estimation in the common denominator form of equation (8.96). Since it is not then possible to constrain the estimate of

[9] An alternative is the different denominator MISO model form but this also requires the estimate of one more parameter.

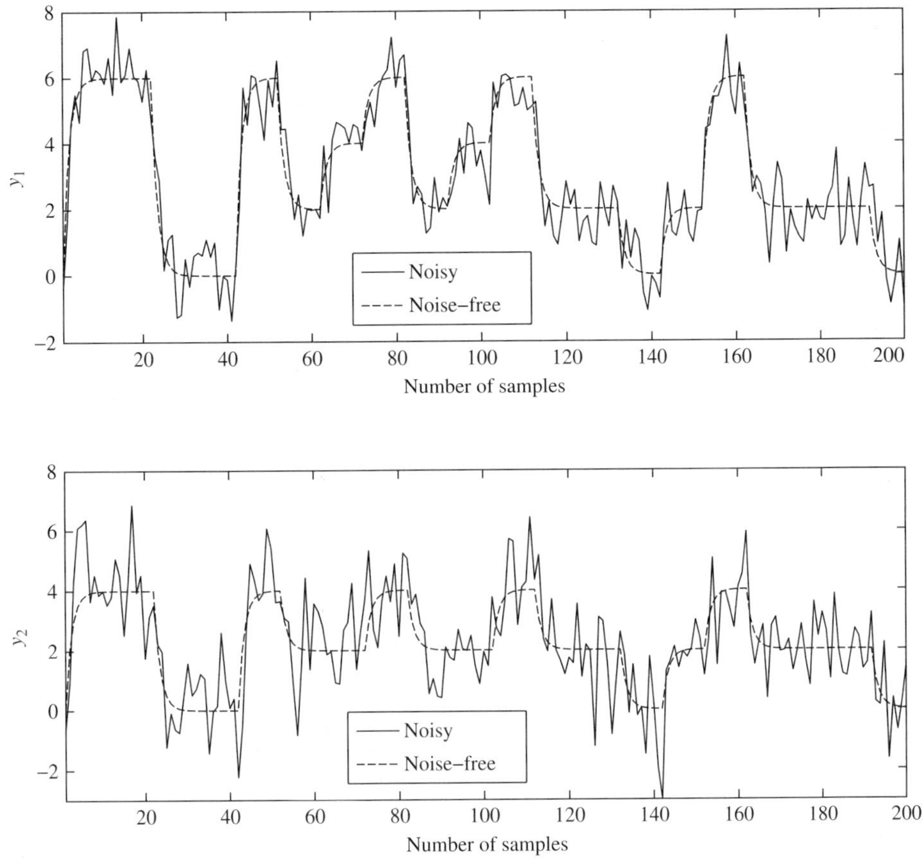

Figure 8.10 Representative section of the data used in Example 8.11

Table 8.5 RIV and PEM results for Example 8.11 MISO Channel 1: $y_1(k)$

Parameter	a_{11}	b_{11}	b_{12}
True	−0.5	1.0	2.0
RIVBJ: mean	−0.49917	1.0047	2.0024
RIVBJ: SD	0.01096	0.027517	0.043935
RIVBJ: single run	−0.4978	1.0299	2.0117
RIVBJ: SE	0.010358	0.026457	0.041748
PEM: mean	−0.50187	0.99844	1.9922
PEM: SD	0.011316	0.028189	0.045492
PEM: single run	−0.49552	1.0435	2.0125
PEM: SE	0.010049	0.02847	0.044774

Table 8.6 RIV and PEM results for Example 8.11 MISO Channel 2: $y_2(k)$

Parameter	a_{21}	b_{21}	b_{22}
True	−0.5	1.0	1.0
RIVBJ: mean	−0.49708	1.0058	1.0079
RIVBJ: SD	0.009498	0.020543	0.021111
RIVBJ: single run	−0.50318	0.99114	0.99291
RIVBJ: SE	0.010349	0.022151	0.022032
PEM: mean	−0.4996	1.0001	1.0025
PEM: SD	0.0090982	0.020326	0.020666
PEM: single run	−0.49383	1.0094	1.0211
PEM: SE	0.010845	0.023259	0.023094

a_2 in (8.97) to be the same as that of a_1, this involves the estimation of four rather than three parameters. The resulting estimates of a_2 in (8.97) are not shown in Table 8.5 and Table 8.6 but, in each case, they are close but not identical to the estimate of a_1.

8.5.2 *The MISO Model with Different Denominator Polynomials*

The MISO model with different denominator polynomials can be estimated using the PEM algorithm in the MATLAB® System Identification Toolbox. However, it presents a more difficult estimation problem in the case of the RIV algorithm type. The RIV algorithm in this case was first suggested by Young and Jakeman (1980) and Jakeman *et al.* (1980), whilst a recent description of its RIVBJ successor is given in section 7.6 of Chapter 7 in Young (2011a). The CAPTAIN routine (Taylor *et al.* 2012) for this algorithm is denoted by rivdd and this is used in Example 8.12.

Example 8.12 Multivariable System with Different Denominators A simple first order example of different denominator model identification and estimation is the following MISO multivariable model:

$$y(k) = \frac{1}{1 - 0.5z^{-1}} u_1(k-1) + \frac{2}{1 - 0.9z^{-1}} u_2(k-1) + \xi(k) \tag{8.98}$$

Here, as in the previous example, $u_1(k)$ and $u_2(k)$ are independent PRBS sequences with a switching period of 10; while $\xi(k)$ is an additive, coloured measurement noise input generated as the output of the following ARMA(1,1) model:

$$\xi(k) = \frac{0.1}{1 - 0.95z^{-1}} e(k) \tag{8.99}$$

where $e(k)$ is a zero mean, white noise input sequence with variance 4.5. The resulting noise is highly coloured, as can be seen in the section of the data shown in Figure 8.11, and the NSR by SD is 0.35.

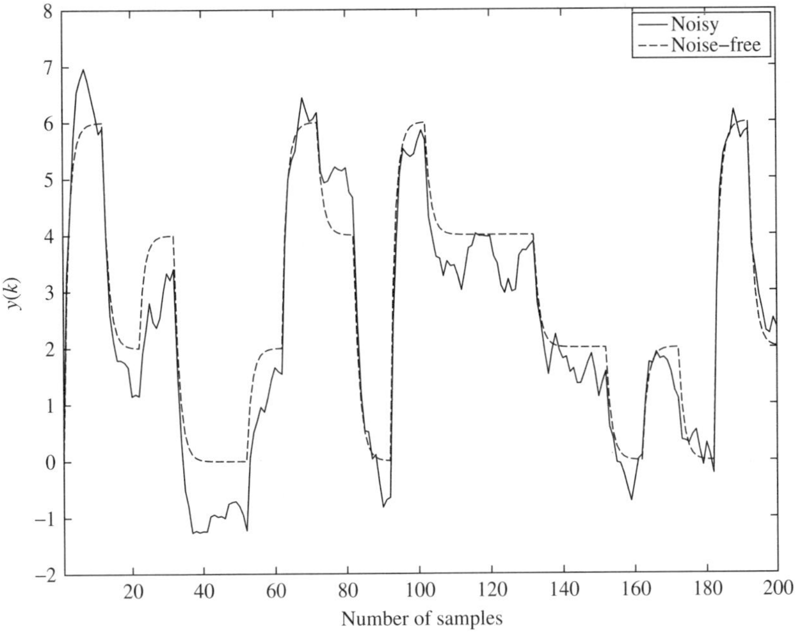

Figure 8.11 A representative section of the noisy output used in Example 8.12 compared with the noise-free output, showing the heavily coloured nature of the noise

MCS results for the RIV and PEM algorithms, based on a 100 realisations, each with a size of 2000 samples, are compared with the results from a single estimation run in Table 8.7. Once again, as in the previous common denominator example, it is clear that the estimates are well defined and that the estimated SEs from the single runs are consistent with the MCS means and SDs. The comparative PEM results in Table 8.8 were obtained using the standard MISO model form with different denominators.

Finally, Figure 8.12a compares the deterministic output of the RIV estimated model with the noise-free output measurement and it is clear that they are hardly distinguishable, with a coefficient of determination $R_T^2 = 0.999$. By contrast, the ARX estimated model has a poor

Table 8.7 RIVBJ results for different denominator MISO model in Example 8.12

Parameters	a_{11}	a_{12}	b_{11}	b_{12}	c_1	d_1
True	−0.5	−0.9	1.0	2.0	−0.95	0.1
RIVBJ: mean	−0.4929	−0.9000	0.9951	2.003	−0.9478	0.1020
RIVBJ: SD	0.0433	0.0029	0.0658	0.0399	0.0073	0.0270
RIVBJ	−0.5059	−0.8996	1.0062	2.0244	−0.95	0.1163
RIVBJ: SE	0.0390	0.0027	0.0587	0.0366	0.0076	0.0241

Table 8.8 PEM results for different denominator MISO model in Example 8.12

PEM: mean	−0.4952	−0.9000	0.9930	2.006	−0.94762	0.1022
PEM: SD	0.0518	0.0034	0.0717	0.0443	0.0072	0.0268
PEM	−0.5152	−0.8987	0.9324	2.0582	−0.9518	0.1203
PEM: SE	0.0434	0.0029	0.0603	0.0385	0.0075	0.0238

fit to the noise-free data and its parameter estimates (not shown) are heavily biased, as would be expected.

8.6 Continuous-Time Models

At first sight, the reader might find it strange to see a section on continuous-time model identification and estimation included in a book on true digital control. However, it is sometimes more convenient to model a system in continuous-time terms since the model can then be converted to a discrete-time form with *any* selected sampling interval Δt. This clearly adds flexibility to the design process since it is often not clear what sampling interval will be best for practical control system implementation and the closed-loop performance needs to be assessed

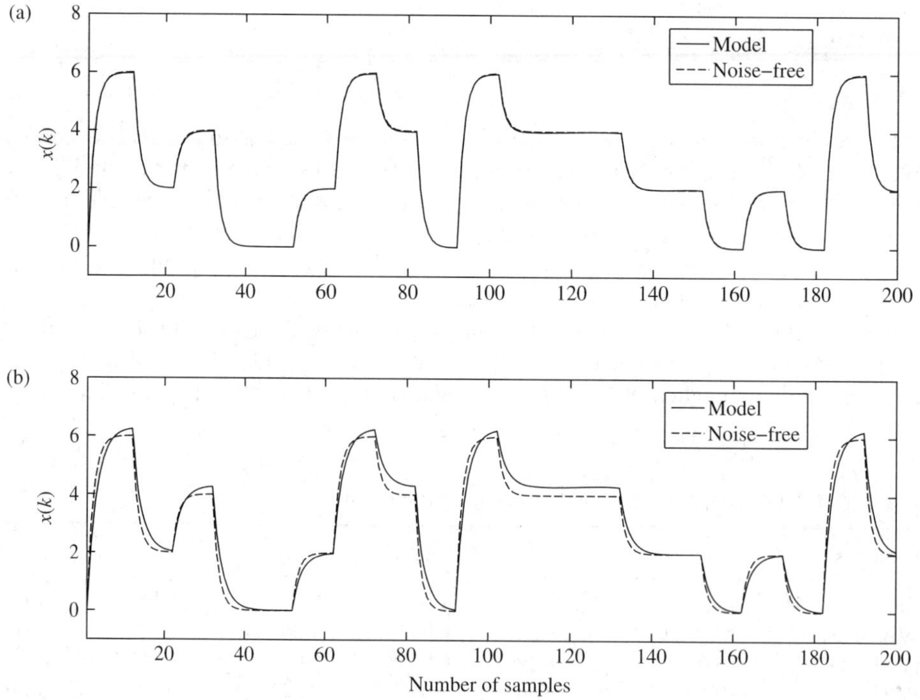

Figure 8.12 Comparison of (a) RIV (solid trace) and (b) ARX (solid trace) estimated model outputs with the noise-free output (dashed trace) for Example 8.12

in this regard. Moreover, continuous-time models have other advantages (see also Garnier and Young 2012):

1. There is only one continuous-time model, regardless of the sampling interval, whereas the parameters of the discrete-time model are a function of Δt.
2. Most models of physical systems are obtained by the application of physical laws (e.g. conservation of mass, energy, etc.) that are normally formulated in continuous-time terms. Because of this, the parameters of the continuous-time model normally have a clear physical significance which is independent of the sampling interval and which is useful when communicating with practitioners who are normally familiar with such parameters and their meaning.
3. If the data are sampled rapidly in relation to the dominant time constants of the system, the discrete-time model estimation can be very poor because the eigenvalues of the discrete-time model will be close to the unit circle in the complex z-plane. As a result, the parameter estimates can be poorly defined[10], particularly if the system is 'stiff', with widely spaced eigenvalues (see Example 8.6 and Example 8.7). By contrast, the continuous-time model estimates are advantaged by rapid sampling.
4. Continuous-time models can be estimated from non-uniformly sampled data and can include 'fractional' time delays where the time delay is not an integral number of sampling intervals (e.g. Young 2006; Ahmed *et al.* 2008).

8.6.1 The SRIV and RIVBJ Algorithms for Continuous-Time Models

The RIV approach to model identification and estimation is, as far as the authors are aware, the only discrete-time TF estimation algorithm that can be applied straightforwardly to continuous-time models, with only fairly minor modifications to the definition of the vectors and the iterative prefiltering procedures. The *Simplified Refined Instrumental Variable method for hybrid Continuous-time models* (SRIVC) was first suggested and evaluated by Young and Jakeman (1980), under the simplifying theoretical assumption of additive white noise. More specifically, in the SISO case, it is concerned with the estimation of the following multi-order continuous-time differential equation model:

$$\frac{d^n x(t)}{dt^n} + \alpha_1 \frac{d^{n-1} x(t)}{dt^{n-1}} + \cdots + \alpha_n x(t) = \beta_0 \frac{d^m u(t-\tau)}{dt^m} + \cdots + \beta_m u(t-\tau) \qquad (8.100)$$

or, using simpler nomenclature where the bracketed superscript denotes the differentiation order:

$$x^{(n)}(t) + \alpha_1 x^{(n-1)}(t) + \cdots + \alpha_n x^{(0)}(t) = \beta_0 u^{(m)}(t-\tau) + \cdots + \beta_m u^{(0)}(t-\tau) \qquad (8.101)$$

Here, the pure time delay τ, in time units, is often assumed to be an integer number related to the sampling time, as in the discrete-time case: that is $\tau = \delta\Delta t$, but this is not essential. In

[10] The data could be sub-sampled by decimation but this clearly involves a loss of information on any higher frequency dynamic modes.

this continuous-time environment, 'fractional' time delays can be introduced if required (see earlier cited references). In TF terms, equation (8.100) or equation (8.101) take the form:

$$x(t) = \frac{B(s)}{A(s)} u(t - \tau) \tag{8.102}$$

where

$$B(s) = \beta_0 s^m + \beta_1 s^{m-1} + \cdots + \beta_m; \quad A(s) = s^n + \alpha_1 s^{n-1} + \cdots + \alpha_n \tag{8.103}$$

and s is the differential operator, i.e. $s^p x(t) = d^p x(t)/dt^p$. It is now assumed that the input signal $u(t),\ t_1 < t < t_N$, is applied to the system and that this input and the output $x(t)$ are sampled at discrete times t_k, not necessarily uniformly spaced.

In the case of uniformly sampled data at a sampling interval Δt, the measured output $y(t_k)$, where $t_k = k\Delta t$, is assumed to be corrupted by an additive measurement noise:

$$y(t_k) = x(t_k) + e(t_k) \quad e(t_k) = \mathcal{N}(0, \sigma^2) \tag{8.104}$$

where the argument t_k indicates that the associated variable is sampled[11] at time t_1 to t_N and, as shown, $e(t_k)$ is a zero mean, normally distributed, white noise sequence with variance σ^2. Combining the continuous-time model equation with the discrete-time observation equation, the complete theoretical model takes the following hybrid form:

$$\begin{aligned} x(t) &= \frac{B(s)}{A(s)} u(t - \tau) && \text{(i)} \\ y(t_k) &= x(t_k) + e(t_k) && \text{(ii)} \end{aligned} \tag{8.105}$$

As in the case of the SRIV algorithm, the statistical model (8.105) is assumed for theoretical purposes and, if its assumptions are correct in any practical situation, then the SRIVC algorithm yields statistically optimal estimates of the model parameters that are both consistent and asymptotically efficient. However, even if the assumptions about the noise $e(t_k)$ are not satisfied, the estimates remain consistent because of the IV implementation. Moreover, practical experience with the SRIVC algorithm shows that the estimates are relatively efficient, that is they normally have relatively low variance.

The more complex RIVCBJ algorithm (Young 2008; Young *et al.* 2008) is based on the following hybrid theoretical model:

$$\begin{aligned} x(t) &= \frac{B(s)}{A(s)} u(t - \tau) && \text{(i)} \\ \xi(t_k) &= \frac{D(z^{-1})}{C(z^{-1})} e(t_k) \quad e(t_k) = \mathcal{N}(0, \sigma^2) && \text{(ii)} \\ y(t_k) &= x(t_k) + \xi(t_k) && \text{(iii)} \end{aligned} \tag{8.106}$$

[11] Note that in the uniform sampling situation, $y(t_k)$ means the same as $y(k)$ and this changed nomenclature for sample variables is utilised simply to emphasise the continuous-time nature of the estimation problem.

where the white additive noise has been replaced by a discrete-time ARMA noise process of the same kind as that assumed previously.

In fact, the RIVCBJ algorithm is very similar to the RIVBJ algorithm: the steps in the algorithm are exactly the same and the only differences are the use of the continuous-time TF model of the system to generate the source of the instrumental variables, and the introduction of hybrid prefilters in the iterative prefiltering procedure. The initial prefiltering by $1/\hat{A}(s)$ is carried out in continuous-time (e.g. exploiting the numerical integration facilities in MATLAB®), while the inverse noise model prefiltering is carried out in discrete-time, based on the sampled output from the continuous-time prefilters.

The continuous-time prefilters now have an additional advantage and role: the inputs to the integrators that appear in the implementaion of the prefilter $1/\hat{A}(s)$ are clearly the time derivatives of their outputs and are precisely the variables that are required to define the nth derivative of the measured, prefiltered output $y_f^{(n)}$ and the vector $\boldsymbol{\phi}^T$ in the estimation model which, in this case, takes the form:

$$y_f^{(n)}(t_k) = \boldsymbol{\phi}^T(t_k)\boldsymbol{\rho}_c + e(t_k) \tag{8.107}$$

where

$$\begin{aligned} \boldsymbol{\phi}^T(t_k) &= \left[-y_f^{(n-1)}(t_k) \cdots - y_f^{(0)}(t_k)\ u_f^{(m)}(t_k - \tau) \cdots u_f^{(0)}(t_k - \tau) \right] \\ \boldsymbol{\rho}_c &= [\alpha_1\ \ldots\ \alpha_n\ \beta_0\ \ldots\ \beta_m]^T \end{aligned} \tag{8.108}$$

Further details of the SRIVC and RIVCBJ algorithms are given in Young (2008, 2011a) and Young *et al.* (2008). They are implemented in the rivcbj and rivcbjid routines in CAPTAIN; while similar routines are available in the CONTSID Toolbox[12]. A continuous-time algorithm for estimating MISO models with different denominator polynomials, using a similar back-fitting procedure to that used for discrete-time systems (section 8.5) was developed by Garnier *et al.* (2007). Other related publications that provide additional information on continuous-time model estimation and its applications are the edited books by Garnier and Wang (2008) and Wang and Garnier (2011), which contain numerous contributions; Young and Garnier (2006) show how these algorithms can be used to model linear and *state-dependent parameter* nonlinear systems in environmental applications; and Laurain *et al.* (2010) also consider how they can be used for modelling related *linear parameter varying* nonlinear systems.

Example 8.13 Continuous-Time Estimation of Hair-Dryer Experimental Data Let us consider again the hair-dryer data from Example 8.8 and see how the RIVC algorithm can be useful in producing multiple discrete-time models at different sampling intervals from a single continuous-time model. In this example, the RIVC estimated [2 3 2 1 0] model is as follows (data sampling interval $\Delta t = 0.08$ s):

$$\begin{aligned} x(t) &= \frac{0.00164s^2 + 0.139s + 25.18}{s^2 + 10.47s + 27.82} u(t - 0.16) \\ y(t_k) &= x(t_k) + \frac{1}{1 - 0.927z^{-1}} e(t_k) \quad \sigma^2 = 0.00147 \end{aligned} \tag{8.109}$$

[12] See http://www.iris.cran.uhp-nancy.fr/contsid/.

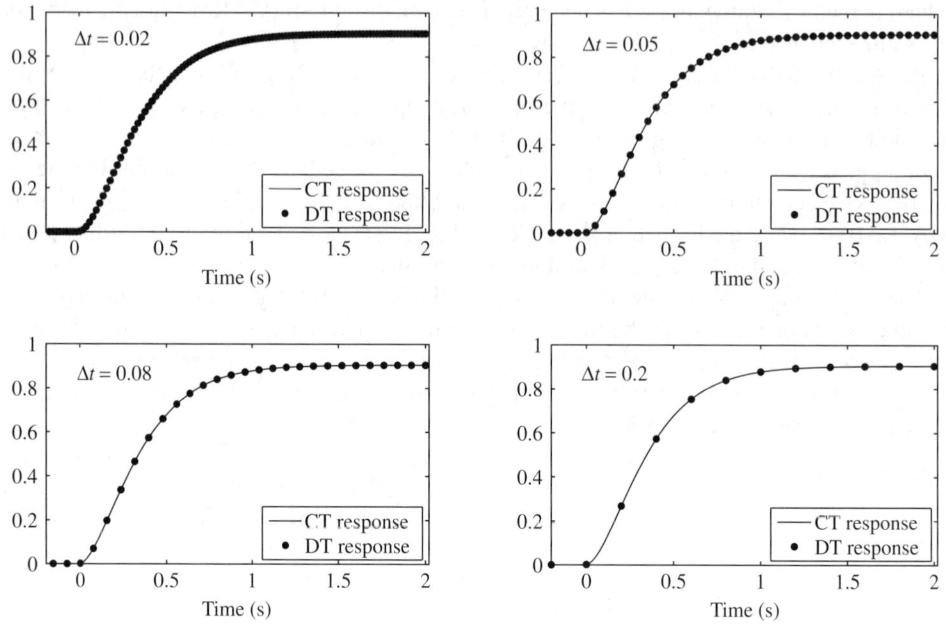

Figure 8.13 Comparison of the four inferred *Discrete-Time* (DT) step responses (dots) with the *Continuous-Time* (CT) step response (solid trace) in Example 8.13

Now, using the continuous-to-discrete time c2d routine in MATLAB®, we are able to generate discrete-time TF models at any specified sampling interval Δt. For example, the results for $\Delta t = 0.02, 0.05, 0.08$ and 0.2 s are[13]:

$$
\begin{aligned}
F_{0.02}(z^{-1}) &= \frac{0.001642 + 0.003928z^{-1} + 0.00351z^{-2}}{1 - 1.801z^{-1} + 0.8111z^{-2}} \\
F_{0.05}(z^{-1}) &= \frac{0.001642 + 0.0286z^{-1} + 0.01849z^{-2}}{1 - 1.539z^{-1} + 0.5925z^{-2}} \\
F_{0.08}(z^{-1}) &= \frac{0.001642 + 0.06539z^{-1} + 0.04053z^{-2}}{1 - 1.314z^{-1} + 0.4329z^{-2}} \\
F_{0.2}(z^{-1}) &= \frac{0.001642 + 0.2653z^{-1} + 0.1196z^{-2}}{1 - 0.6962z^{-1} + 0.1233z^{-2}}
\end{aligned}
\tag{8.110}
$$

The step responses of each TF are compared with the step response of the continuous-time model (8.109) in Figure 8.13: it is clear that, at each sampling instant, the response of each discrete-time model corresponds exactly with the continuous-time response, as expected. In this case, the most appropriate sampling interval for discrete-time control design is probably that of the data used to estimate the continuous-time model ($\Delta t = 0.08$ s), where it will be noted that the model parameters obtained here are very similar to those obtained by direct discrete-time model estimation in Example 8.8. This is not surprising because the data were

[13] Note that, for simplicity, the pure time delay in the continuous-time model (8.109) was omitted in this analysis.

obtained through a planned laboratory experiment on the dryer, but this may not always be the case in practical situations.

A typical scenario for utilising a continuous-time model within digital control system design would be when data are sampled very rapidly, as is often the case with modern data acquisition systems. These data can be used to identify and estimate a continuous-time model and the derived discrete-time models, at different sampling intervals, then provide the basis for control system design exercises, normally via simulation modelling. Here, the discrete-time model to be used for final control system design and implementation is selected on the basis of the best overall closed-loop performance, as evaluated in terms of factors such as input signal and response characteristics, robustness to uncertainty on the model parameters and multi-objective performance.

8.6.2 Estimation of δ-Operator Models

An alternative to continuous-time TF model estimation is to consider a δ-operator model (Middleton and Goodwin 1990). As discussed in Chapter 9, the δ-operator is the discrete-time, sampled data equivalent of the differential operator s. The RIVBJ approach to the estimation of such δ-operator TF models follows directly from the RIVCBJ method outlined above and was suggested by Young *et al.* (1991). However, the model can be obtained from the RIVCBJ estimated continuous-time model using a continuous-time to delta conversion, such as the c2del routine developed by I. Kaspura in Goodwin and Middleton's original δ-operator Toolbox for MATLAB®. Chapter 9 considers PIP control system design based on such δ-operator models. The advantage of the δ-operator approach is that it retains the completely digital approach to model estimation and control system design, while being superior to the standard discrete-time approach when the data are sampled rapidly, as discussed at length by Middleton and Goodwin (1990). Its disadvantage is that the parameters are not quite the same as those of the continuous-time model, except when the sampling rate is very fast, so that their physical interpretation is not as clear as in the completely continuous-time model case.

8.7 Identification and Estimation in the Closed-Loop

The identification and estimation of TF models in a closed-loop situation has received a lot of attention in the control systems literature (see e.g. Söderström and Stoica 1989; Verhaegen 1993; Van den Hof 1998; Ljung 1999; Gilson and Van den Hof 2005). Provided there is an external command input signal, simple, sub-optimal transfer function estimation within a closed automatic control loop has always been straightforward when using IV estimation methodology (see e.g. Young 1970). However, the more recent RIVBJ and RIVCBJ algorithms, as discussed in previous sections of this chapter, provide a stimulus to the development of statistically optimal methods for closed-loop estimation and a number of possible solutions are discussed by Gilson *et al.* (2008, 2009) within a continuous-time setting.

These latest optimal RIVBJ procedures are fairly complicated and so, in the spirit of the present practically orientated book, we will consider here a new and particularly simple generalised RIV method for estimating discrete- and continuous-time TF models enclosed within a feedback control system. This 'three-stage' method (Young 2011a) derives from a simple two-stage algorithm (Young 2008; Young *et al.* 2009) that yields consistent, but statistically inefficient, parameter estimates. The additional third stage allows for statistically efficient

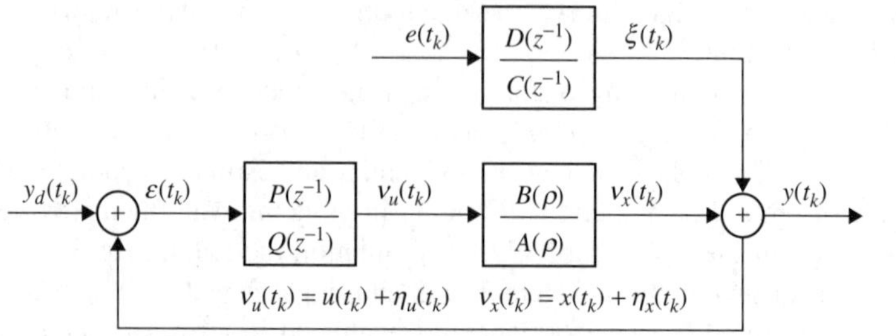

Figure 8.14 Typical closed-loop system: GBJ model (8.111)

estimation of the enclosed TF model parameters *when the system within the closed-loop is stable*. The attraction of this new approach is its relative simplicity: in particular, the resulting closed-loop RIVBJ and closed-loop RIVCBJ algorithms, for discrete-time and continuous-time models, are straightforward to implement since they use the existing estimation algorithms described in previous sections of this chapter, as implemented by the rivbj and rivcbj routines in the CAPTAIN Toolbox. As a result, the coding of the new closed-loop routines in CAPTAIN is straightforward, requiring only three calls to these existing algorithms.

8.7.1 The Generalised Box–Jenkins Model in a Closed-Loop Context

Consider the feedback control system shown in Figure 8.14.

The *Generalised Box–Jenkins* (GBJ) model then takes the following form:

$$\begin{aligned} v_x(t_k) &= G(\rho)v_u(t_k) \\ y(t_k) &= v_x(t_k) + \xi(t_k) \\ \xi(t_k) &= H(z^{-1})e(t_k) \qquad e(t_k) = \mathcal{N}(0, \sigma^2) \end{aligned} \tag{8.111}$$

The system and ARMA noise TFs, $G(\rho)$ and $H(z^{-1})$, are defined as the following ratios of rational polynomials in a generalised ρ operator[14] and z^{-1}, respectively:

$$G(\rho) = \frac{B(\rho)}{A(\rho)}; \quad H(z^{-1}) = \frac{D(z^{-1})}{C(z^{-1})} \tag{8.112}$$

More specifically,

$$\begin{aligned} A(\rho) &= 1 + a_1\rho + a_2\rho^2 + \cdots + a_n\rho^n \\ B(\rho) &= b_0 + b_1\rho + b_2\rho^2 + \cdots + b_m\rho^m \\ C(z^{-1}) &= 1 + c_1z^{-1} + c_2z^{-2} + \cdots + c_pz^{-p} \\ D(z^{-1}) &= 1 + d_1z^{-1} + d_2z^{-2} + \cdots + d_qz^{-q} \end{aligned} \tag{8.113}$$

The generalised operator ρ represents the backward shift operator z^{-1} in the discrete-time case and the inverse of the differential operator s^{-1}, where $s = d/dt$, in the hybrid continuous-time case. Referring to Figure 8.14, $y_d(t_k)$ is the command input to the closed-loop system;

[14] Do not confuse the lower case ρ used here with the boldface $\boldsymbol{\rho}$ used to denote the system model parameter vector: ρ is used here because it is employed as a generalised operator by Middleton and Goodwin (1990).

$\varepsilon(t_k) = y_d(t_k) - y(t_k)$ is the control system error; and the forward path controller is denoted by $P(z^{-1})/Q(z^{-1})$. The control input to the system, $v_u(t_k)$, is generated by the controller and is affected by additive circulatory noise $\zeta_u(t_k)$, with $u(t_k)$ denoting the underlying 'noise-free' control input to the system that would be measured at this point if $\xi(t_k) = \zeta_u(t_k) = 0 \; \forall \; t_k$. The output from the system $v_x(t_k)$ is also affected by additive circulatory noise $\zeta_x(t_k)$, with the noise-free output $x(t_k)$ defined in a similar manner to the noise-free input. Finally, the noisy measured output $y(t_k)$ is the sum of $v_x(t_k)$ and the additive ARMA noise $\xi(t_k)$, while $e(t_k)$ is the zero mean, normally distributed white noise source to the ARMA noise model. All of these signals are sampled at a uniform sampling interval Δt.

Note that the ARMA noise model is always considered in discrete-time terms and that ρ could also represent the δ-operator mentioned in the previous section. It must also be stressed that Figure 8.14 is purely illustrative: the closed-loop identification and estimation procedures outlined below can be applied to any standard control structure.

The primary aim of this section is to show how the various RIV algorithms described in previous sections of this chapter can be exploited for closed-loop TF model identification and estimation. In order to simplify the presentation, however, the acronyms GRIVBJ and GSRIV are used in unified situations where both discrete- and continuous-time estimation is being considered, so that all of the SRIV/RIVBJ and SRIVC/RIVCBJ acronyms introduced earlier apply in the context of the model (8.111).

Section 8.7.2 outlines two very simple methods of closed-loop model estimation that exploit the GSRIV algorithms and yield consistent, asymptotically unbiased estimates of the parameters in both discrete or hybrid continuous-time TF models of the enclosed system. These simple, two-stage algorithms can be applied reasonably successfully if the noise $\xi(t_k)$ does not comply with the theoretical assumptions; and even if the enclosed system is inherently unstable, although convergence cannot be guaranteed in this unstable situation. Section 8.7.3 then goes on to describe how one of these simple algorithms can be enhanced by the addition of a third stage to induce asymptotic efficiency and ensure that the estimates have desirable minimum variance characteristics in the case where $\xi(t_k)$ can be described by an ARMA model, *but only when the enclosed system is stable.*

Note that an important aspect of all these methods, not referred to specifically in the following descriptions of the estimation procedures, is the identification of appropriate orders for the TF model polynomials in the model (8.111). However, these would be identified normally using standard GRIVBJ structure identification statistics available from the rivbjid and rivcbjid routines in the CAPTAIN Toolbox (section 8.4).

8.7.2 *Two-Stage Closed-Loop Estimation*

Provided the command input $y_d(t_k)$ is free of noise and persistently exciting (see e.g. Young 2011a) it can be used as a source on instrumental variables. Note that neither $\varepsilon(t_k) = y_d(t_k) - y(t_k)$ nor $v_u(t_k)$ could be used as the source of the IVs because both are contaminated by the circulatory noise within the closed-loop system. Now, because both $v_u(t_k)$ and $y(t_k)$ are available for measurement, in addition to $y_d(t_k)$, two rather obvious approaches to closed-loop estimation are possible:

1. **Method 1.** Estimate the parameters of the TF model between between $y_d(t_k)$ and the control input $v_u(t_k)$ using the appropriate GSRIV algorithm. Note that $v_u(t_k)$ is a function of $y(t_k)$ and so is affected by the component of the noise circulating around the closed-loop at

this location, denoted by $\eta_u(t_k)$, i.e. $v_u(t_k) = u(t_k) + \eta_u(t_k)$, where $u(t_k)$ is the underlying noise-free control input to the system (see Figure 8.14 and the associated definition of the variables). As a result, the deterministic output of this estimated TF model provides a good but sub-optimal estimate $\hat{u}(t_k)$ of the noise-free input $u(t_k)$ to the enclosed system. Hence, either the GSRIV or the GRIVBJ algorithm can be used again to estimate the required TF between $\hat{u}(t_k)$ and the noisy $y(t_k)$. This is the two-stage approach first suggested by Young (2008): see also Young *et al.* (2009)[15].

2. **Method 2.** Estimate the parameters of the TF model for the whole closed-loop system between $y_d(t_k)$ and the measured, noisy output $y(t_k)$. The deterministic output of this model then provides a good estimate $\hat{x}(t_k)$ of the noise-free output from the system and the appropriate GSRIV algorithm can be used again, this time in order to estimate the TF between the two estimated variables $\hat{u}(t_k)$ obtained from Method 1, and $\hat{x}(t_k)$. This approach is less satisfying than the first method in statistical terms because the final estimation involves two estimated noise-free variables, without direct reference to the measured output $y(t_k)$.

8.7.3 *Three-Stage Closed-Loop Estimation*

The three stages of the estimation algorithm are as follows, where it will be noted that the first two stages are very similar to the two-stage Method 1, outlined in Section 8.7.2, except that full GRIVBJ, rather than GSRIV algorithms are utilised throughout:

1. **Stage 1.** Estimate the TF between the command input $y_d(t_k)$ and the noisy control input $v_u(t_k)$ using the appropriate GRIVBJ algorithm, and generate an estimate $\hat{u}(t_k)$ of the underlying noise-free control input $u(t_k)$ using this model.
2. **Stage 2.** Use the appropriate GRIVBJ algorithm to obtain initial, two stage estimates $\hat{A}(\rho)$ and $\hat{B}(\rho)$ of the system TF model polynomials $A(\rho)$ and $B(\rho)$, respectively, based on the estimated noise-free control input signal $\hat{u}(t_k)$ obtained in Stage 1 and the noisy measured output signal $y(t_k)$. Note that (Figure 8.14), $y(t_k) = v_x(t_k) + \xi(t_k) = x(t_k) + \eta_x(t_k) + \xi(t_k)$ where $x(t_k)$ is the underlying 'noise-free' output of the system that would be measured at this point if $\xi(t_k) = \eta_u(t_k) = 0 \; \forall \; t_k$. Note also that, because of this, $y(t_k) - \eta_x(t_k) = x(t_k) + \xi(t_k)$, which is referred to below.
3. **Stage 3.** Compute the estimate $\hat{\eta}_u(t_k) = u(t_k) - \hat{u}(t_k)$ of the circulatory noise component of the control input signal, $\eta_u(t_k)$, and transfer this to the output of the system using the system model obtained in Stage 2, i.e.

$$\hat{\eta}_x(t_k) = \frac{\hat{B}(\rho)}{\hat{A}(\rho)}\hat{\eta}_u(t_k) \tag{8.114}$$

where it provides an estimate of the component of the circulatory noise at the output of the system, $\eta_x(t_k)$, that derives from $\zeta_u(t_k)$. Consequently, if this estimate is subtracted from the measured output it yields an estimate $\hat{y}(t_k) = y(t_k) - \hat{\eta}_x(t_k)$ of the output signal *that does not include the circulatory noise component from the closed-loop*. It is, therefore, an estimate of the noise-free output $x(t_k)$ plus only the additive noise $\xi(t_k)$ [cf. equation

[15] This is conceptually similar to the two-stage algorithm suggested by Van den Hof and Schrama (1993) but they used output error estimation of an FIR model, rather than SRIV.

(8.111)]. As a result, the data set $\{\hat{u}(t_k)\ ;\ \hat{y}(t_k)\}$ provides an estimate of the data set *that would have been obtained if the system was being estimated in the open-loop situation.* Finally, therefore, use the appropriate open-loop GRIVBJ algorithm for a second time to re-estimate the system model based on this constructed data set.

In computational terms, this three-stage procedure is straightforward to implement because it makes use of estimation routines rivbj or rivcbj already available in CAPTAIN.

Example 8.14 Control of CO_2 in Carbon-12 Tracer Experiments Let us think again about the control of the carbon dioxide ($^{12}CO_2$) level in carbon tracer experiments, as previously considered in Chapter 5 and Chapter 6 (Example 5.9, Example 6.3 and Example 6.4). The model of the system was considered as an integrator in Chapter 5 but as a first order system with a long time constant in Chapter 6. The latter model proved sufficient to design a PIP controller that, when implemented, produced reasonable closed-loop behaviour. An interesting question, therefore, is given the measured closed-loop signals, does closed-loop analysis confirm the initial assumptions about the system model?

In completely objective identification and estimation terms, the answer to this question is in the negative: the two-stage analysis suggests that the system is second order, with two real eigenvalues having associated time constants of 523 and 92.5 s. Two-stage estimation is used here because the noise is rather odd, as a result of the pulse-width modulated control input. Although it can be modelled as an AR(5) process, the three-stage parameter estimates are clearly not as good as those obtained by the more robust two-stage algorithm. In particular, the model takes the form ($\Delta t = 5$ s):

$$
\begin{aligned}
y(k) &= \frac{b_0 + b_1 z^{-1}}{1 + a_1 z^{-1} + b_2 z^{-2}} u(k-4) + \xi(k) \\
\xi(k) &= \frac{1}{1 + c_1 z^{-1} + c_2 z^{-2} + c_3 z^{-3} + c_4 z^{-4} + c_5 z^{-5}} e(k); \qquad e(k) = \mathcal{N}(0, \sigma^2)
\end{aligned}
\tag{8.115}
$$

where the estimated parameters are as follows, with the estimated SEs shown in parentheses:

$$
\begin{aligned}
\hat{a}_1 &= -1.9379\,(0.0048);\ \hat{a}_2 = 0.9384\,(0.0048);\ \hat{b}_0 = 0.005953\,(0.00014) \\
\hat{b}_1 &= -0.005771\,(0.00013);\ \hat{c}_1 = -0.499\,(0.035);\ \hat{c}_2 = -0.036\,(0.039) \\
\hat{c}_3 &= -0.042\,(0.039);\ \hat{c}_4 = -0.0068\,(0.040);\ \hat{c}_5 = 0.120\,(0.035);\ \hat{\sigma}^2 = 9.97 \times 10^{-7}
\end{aligned}
$$

Given the high $R_T^2 = 0.996$, it is not surprising that the estimated model outputs explain both the measured output and the control input of the system very well, as shown in Figure 8.15. Note that the above parameter estimates are cited to four significant figures because the time contant values and any decomposition of the TF model (for instance to a parallel form obtained by partial fraction expansion using the MATLAB® routine residuez) are sensitive to these values.

Fortunately, this uncertainty only affects the estimation of these derived parameters: the uncertainty on the model parameters and the system as a whole is very small: see, for example, the three times SE uncertainty bounds on the frequency response for the second order model shown in Figure 8.16. One word of caution is necessary, however, since these uncertainty

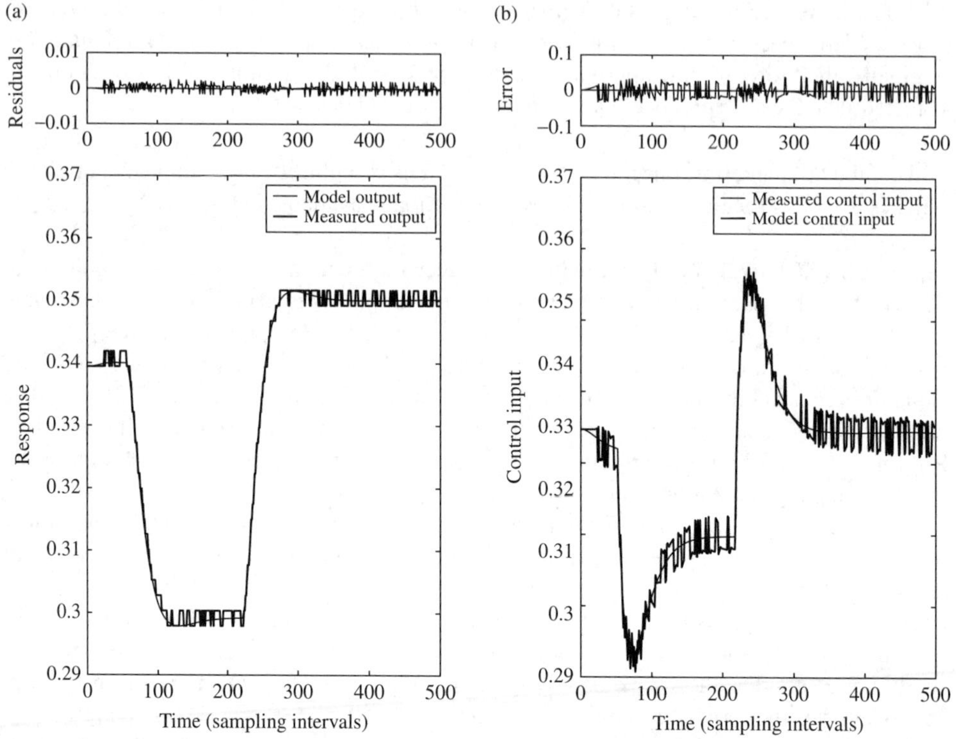

Figure 8.15 Comparison of model and measured responses for Example 8.14. (a) Output responses and (b) control input responses. The residuals in (a) are the final stochastic residuals $\hat{e}(k)$

bounds are based on the statistical assumptions about the nature of the AR noise process and these are contravened to some extent in this case. For instance, the estimated final residuals $\hat{e}(k)$ are small, zero mean, and serially uncorrelated but, as we can see from the error plot at the top of Figure 8.15a, they do not have a normal distribution. Indeed, as pointed out above, one might question the use of an AR model in this case, where the noise has special characteristics caused by the pulse-width-modulation control method implemented in the plant physiology research laboratory. But these are the kind of difficult ambiguities that can beset our real world, which does not always conform to the assumptions of theorists! For practical purposes in this case, the residual error is small and one can have reasonable confidence in the model, despite its violation of the assumptions.

As noted above, PIP control of this system represents one of the early practical examples of the TDC approach (Taylor *et al.* 1996) but was based on a first order model. If the above analysis is repeated and the model is constrained to be first order then the system model is estimated as follows:

$$y(k) = \frac{0.00439}{1 - 0.986z^{-1}}u(k-4) + \xi(k) \tag{8.116}$$

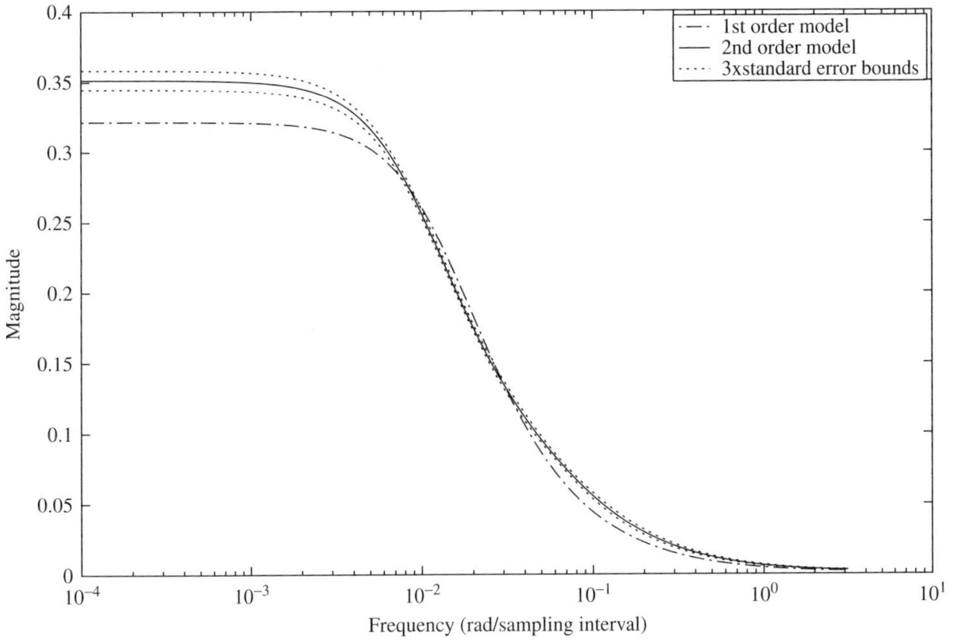

Figure 8.16 Magnitude frequency response of the estimated first and second order models with the three times SE bounds for the latter shown as dotted lines

where the noise $\xi(k)$ is now identified as an AR(11) process. However, the explanation of the data is visibly poorer than that shown in Figure 8.15 and on these grounds the model would be rejected. Despite this, when the frequency response characteristics of this model and the second order model are compared, as shown in Figure 8.16, they are quite similar, except at low frequencies. A comparison with the model (6.20) used for PIP design in Chapter 6 shows similar results, although the low frequency magnitude is smaller still. Consequently, the main differences between the various models are in their steady-state gains; differences that can be accommodated for in the servomechanism PIP design procedure, where the control weightings are always adjusted to provide the required closed-loop response characteristics. Of course, given these closed-loop estimation results, the robustness of the closed-loop response to these kind of differences should be checked by MCS analysis of the kind considered in Chapter 4 and Chapter 6.

Finally, continuous-time estimation yields similar but not identical results. The estimated model, in factorised form, is as follows:

$$y(t) = \frac{52.5\text{s} + 0.344}{(1 + 488s)(1 + 89.5s)} u(t - 20) \tag{8.117}$$

which reveals straightaway that the time constants are 488 and 89.5 s and the steady-state gain is 0.344. Earlier, in section 8.4, it was stressed that, in DBM modelling, the model should be credible when interpreted in physically meaningful terms and this continuous-time TF provides an immediate interpretation of the system dynamics for consideration in such terms. In this regard, the Principle Investigator who managed the research that generated the data used in this example was Dr Peter Minchin, then of the Department of Scientific and Industrial Research in New Zealand. Dr Minchin found the present results interesting and confirmed that the system could be of the identified second order form. Paraphrasing his comments (P.C. Young, personal communication, 2012):

> There was a water bubbler system containing approximately one litre of water in the closed-loop, in order to maintain a dew point of 4°C in the circulated air and avoiding any increase in the dew point caused by the leaf. So when a correction pulse of CO_2 was introduced, the bubbler would have taken some time to come to a new CO_2 equilibrium, which is probably the cause of the long time constant. The short time constant is likely to be that associated with the gas loop. Also, the $^{12}CO_2$ used in the experiments must be conserved and the closed-loop into which this was injected had to have a route for the volume increase of gas to be released, and this would have resulted in some loss of CO_2. This leak, controlled to maintain the pressure of the leaf chamber at just below atmospheric, would result in loss of CO_2 from the loop. Also the injected CO_2 was not pure, but a CO_2/air mix. Given these considerations the estimated gain of less than unity is not surprising.

8.7.4 Unstable Systems

One major reason why one would wish to identify and estimate a system within a closed-loop is when the system is open-loop unstable, so that open-loop estimation is difficult, if not impossible. However, the three-stage estimation procedure described in the previous subsection includes explicit simulation of the estimated model in the form of the iteratively updated 'auxiliary model' and utilisation of the estimated system denominator polynomial $\hat{A}(\rho)$ in the iteratively updated prefilters. In the normal CAPTAIN Toolbox implementation of the GRIVBJ-type algorithms for open-loop systems, therefore, the auxiliary model and prefilter denominator polynomials are stabilised at each iteration in order to avoid any (rare) transient violations of stability during the iterative updating process (using the MATLAB® routines polystab for discrete-time models and polystc for continuous-time models). Despite these precautions, the full three-stage algorithm fails to produce satisfactory estimation results when applied to open-loop unstable systems, even when they are contained within a stable closed-loop. Fortunately, however, the simple two-stage methods function satisfactorily in this closed-loop environment: comprehensive MCS analysis (Young 2011a) suggests that they normally yield consistent, asymptotically unbiased parameter estimates with acceptable confidence bounds, provided sufficient data are available from the closed-loop experiments when the data are noisy.

8.8 Concluding Remarks

This chapter has presented most of the background information to TF model identification and estimation that is required for TDC design. However, some other important concepts, such as identifiability and the required nature of input signals that will ensure model identifiability have not been discussed. A much more complete coverage of this topic is available in the second author's recent book (Young 2011a), to which the reader is directed for these and other details, particularly in the areas of recursive and closed-loop estimation. The present chapter has also concentrated on algorithmic details and the practical application of these algorithms that will be of most use to the practitioner, rather than the theoretical fundamentals of optimal 'PEM'-type methods, which are covered well in books such as by Ljung (1987, 1999). However, note that this latter book does not consider the full RIVBJ methods specifically, even though their estimates have the same statistical properties as the PEM estimates obtained by gradient optimisation procedures, so it is useful mainly in a more general theoretical sense.

References

Ahmed, S., Huang, B. and Shah, S.L. (2008) Process parameter and delay estimation from non–uniformly sampled data. In H. Garnier and L. Wang (Eds), *Identification of Continuous-Time Models from Sampled Data*, Springer-Verlag, London.

Akaike, H. (1974) A new look at statistical model identification, *IEEE Transactions on Automatic Control*, **19**, pp. 716–723.

Åström, K.J. and Bohlin, T. (1966) Numerical identification of linear dynamic systems from normal operating records. In P.H. Hammond (Ed.), *Theory of Self Adaptive Control Systems*, Plenum Press, New York.

Åström, K.J. and Eykhoff, P. (1971) System identification – a survey, *Automatica*, **7**, pp. 123–162.

Bodewig, E. (1956) *Matrix Calculus*, North Holland, Amsterdam.

Box, G.E.P. and Jenkins, G.M. (1970) *Time Series Analysis Forecasting and Control*, Holden-Day, San Francisco.

Clarke, D.W. (1967) Generalised least squares estimation of the parameters of a dynamic model. *Proceedings IFAC Congress*, Paper 3.17.

Durbin, J. (1954) Errors in variables. *Review of the International Statistical Institute*, **22**, pp. 23–32.

Eykhoff, P. (1974) *System Identification – Parameter and State Estimation*, John Wiley & Sons, Ltd, Chichester.

Garnier, H., Gilson, M., Young, P.C. and Huselstein, E. (2007) An optimal IV technique for identifying continuous-time transfer function model of multiple input systems, *Control Engineering Practice*, **15**, pp. 471–486.

Garnier, H. and Wang, L. (Eds) (2008) *Identification of Continuous–Time Models from Sampled data*, Springer-Verlag, London.

Garnier, H. and Young, P.C. (2012) What does continuous-time model identification have to offer? *Proceedings 14th IFAC Symposium on System Identification SYSID12*, pp. 810–815.

Gilson, M., Garnier, H., Young, P.C. and Van den Hof, P. (2008) Instrumental variable methods for closed-loop continuous-time model identification. In H. Garnier and L. Wang (Eds), *Identification of Continuous-Time Models from Sampled Data*, Springer-Verlag, London, pp. 133–160.

Gilson, M., Garnier, H., Young, P.C. and Van den Hof, P. (2009) Refined instrumental variable methods for closed-loop system identification. *Proceedings 15th IFAC Symposium on System Identification SYSID09*, pp. 284–289.

Gilson, M. and Van den Hof, P. (2005) Instrumental variable methods for closed–loop system identification, *Automatica*, **41**, pp. 241–249.

Goodwin, G.C. and Payne, R.P. (1977) *Dynamic System Identification. Experiment Design and Data Analysis*, Academic Press, New York.

Hannan, E.J. and Quinn, B.G. (1979) The determination of the order of an autoregression. *Journal of Royal Statistical Society, Series B*, **41**, pp. 190–195.

Hastings-James, R. and Sage, M.W. (1969). Recursive generalised least squares procedure for on–line identification of process parameters. *Prococeedings of the Institute of Electrical Engineers*, **116**, pp. 2057–2062.

Ho, Y.C. (1962) On the stochastic approximation method and optimal filtering theory, *Journal of Mathematical Analysis and Applications*, **6**, p. 152.

Jakeman, A.J., Steele, L.P. and Young, P.C. (1980) Instrumental variable algorithms for multiple input systems described by multiple transfer functions. *IEEE Transactions on Systems, Man, and Cybernetics*, **SMC-10**, pp. 593–602.

Jakeman, A.J. and Young, P.C. (1979) Refined instrumental variable methods of time-series analysis: Part II, multivariable systems. *International Journal of Control*, **29**, pp. 621–644.

Jazwinski, A.H. (1970) *Stochastic Processes and Filtering Theory*, Academic Press, San Diego.

Johnston, J. (1963) *Econometric Methods*, McGraw-Hill, New York.

Kendall, M.G. and Stuart, A. (1961) *The Advanced Theory of Statistics*, volume 2, Griffin, London.

Landau, I.D. (1976) Unbiased recursive estimation using model reference adaptive techniques. *IEEE Transactions on Automatic Control*, **AC-21**, pp. 194–202.

Laurain, V., Toth, R., Gilson, M. and Garnier, H. (2010) Refined instrumental variable methods for identification of LPV Box–Jenkins models. *Automatica*, **46**, pp. 959–967.

Lee, R.C.K. (1964) *Optimal Identification, Estimation and Control*, MIT Press, Cambridge, MA.

Ljung, L. (1979) Asymptotic behaviour of the extended Kalman filter as a parameter estimator for linear systems, *IEEE Transactions on Automatic Control*, **AC-24**, pp. 36–50.

Ljung, L. (1987) *System Identification. Theory for the User*, Prentice Hall, Englewood Cliffs, NJ.

Ljung, L. (1999) *System Identification. Theory for the User*, Second Edition, Prentice Hall, Upper Saddle River, NJ.

Ljung, L. and Söderström, T. (1983) *Theory and Practice of Recursive Identification*, MIT Press, Cambridge, MA.

Mayne, D.Q. (1963) Optimal nonstationary estimation of the parameters of a dynamic system, *Journal of Electronics and Control*, **14**, pp. 101–112.

Middleton, R.H. and Goodwin, G.C. (1990) *Digital Control and Estimation – A Unified Approach*, Prentice Hall, Englewood Cliffs, NJ.

Norton, J.P. (1986) *An Introduction to Identification*, Academic Press, New York.

Panuska, V. (1969) An adaptive recursive least squares identification algorithm. *Proceedings 8th IEEE Symposium on Adaptive Processes*, Paper 6e.

Pierce, D.A. (1972) Least squares estimation in dynamic disturbance time-series models. *Biometrika*, **5**, pp. 73–78.

Price, L., Young, P., Berckmans, D., Janssens, K. and Taylor, J. (1999) Data-Based Mechanistic Modelling (DBM) and control of mass and energy transfer in agricultural buildings, *Annual Reviews in Control*, **23**, pp. 71–82.

Schwarz, G. (1978) Estimating the dimension of a model. *Annals of Statistics*, **6**, pp. 461–464.

Shaban, E.M. (2006) *Nonlinear Control for Construction Robots using State Dependent Parameter Models*, PhD thesis, Engineering Department, Lancaster University.

Shaban, E.M., Ako, S., Taylor, C.J. and Seward, D.W. (2008) Development of an automated verticality alignment system for a vibro-lance, *Automation in Construction*, **17**, pp. 645–655.

Shibata, R. (1985) Various model selection techniques in time series analysis. In E.J. Hannan, P.R. Krishnaiah and M. Rao (Eds), *Handbook of Statistics 5: Time Series in the Time Domain*, North-Holland, Amsterdam, pp. 179–187.

Söderström, T., Ljung, L. and Gustavsson, I. (1974) A comparative study of recursive identification methods. *Technical Report TR 7308,* Lund Institute of Technology, Division of Automatic Control.

Söderström, T. and Stoica, P. (1989) *System Identification*, Series in Systems and Control Engineering, Prentice Hall, New York.

Talmon, J.L. and van den Boom, A.J.W. (1973) On the estimation of the transfer function parameters of process and noise dynamics using a single stage estimator. In P. Eykhoff (Ed.), Preprints of the *3rd IFAC Symposium on Identification and System Parameter Estimation*, 12–15 June, The Hague–Delft, North-Holland, Amsterdam, pp. 711–720.

Taylor, C.J. (2004) Environmental test chamber for the support of learning and teaching in intelligent control, *International Journal of Electrical Engineering Education*, **41**, pp. 375–387.

Taylor, C.J., Lees, M.J., Young, P.C. and Minchin, P.E.H. (1996) True digital control of carbon dioxide in agricultural crop growth experiments, *International Federation of Automatic Control 13th Triennial World Congress* (IFAC-96), 30 June–5 July, San Francisco, Elsevier, Vol. B, pp. 405–410.

Taylor, C.J., Leigh, P.A., Chotai, A., Young, P.C., Vranken, E. and Berckmans, D. (2004a) Cost effective combined axial fan and throttling valve control of ventilation rate, *IEE Proceedings: Control Theory and Applications*, **151**, pp. 577–584.

Taylor, C.J., Leigh, P., Price, L., Young, P.C., Berckmans, D. and Vranken, E. (2004b) Proportional-Integral-Plus (PIP) control of ventilation rate in agricultural buildings, *Control Engineering Practice*, **12**, pp. 225–233.

Taylor, C.J., Pedregal, D.J., Young, P.C. and Tych, W. (2007a) Environmental time series analysis and forecasting with the Captain toolbox, *Environmental Modelling and Software*, **22**, pp. 797–814.

Taylor, C.J., Shaban, E.M., Stables, M.A. and Ako, S. (2007b) Proportional-Integral-Plus (PIP) control applications of state dependent parameter models, *IMECHE Proceedings: Systems and Control Engineering*, **221**, pp. 1019–1032.

Taylor, C.J., Young, P.C. and Cross, P. (2012) Practical experience with unified discrete and continuous-time, multi-input identification for control system design, *16th IFAC Symposium on System Identification* (SYSID-2012), Paper 123.

Van den Hof, P.M.J. (1998) Closed-loop issues in system identification, *Annual Reviews in Control*, **22**, 173–186.

Van den Hof, P.M.J. and Schrama, R.J.P. (1993) An indirect method for transfer function estimation from closed loop data, *Automatica*, **29**, pp. 1523–1527.

Verhaegen, M. (1993) Application of a subspace model identification technique to identify LTI systems operating in closed-loop, *Automatica*, **29**, pp. 1027–1040.

Wang, L. and Garnier, H. (Eds) (2011) *System Identification, Environmental Modelling, and Control System Design*, Springer-Verlag, London.

Wellstead, P.E. (1978) An instrumental product moment test for model order estimation. *Automatica*, **14**, pp. 89–91.

Wellstead, P.E. and Zarrop, M.B. (1991) *Self-Tuning Systems: Control and Signal Processing*, John Wiley & Sons, Ltd, New York.

Wong, K.Y. and Polak, E. (1967) Identification of linear discrete-time systems using the instrumental variable approach. *IEEE Transactions on Automatic Control*, **AC-12**, pp. 707–718.

Young, P.C. (1965) Process parameter estimation and self-adaptive control. In P.H. Hammond (Ed.), *Theory of Self Adaptive Control Systems*, Plenum Press, New York, pp. 118–140.

Young, P.C. (1968) The use of linear regression and related procedures for the identification of dynamic processes. *Proceedings 7th IEEE Symposium on Adaptive Processes,* pp. 501–505.

Young, P.C. (1969) Applying parameter estimation to dynamic systems: Part I – theory. *Control Engineering*, **16**, pp. 119–125.

Young, P.C. (1970) An instrumental variable method for real–time identification of a noisy process, *Automatica*, **6**, pp. 271–287.

Young, P.C. (1974) Recursive approaches to time-series analysis. *Bulletin of the Institute of Mathematics and its Applications*, **10**, pp. 209–224.

Young, P.C. (1976) Some observations on instrumental variable methods of time-series analysis. *International Journal of Control*, **23**, pp. 593–612.

Young, P.C. (1981) Parameter estimation for continuous-time models – a survey. *Automatica*, **17**, pp. 23–39.

Young, P.C. (1984) *Recursive Estimation and Time–Series Analysis: An Introduction*, Springer-Verlag, Berlin.

Young, P.C. (1989) Recursive estimation, forecasting and adaptive control. In C.T. Leondes (Ed.), *Control and Dynamic Systems*, Academic Press, San Diego, pp. 119–166.

Young, P.C. (1998) Data-based mechanistic modelling of environmental, ecological, economic and engineering systems, *Environmental Modelling and Software*, **13**, pp. 105–122.

Young, P.C. (2006) An instrumental variable approach to ARMA model identification and estimation. *Proceedings 14th IFAC Symposium on System Identification SYSID06,* pp. 410–415.

Young, P.C. (2008) The refined instrumental variable method: unified estimation of discrete and continuous-time transfer function models. *Journal Européen des Systèmes Automatisés*, **42**, pp. 149–179.

Young, P.C. (2011a) *Recursive Estimation and Time-Series Analysis: An Introduction for the Student and Practitioner*, Springer-Verlag, Berlin.

Young, P.C. (2011b) Data-based mechanistic modelling: natural philosophy revisited? In L. Wang and H. Garnier (Eds), *System Identification, Environmetric Modelling and Control*, Springer-Verlag, Berlin, pp. 321–340.

Young, P.C., Chotai, A. and Tych, W. (1991) Identification, estimation and control of continuous-time systems described by delta operator models. In N. Sinha and G. Rao (Eds), *Identification of Continuous-Time Systems*, Kluwer, Dordrecht, pp. 363–418.

Young, P.C. and Garnier, H. (2006) Identification and estimation of continuous-time, data-based mechanistic models for environmental systems. *Environmental Modelling & Software*, **21**, pp. 1055–1072.

Young, P.C., Garnier, H. and Gilson, M. (2008) Refined instrumental variable identification of continuous-time hybrid Box–Jenkins models. In H. Garnier and L. Wang (Eds), *Identification of Continuous-Time Models from Sampled Data*, Springer-Verlag, London, pp. 91–131.

Young, P.C., Garnier, H. and Gilson, M. (2009) Simple refined IV methods of closed-loop system identification. *Proceedings 15th IFAC Symposium on System Identification SYSID09*, pp. 1151–1156.

Young, P.C. and Jakeman, A.J. (1979) Refined instrumental variable methods of time-series analysis: Part I, SISO systems. *International Journal of Control*, **29**, pp. 1–30.

Young, P.C. and Jakeman, A.J. (1980) Refined instrumental variable methods of time-series analysis: Part III, extensions. *International Journal of Control*, **31**, pp. 741–764.

Young, P.C., Jakeman, A.J. and McMurtrie, R. (1980) An instrumental variable method for model order identification. *Automatica*, **16**, pp. 281–296.

Young, P.C. and Lees, M.J. (1993) The active mixing volume: a new concept in modelling environmental systems. In V. Barnett and K. Turkman (Eds), *Statistics for the Environment*, John Wiley & Sons, Ltd, Chichester, pp. 3–43.

9

Additional Topics

In this book, we are concerned mainly with *True Digital Control* (TDC) based on discrete-time *Transfer Function* (TF) models of dynamic systems. In Chapter 8, however, we discussed the identification and estimation of continuous-time TF models, pointing out that many systems were more easily considered in continuous-time terms at the conceptual level and that a continuous-time TF model could be converted into discrete-time for the purposes of TDC design. An alternative approach is to exploit the δ-operator TF model, which is the discrete-time equivalent of the continuous-time model. In recent times, the main impetus for the use of δ-operator TF models in control system design has come from the book by Middleton and Goodwin (1990), to which the reader is referred for background information. In section 9.1, we will simply introduce the δ-operator TF model; show how it can be presented in *Non-Minimal State Space* (NMSS) form; and develop the δ-operator *Proportional-Integral-Plus* (PIP) control algorithm.

The recursive nature of all the *Refined Instrumental Variable* (RIV) algorithms discussed in Chapter 8, also allows for the development of *Time Variable Parameter* (TVP) versions that can be used for real-time applications, such as self-tuning or adaptive control, i.e. in those situations where the model parameters are changing relatively slowly over time. Here, the algorithm updates the parameter estimates as the input–output data are acquired and the latest estimates are used to update the control system design on a continuing basis. However, if these changes in the parameters are much more rapidly changing as functions of the state or input variables (i.e. they actually constitute stochastic state variables), then the system is truly nonlinear and likely to exhibit severe nonlinear behaviour. Normally, this cannot be approximated in a simple TVP manner; in which case, recourse must be made to alternative *State-Dependent Parameter* (SDP) modelling methods. Section 9.2 and section 9.3 provide a review of recursive TVP and SDP estimation but a much more comprehensive treatment is available in Young (2011). The reader is also referred to recent research on the wider application of SDP modelling within a PIP design context (Taylor *et al.* 2007b, 2009, 2011).

Finally in relation to SDP models, note that the terms *Linear Parameter Varying* (LPV) and *NonLinear Parameter Varying* (NLPV) are often used in the systems and control literature to describe SDP-type models. However LPV, in particular, is a quite misleading term since SDP

True Digital Control: Statistical Modelling and Non-Minimal State Space Design, First Edition.
C. James Taylor, Peter C. Young and Arun Chotai.

models are truly nonlinear dynamic systems and, as we shall see, can even exhibit chaotic dynamic behaviour.

9.1 The δ-Operator Model and PIP Control

The δ-operator was first utilised in a PIP design context by Young *et al.* (1991). Later papers formalised this approach for both *Single-Input, Single-Output* (SISO; Young *et al.* 1998) and multivariable (Chotai *et al.* 1998) systems (see also McKenna 1997). An outline of the modelling and PIP design approaches are discussed in these references, with the present chapter concentrating on the following nth order, SISO system, whose input and output signals are sampled regularly with a sampling interval of Δt time units. The discrete differential or δ-operator TF model of this system takes the form:

$$y(k) = \frac{B(\delta)}{A(\delta)}u(k) \tag{9.1}$$

where $y(k)$ and $u(k)$ denote, respectively, the sampled output and input signals at the kth sampling instant; while $A(\delta)$ and $B(\delta)$ are the following polynomials in the δ-operator:

$$A(\delta) = \delta^n + a_1\delta^{n-1} + \cdots + a_n; \quad B(\delta) = b_1\delta^{n-1} + b_2\delta^{n-2} + \cdots + b_n \tag{9.2}$$

Here, the δ-operator is defined as follows in terms of the forward shift operator z:

$$\delta = \frac{z-1}{\Delta t} \quad \text{i.e.} \quad \delta y(k) = \frac{y(k+1) - y(k)}{\Delta t} \tag{9.3}$$

In general, no prior assumptions are made about the nature of the TF, which may be marginally stable, unstable, or possess non-minimum phase characteristics. The order of the numerator polynomial $B(\delta)$ is set to n–1 to ensure that the TF is proper. However, it can be of less dimension than this if identification and estimation analysis shows that it is more appropriate. Any pure time delay in the system can be handled in various ways: e.g. by introducing additional poles at $-1/\Delta t$ to accommodate the time delay; or, in the case of long time delays, by introducing the δ form of the Smith Predictor using a similar approach to that described in section 6.3 (see also e.g. Taylor *et al.* 1998).

It is clear that the δ-operator is the discrete-time, sampled data equivalent of the differential operator $s = d/dt$ considered in Chapter 8. One attraction of the δ-operator model (9.1) is that it can be applied to a wide range of discrete-time systems, from sampled data systems with coarse sampling intervals to rapidly sampled, near continuous-time systems. For example, it is easy to see that the unit circle in the complex z-plane (see section 2.2) maps to a circle with centre $-1/\Delta t$ and radius $1/\Delta t$ in the complex δ-plane; so that, as $\Delta t \rightarrow 0$, this circular stability region is transformed to the left half of the complex s-plane. For very rapidly sampled systems, therefore, the δ-operator model can be considered in almost continuous-time terms, with the pole positions in the δ-plane close to those of the *equivalent* continuous-time system in the s-plane. As such, the δ-operator model provides a rather natural digital representation for a rapidly sampled, continuous-time system; one which avoids the approximate digitisation of continuous-time designs and provides a direct basis for TDC system design. However, it should be realised that, unless the sampling rate is very high, the parameters of the δ-operator

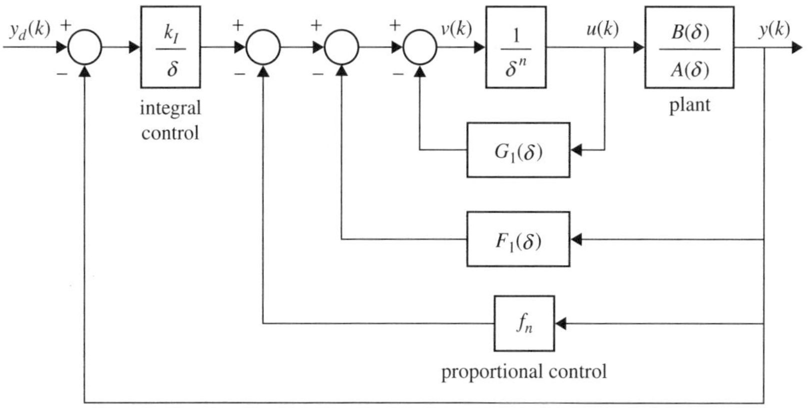

Figure 9.1 The δ-operator PIP servomechanism system (see section 9.1.3 for control polynomials)

model will not be the same as those of the continuous-time model, so that they do not have the same physically meaningful interpretation as the parameters of the continuous-time model.

9.1.1 The δ-operator NMSS Representation

The most obvious NMSS representation of the TF model (9.1) is formulated directly in terms of the discrete-time δ derivatives of the output and input signals and takes the form [cf. equations (5.13)]:

$$\delta \boldsymbol{x}(k) = \boldsymbol{F}\boldsymbol{x}(k) + \boldsymbol{g}v(k) + \boldsymbol{d}y_d(k) \tag{9.4}$$

where $v(k)$ is an intermediate control variable: see below and Figure 9.1. The system transition matrix F, input vector $\boldsymbol{g}$ and command input vector $\boldsymbol{d}$ are defined as follows:

$$\boldsymbol{F} = \begin{bmatrix} -a_1 & -a_2 & \cdots & -a_{n-1} & -a_n & b_1 & b_2 & \cdots & b_{n-1} & b_n & 0 \\ 1 & 0 & \cdots & 0 & 0 & 0 & 0 & \cdots & 0 & 0 & 0 \\ 0 & 1 & \cdots & 0 & 0 & 0 & 0 & \cdots & 0 & 0 & 0 \\ \vdots & \vdots & \ddots & \vdots & \vdots & \vdots & \vdots & \ddots & \vdots & \vdots & \vdots \\ 0 & 0 & \cdots & 1 & 0 & 0 & 0 & \cdots & 0 & 0 & 0 \\ 0 & 0 & \cdots & 0 & 0 & 0 & 0 & \cdots & 0 & 0 & 0 \\ 0 & 0 & \cdots & 0 & 0 & 1 & 0 & \cdots & 0 & 0 & 0 \\ 0 & 0 & \cdots & 0 & 0 & 0 & 1 & \cdots & 0 & 0 & 0 \\ \vdots & \vdots & \ddots & \vdots & \vdots & \vdots & \vdots & \ddots & \vdots & \vdots & \vdots \\ 0 & 0 & \cdots & 0 & 0 & 0 & 0 & \cdots & 1 & 0 & 0 \\ 0 & 0 & \cdots & 0 & -1 & 0 & 0 & \cdots & 0 & 0 & 0 \end{bmatrix} \tag{9.5}$$

$$\boldsymbol{d} = \begin{bmatrix} 0 & 0 & \cdots & 0 & 0 & 0 & 0 & \cdots & 0 & 0 & 1 \end{bmatrix}^T \tag{9.6}$$

$$\boldsymbol{g} = \begin{bmatrix} 0 & 0 & \cdots & 0 & 0 & 1 & 0 & \cdots & 0 & 0 & 0 \end{bmatrix}^T \tag{9.7}$$

In this formulation, the control variable $v(k)$ is defined as the nth δ differential of the actual control input $u(k)$, i.e.

$$v(k) = \delta^n u(k) \tag{9.8}$$

and the non-minimal state vector $\boldsymbol{x}(k)$ is defined as:

$$\boldsymbol{x}(k) = \begin{bmatrix} \delta^{n-1}y(k) & \delta^{n-2}y(k) & \cdots & \delta y(k) & y(k) & \delta^{n-1}u(k) & \cdots & u(k) & z(k) \end{bmatrix}^T \tag{9.9}$$

In these equations, $z(k)$ will be recognised from earlier chapters as an additional *integral-of-error* state of the form:

$$z(k) = \delta^{-1}\{y_d(k) - y(k)\} \tag{9.10}$$

in which $y_d(k)$ is the reference or command input to the servomechanism system.

9.1.2 Characteristic Polynomial and Controllability

The open-loop characteristic polynomial of the NMSS representation (9.4) can be written as [cf. equation (5.17)]:

$$|\lambda \boldsymbol{I} - \boldsymbol{F}| = \lambda \left| \begin{array}{ccccc|ccccc} \lambda + a_1 & a_2 & \cdots & a_{n-1} & a_n & -b_1 & -b_2 & \cdots & -b_{n-1} & -b_n \\ -1 & \lambda & \cdots & 0 & 0 & 0 & 0 & \cdots & 0 & 0 \\ 0 & -1 & \cdots & 0 & 0 & 0 & 0 & \cdots & 0 & 0 \\ \vdots & \vdots & \ddots & \vdots & \vdots & \vdots & \vdots & \ddots & \vdots & \vdots \\ 0 & 0 & \cdots & -1 & \lambda & 0 & 0 & \cdots & 0 & 0 \\ \hline 0 & 0 & \cdots & 0 & 0 & \lambda & 0 & \cdots & 0 & 0 \\ 0 & 0 & \cdots & 0 & 0 & -1 & \lambda & \cdots & 0 & 0 \\ 0 & 0 & \cdots & 0 & 0 & 0 & -1 & \cdots & 0 & 0 \\ \vdots & \vdots & \ddots & \vdots & \vdots & \vdots & \vdots & \ddots & \vdots & \vdots \\ 0 & 0 & \cdots & 0 & 0 & 0 & 0 & \cdots & -1 & \lambda \end{array} \right| \tag{9.11}$$

As a result, the open-loop characteristic polynomial is given by:

$$|\lambda \boldsymbol{I} - \boldsymbol{F}| = \lambda^{n+1} A(\lambda) \tag{9.12}$$

which is a product of the characteristic polynomial of the minimal state space representation, i.e. $A(\lambda)$, and a term λ^{n+1} due to the introduction of the additional states.

If the NMSS model (9.4) is to be used as the basis for the design of *State Variable Feedback* (SVF) control systems, such as closed-loop pole assignment or *Linear Quadratic* (LQ) optimal control, it is important to evaluate the conditions for controllability of this model. These are provided by Theorem 9.1.

Theorem 9.1 Controllability of the δ-operator NMSS Model Given a SISO discrete-time δ-operator system described by (9.1), the NMSS representation (9.4), as described by the pair $[\boldsymbol{F}, \boldsymbol{g}]$, is completely controllable if, and only if, the following conditions are satisfied:

(i) the polynomials $A(\delta)$ and $B(\delta)$ are coprime;
(ii) $b_n \neq 0$.

The proof of this theorem is given by Young *et al.* (1998).

The conditions in Theorem 9.1 have obvious physical interpretations. The coprimeness condition is equivalent to the normal requirement that the δ-operator TF (9.1) should have no pole-zero cancellations. The second condition avoids the presence of a zero at the origin in the complex δ-plane, which would cancel with the pole associated with the inherent integral action. The controllability conditions of the above theorem are, of course, equivalent to the normal requirement that the controllability matrix associated with the NMSS representation, i.e.

$$\boldsymbol{S}_1 = [\boldsymbol{g} \quad \boldsymbol{F}\boldsymbol{g} \quad \boldsymbol{F}^2\boldsymbol{g} \cdots \boldsymbol{F}^{2n}\boldsymbol{g}] \tag{9.13}$$

has full rank $2n + 1$.

9.1.3 The δ-Operator PIP Control Law

In the context of the NMSS model (9.4), the automatic control objective is to design a SVF control law [cf. equation (5.32)]:

$$\begin{aligned} v(k) &= -\boldsymbol{k}^T \boldsymbol{x}(k) \\ &= -f_1 \delta^{n-1} y(k) - f_2 \delta^{n-2} y(k) - \cdots - f_n y(k) \\ &\quad -g_1 \delta^{n-1} u(k) - \cdots - g_n u(k) + k_I z(k) \end{aligned} \tag{9.14}$$

such that either the closed-loop poles are at pre-assigned positions in the complex δ-plane; or the system is optimised in some manner; for example, in an LQ sense. Here, the SVF gain vector $\boldsymbol{k}$ is defined as:

$$\boldsymbol{k} = [f_1 \quad f_2 \quad \ldots \quad f_n \quad g_1 \quad \ldots \quad g_n \quad -k_I]^T \tag{9.15}$$

The closed-loop system block diagram obtained directly from the SVF control law (9.14) takes the form shown in Figure 9.1 (cf. Figure 5.3).

The control polynomials in Figure 9.1 are defined as follows:

$$F_1(\delta) = f_1 \delta^{n-1} + \cdots + f_{n-1}\delta; \quad G_1(\delta) = g_1 \delta^{n-1} + \cdots + g_n \tag{9.16}$$

This shows that, as in the standard discrete-time PIP control design, the controller designed in this manner has a basic structure similar to a conventional PI controller with inherent proportional and integral-of-error feedback terms; but, for systems higher than first order,

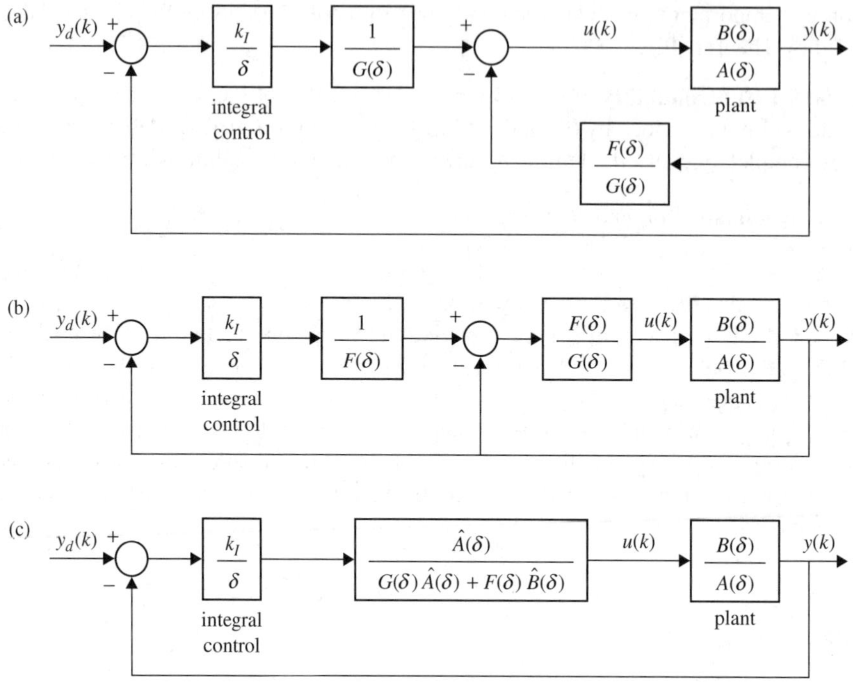

Figure 9.2 Alternative realisable implementations of the system shown in Figure 9.1. (a) $F(\delta)/G(\delta)$ in the feedback path; (b) $F(\delta)/G(\delta)$ in the forward path; (c) Unity gain feedback with a single pre-compensation filter

these basic control actions are enhanced by additional feedback and forward path compensation filters, i.e. $F_1(\delta)$ and $G_1(\delta)$, which exploit the information on the higher order system dynamics that is provided by the TF model and so introduce a general form of derivative action.

9.1.4 Implementation Structures for δ-Operator PIP Control

One of the major attractions of the z^{-1} operator PIP control system designs discussed in previous chapters is that they are easy to implement in practice because the NMSS vector only involves the present and past sampled values of the system output and the past values of its input, all of which are available for direct utilisation in any digital computer implementation of the control system. In contrast, the δ-operator control law involves the discrete δ derivatives of the input and output signals which, as in the case of the continuous-time derivatives required for continuous-time TF model estimation, are not available for direct measurement and cannot be generated directly because of the need for multiple δ differentiation with its associated, unacceptable, noise amplification. This is, of course, a similar situation to that discussed in connection with the estimation of continuous-time TF models in Chapter 8. And the solution is also similar: we look for a way in which realisable prefiltered derivatives that do not amplify high frequency noise can be generated and used in the PIP implementation. The most obvious way to introduce a suitable prefilter into the PIP control system is shown in Figure 9.2, where the two filters:

$$F(\delta) = F_1(\delta) + f_n; \quad G(\delta) = G_1(\delta) + \delta^n \tag{9.17}$$

are combined into a single filter positioned in either the feedback or forward paths of the control system. This filter takes the following general form:

$$\frac{F(\delta)}{G(\delta)} = \frac{f_1 \delta^{n-1} + \cdots + f_{n-1}\delta + f_n}{\delta^n + g_1 \delta^{n-1} + \cdots + g_n} \tag{9.18}$$

In Figure 9.2a, the filter is incorporated into the inner feedback loop; while in Figure 9.2b, it appears in the forward path (cf. Figure 6.1).

The alternative full forward path is shown in Figure 9.2c (cf. Figure 6.2). Here, the forward path filter reveals the *implicit* cancellation of the system denominator polynomial by the PIP controller when there is no plant–model mismatch, i.e. $\hat{A}(\delta) = A(\delta)$, where $\hat{A}(\delta)$ represents the model estimate of $A(\delta)$. As we shall see in Example 9.1, the sensitivity to plant–model mismatch and noise rejection properties are quite different for all three forms of the PIP controller and so the implemented form of the resultant PIP control system becomes a design consideration.

More generally, the design method for δ-operator PIP control of SISO systems follows the same general approach as that used in the standard discrete-time TF model case of Chapter 5, but with the δ-operator equivalents of the various design steps replacing the standard ones. The pole assignment and optimal LQ calculations are outlined below.

9.1.5 Pole Assignment δ-Operator PIP Design

If the *Closed-Loop* (CL) and *Open-Loop* (OL) characteristic polynomials of the PIP system are defined as follows:

$$CL(\delta) = \det\left\{\delta \boldsymbol{I} - \boldsymbol{F} + \boldsymbol{g}\boldsymbol{k}^T\right\} \tag{9.19}$$

$$OL(\delta) = \det\{\delta \boldsymbol{I} - \boldsymbol{F}\} = \delta^{n+1} A(\delta) = \delta^{2n+1} + \sum_{i=1}^{2n+1} a_i \delta^{2n+1-i} \tag{9.20}$$

where $a_i = 0$ for $i > n$, then, by simple algebraic manipulation, it is easy to show that the two polynomials are related to the SVF vector $\boldsymbol{k}$ by:

$$\boldsymbol{k}^T \boldsymbol{q}(\delta) = CL(\delta) - OL(\delta) \tag{9.21}$$

where $\boldsymbol{q}(\delta) = \mathrm{Adj}\,\{\delta \boldsymbol{I} - \boldsymbol{F}\}\,\boldsymbol{g}$. Thus, knowing $\boldsymbol{q}(\delta)$ and $OL(\delta)$, it is possible to solve (9.21) for the feedback gain vector $\boldsymbol{k}$ that ensures any desired closed-loop polynomial $CL(\delta)$, provided only that the pair $[\boldsymbol{F}, \boldsymbol{g}]$ is completely controllable.

To gain greater theoretical insight into the nature of equation (9.21), however, we can substitute from (9.20) into (9.21) and replace $CL(\delta)$ by the desired closed-loop characteristic polynomial $d(\delta)$, where:

$$d(\delta) = \delta^{2n+1} + \sum_{i=1}^{2n+1} d_i \delta^{2n+1-i} \tag{9.22}$$

It is then straightforward to obtain the following equation for the computation of the SVF gain vector $\boldsymbol{k}$ which will ensure that the closed-loop characteristic polynomial is as specified by $d(\delta)$ and, therefore, that the closed-loop poles are at their required locations in the complex δ-plane:

$$\boldsymbol{M}\boldsymbol{S}_1^T \boldsymbol{k} = \boldsymbol{d} - \boldsymbol{p} \tag{9.23}$$

In this equation, $\boldsymbol{d}$ and $\boldsymbol{p}$ are the vectors of coefficients of the desired closed- and open-loop characteristic polynomials of the NMSS system, respectively, i.e.

$$\begin{aligned} \boldsymbol{d}^T &= \begin{bmatrix} d_1 & d_2 & d_3 & \cdots & d_n & d_{n+1} & \cdots & d_{2n} & d_{2n+1} \end{bmatrix} \\ \boldsymbol{p}^T &= \begin{bmatrix} a_1 & a_2 & a_3 & \cdots & a_n & 0 & \cdots & 0 & 0 \end{bmatrix} \end{aligned} \tag{9.24}$$

while $\boldsymbol{S}_1$ is the controllability matrix (9.13) and the matrix $\boldsymbol{M}$ is defined as

$$\boldsymbol{M} = \begin{bmatrix} 1 & 0 & 0 & \cdots & 0 & 0 & \cdots & 0 & 0 \\ a_1 & 1 & 0 & \cdots & 0 & 0 & \cdots & 0 & 0 \\ a_2 & a_1 & 1 & \cdots & 0 & 0 & \cdots & 0 & 0 \\ \vdots & \vdots & \vdots & \ddots & \vdots & \vdots & \ddots & \vdots & \vdots \\ a_n & a_{n-1} & a_{n-2} & \cdots & 1 & 0 & \cdots & 0 & 0 \\ 0 & a_n & a_{n-1} & \cdots & a_1 & 1 & \cdots & 0 & 0 \\ \vdots & \vdots & \ddots & \ddots & \vdots & \vdots & \ddots & \vdots & \vdots \\ 0 & 0 & 0 & a_n & a_{n-1} & a_{n-2} & \cdots & 1 & 0 \\ 0 & 0 & 0 & \cdots & a_n & a_{n-1} & \cdots & a_1 & 1 \end{bmatrix} \tag{9.25}$$

As noted in the case of standard z-operator PIP pole assignment (section 5.3), this derivation is particularly useful in theoretical terms because it reveals very clearly that a solution to the simultaneous equations exists *if and only if* the system is controllable, i.e. if the controllability matrix $\boldsymbol{S}_1$ is of full rank $(2n + 1)$.

9.1.6 *Linear Quadratic Optimal δ-Operator PIP Design*

In the case of LQ optimal control, the aim is to design a feedback gain vector $\boldsymbol{k}$ that will minimise the following quadratic cost function [cf. equation (5.75)]:

$$J = \sum_{i=0}^{\infty} \boldsymbol{x}(i)^T \boldsymbol{Q} \boldsymbol{x}(i) + r v(i)^2 \tag{9.26}$$

where, because of the NMSS formulation, $\boldsymbol{Q}$ is a $(2n + 1 \times 2n + 1)$ symmetric, positive semi-definite matrix; and r is a positive scalar. In this $\delta-$operator context, the optimum feedback gain vector $\boldsymbol{k}$ is given by (Middleton and Goodwin 1990):

$$\boldsymbol{k}^T = (r + \Delta t \boldsymbol{g}^T \boldsymbol{P} \boldsymbol{g})^{-1} \boldsymbol{g}^T \boldsymbol{P}(\boldsymbol{I} + \boldsymbol{F}\Delta t) \tag{9.27}$$

where Δt is the sampling rate and the matrix $\boldsymbol{P}$ is the steady-state solution of the following algebraic matrix Riccati equation:

$$\boldsymbol{Q} + \boldsymbol{F}^T \boldsymbol{P} + \boldsymbol{P}\boldsymbol{F} + \Delta t \boldsymbol{F}^T \boldsymbol{P}\boldsymbol{F} - \boldsymbol{k}(r + \Delta t \boldsymbol{g}^T \boldsymbol{P}\boldsymbol{g})\boldsymbol{k}^T = 0 \tag{9.28}$$

The optimal SVF control law then takes the form:

$$v(k) = -\boldsymbol{k}^T \boldsymbol{x}(k) = -(r + \Delta t \boldsymbol{g}^T \boldsymbol{P}\boldsymbol{g})^{-1} \boldsymbol{g}^T \boldsymbol{P}(\boldsymbol{I} + \boldsymbol{F}\Delta t)\boldsymbol{x}(k) \tag{9.29}$$

In general, this LQ design appears to provide a more satisfactory PIP controller than the pole assignment alternative, normally yielding higher phase and gain margins, and control which appears more robust to uncertainty in the model parameters[1]. However, the pole assignment design reveals more clearly the nature of the design and so this is used in Example 9.1.

Example 9.1 Proportional-Integral-Plus Design for a Non-Minimum Phase Double Integrator System In this section, we consider a simulation example in the form of a non-minimum phase double integrator system that demonstrates the utility of the δ-operator PIP design method. A practical example concerned with the control of a laboratory scale coupled drives rig is described in Young *et al.* (1998). As pointed out in Chapter 8, in a practical situation where suitable input–output, sampled data are available, the identification and estimation of the δ-operator models from data can be accomplished by either directly using a RIV δ-operator estimation algorithm (Young *et al.* 1991), or by conversion of the RIVC estimated continuous-time model to δ-operator form (e.g. in MATLAB®[2] using the c2del routine developed by I. Kaspura in Goodwin and Middleton's original δ-operator Toolbox). Since the system here is a double integrator and so open-loop unstable, this would require the design of an initial stabilising controller and identification within the resulting closed-loop, using a method such as that described in section 8.7.

The simulated system is described by the following δ-operator TF model:

$$y(k) = \frac{-0.5\delta + 1}{\delta^2} u(k) \tag{9.30}$$

[1] This statement refers to a typical 'hand-tuned' design in which the poles are adjusted in simulation by trial and error; of course, any given LQ design can always be reproduced by pole assignment using the LQ closed-loop pole positions.

[2] MATLAB®, The MathWorks Inc., Natick, MA, USA.

where the sampling interval $\Delta t = 0.05$ time units. The NMSS δ-operator model for this system takes the form:

$$\delta \boldsymbol{x}(k) = \begin{bmatrix} 0 & 0 & -0.5 & 1 & 0 \\ 1 & 0 & 0 & 0 & 0 \\ 0 & 0 & 0 & 0 & 0 \\ 0 & 0 & 1 & 0 & 0 \\ 0 & -1 & 0 & 0 & 0 \end{bmatrix} \boldsymbol{x}(k) + \begin{bmatrix} 0 \\ 0 \\ 1 \\ 0 \\ 0 \end{bmatrix} v(k) + \begin{bmatrix} 0 \\ 0 \\ 0 \\ 0 \\ 1 \end{bmatrix} y_d(k) \tag{9.31}$$

in which the NMSS state vector, $\boldsymbol{x}(k)$ is defined as:

$$\boldsymbol{x}(k) = \begin{bmatrix} \delta y(k) & y(k) & \delta u(k) & u(k) & z(k) \end{bmatrix}^T \tag{9.32}$$

The control input to the system $u(k)$ is related to the intermediate variable $v(k)$ by the equation:

$$u(k) = \frac{1}{\delta^2} v(k) \tag{9.33}$$

where $v(k) = -[\, f_1 \ f_2 \ g_1 \ g_2 \ -k_I \,]\boldsymbol{x}(k)$ is the SVF control law (9.14).

Since the NMSS order is 5, the closed-loop system will be fifth order. The desired closed-loop characteristic polynomial $d(\delta)$ is selected so that all five closed-loop poles are at –4 in the complex δ–plane, i.e.

$$d(\delta) = (\delta + 4)^5 = \delta^5 + 20\delta^4 + 160\delta^3 + 640\delta^2 + 1280\delta + 1024 \tag{9.34}$$

These pole positions have been arbitrarily chosen for the purposes of this tutorial simulation example. The vectors of the coefficients of the desired closed-loop and the open-loop characteristic polynomials, $\boldsymbol{d}$ and $\boldsymbol{p}$ are then defined as:

$$\begin{aligned} \boldsymbol{d}^T &= [\,20 \quad 160 \quad 640 \quad 1280 \quad 1024\,] \\ \boldsymbol{p}^T &= [\,0 \quad 0 \quad 0 \quad 0 \quad 0\,] \end{aligned} \tag{9.35}$$

The controllability matrix $\boldsymbol{S}_1$ (9.13) in this case is given by:

$$\boldsymbol{S}_1 = [\boldsymbol{g} \quad \boldsymbol{F}\boldsymbol{g} \quad \boldsymbol{F}^2\boldsymbol{g} \quad \boldsymbol{F}^3\boldsymbol{g} \quad \boldsymbol{F}^4\boldsymbol{g}] = \begin{bmatrix} 0 & -0.5 & 1 & 0 & 0 \\ 0 & 0 & -0.5 & 1 & 0 \\ 1 & 0 & 0 & 0 & 0 \\ 0 & 1 & 0 & 0 & 0 \\ 0 & 0 & 0 & 0.5 & -1 \end{bmatrix} \tag{9.36}$$

and the matrix $\boldsymbol{M}$ (9.25) is the identity matrix. As a result, the vector-matrix equation for the SVF control gains (9.23) takes the form:

$$\boldsymbol{M}\boldsymbol{S}_1^T\boldsymbol{k} = \boldsymbol{d} - \boldsymbol{p} = \begin{bmatrix} 0 & 0 & 1 & 0 & 0 \\ -0.5 & 0 & 0 & 1 & 0 \\ 1 & -0.5 & 0 & 0 & 0 \\ 0 & 1 & 0 & 0 & 0.5 \\ 0 & 0 & 0 & 0 & -1 \end{bmatrix} \begin{bmatrix} f_1 \\ f_2 \\ g_1 \\ g_2 \\ -k_I \end{bmatrix} = \begin{bmatrix} 20 \\ 160 \\ 640 \\ 1280 \\ 1024 \end{bmatrix} - \begin{bmatrix} 0 \\ 0 \\ 0 \\ 0 \\ 0 \end{bmatrix} \tag{9.37}$$

which, since the system is controllable, has the unique solution:

$$f_1 = 1536;\ f_2 = 1792;\ g_1 = 20;\ g_2 = 928;\ k_I = 1024 \tag{9.38}$$

Example 9.2 Simulation Experiments for Non-Minimum Phase Double Integrator Using the control gains (9.38) and system model (9.30), all of the three realisable forms of the δ-operator PIP controller (Figure 9.2) produce the same closed-loop response, as shown in Figure 9.3. This is because the closed-loop TF between the command input and the output is exactly the same in each case. The differences between the various PIP control implementations only become evident when either a load disturbance is applied to the plant, or when there is a mismatch between the characteristics of the model used to calculate the gains and the actual system [i.e. when $\hat{A}(\delta) \neq A(\delta)$ or $\hat{B}(\delta) \neq B(\delta)$, where $\hat{A}(\delta)$ and $\hat{B}(\delta)$ are the estimated model polynomials]. In such cases, the advantages and disadvantages of each control structure are similar to those discussed in section 6.1 (i.e. using the feedback and forward path forms of the standard PIP controller).

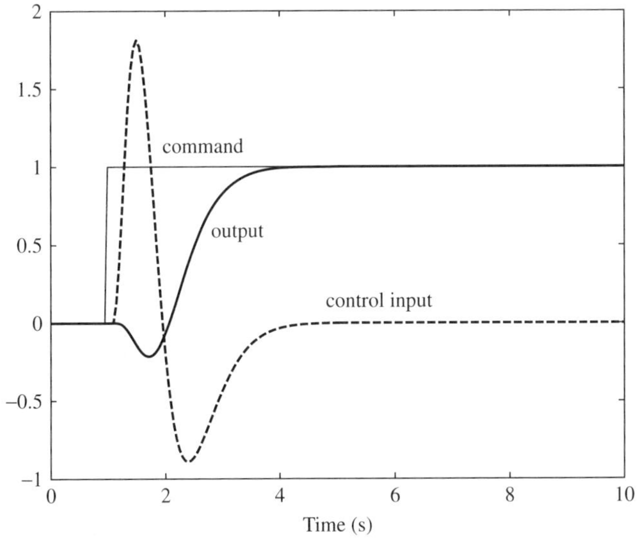

Figure 9.3 Closed-loop response for a unit step in the command level (thin trace) for δ-operator PIP control of non-minimum phase double integrator system in Example 9.1, with desired closed-loop poles all set to –4 in the complex δ-plane, showing the output (bold solid trace) and control input (dashed trace)

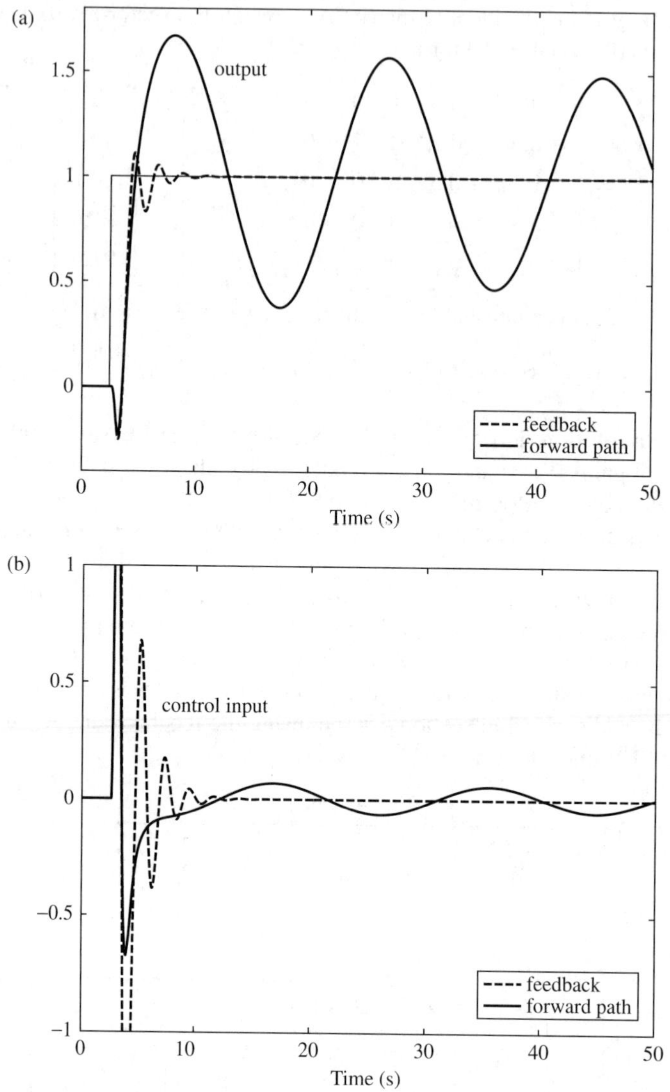

Figure 9.4 Closed-loop response for a unit step in the command level (thin trace) for δ-operator PIP control of non-minimum phase double integrator system in Example 9.1, with a mismatch between the model and system denominator polynomials. (a) Output response obtained using the feedback (dashed trace) and forward path (bold solid trace) forms; (b) Control input signals

For the main two control implementations, i.e. Figure 9.2a and c, the responses when there is a mismatch between the model and plant *denominator* polynomials are illustrated in Figure 9.4. Here, the model denominator polynomial is set to $\hat{A}(\delta) = \delta^2 + 0.2\delta + 0.09$ rather than the true polynomial $A(\delta) = \delta^2$: hence the design is based on open-loop poles that are stable and located at $-0.1 \pm 0.2828j$ in the complex δ-plane, whereas the true system poles are at the origin. The step responses in Figure 9.4 show that, whilst the output generated by the standard PIP feedback form of Figure 9.2a becomes somewhat oscillatory in the presence of this mismatch, the oscillations are reasonably damped, in contrast to those from the pre-compensator implementation of Figure 9.2c, which is characterised by very poorly damped and unacceptable oscillations.

Figure 9.5 shows that the performances of the same two PIP control implementations are reversed when there is mismatch between the model and actual *numerator* polynomials in the system TF. In contrast to the results in Figure 9.4, the sensitivity of the forward path pre-compensation form to model mismatch in $B(\delta)$ is quite small, resulting in a small overshoot

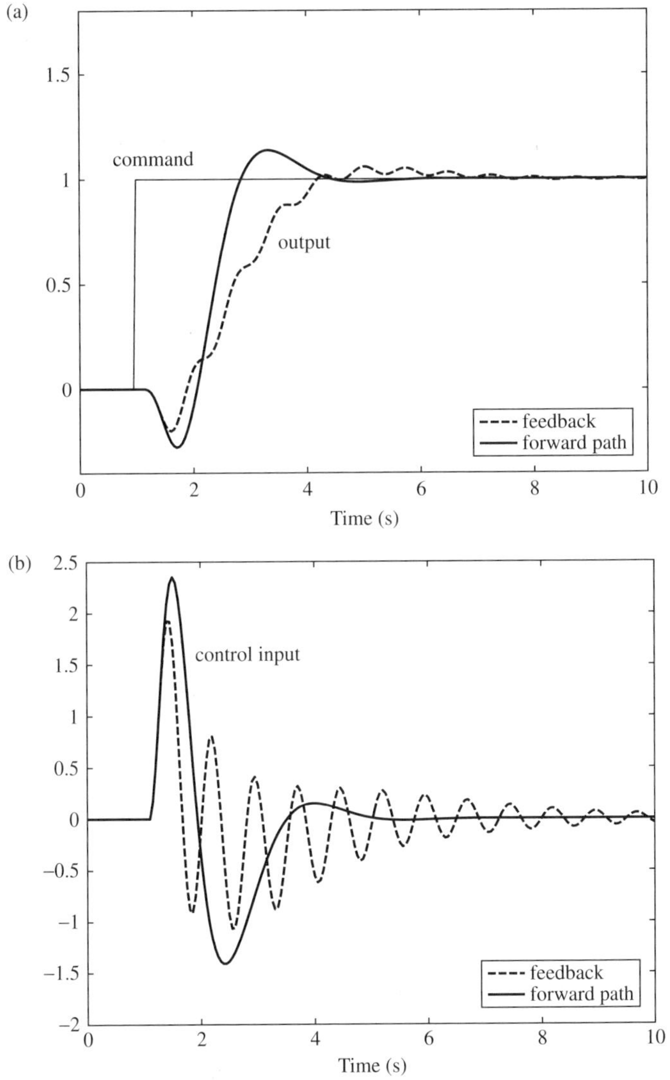

Figure 9.5 Closed-loop response for a unit step in the command level (thin trace) for δ-operator PIP control of non-minimum phase double integrator system in Example 9.1, with a mismatch between the model and system numerator polynomials. (a) Output response obtained using the feedback (dashed trace) and forward path (bold solid trace) forms; (b) Control input signals

above the desired output; whereas the standard feedback implementation response is now sluggish with a pronounced, lowly damped oscillation, which is particularly noticeable on the control input. In this example $\hat{B}(\delta) = -0.5\delta + 0.77$ rather than the actual system numerator polynomial $B(\delta) = -0.5\delta + 1$.

The responses of the two PIP control implementations to the addition of a step load disturbance of magnitude 0.1 after 5 time units are shown in Figure 9.6. Again, the forward

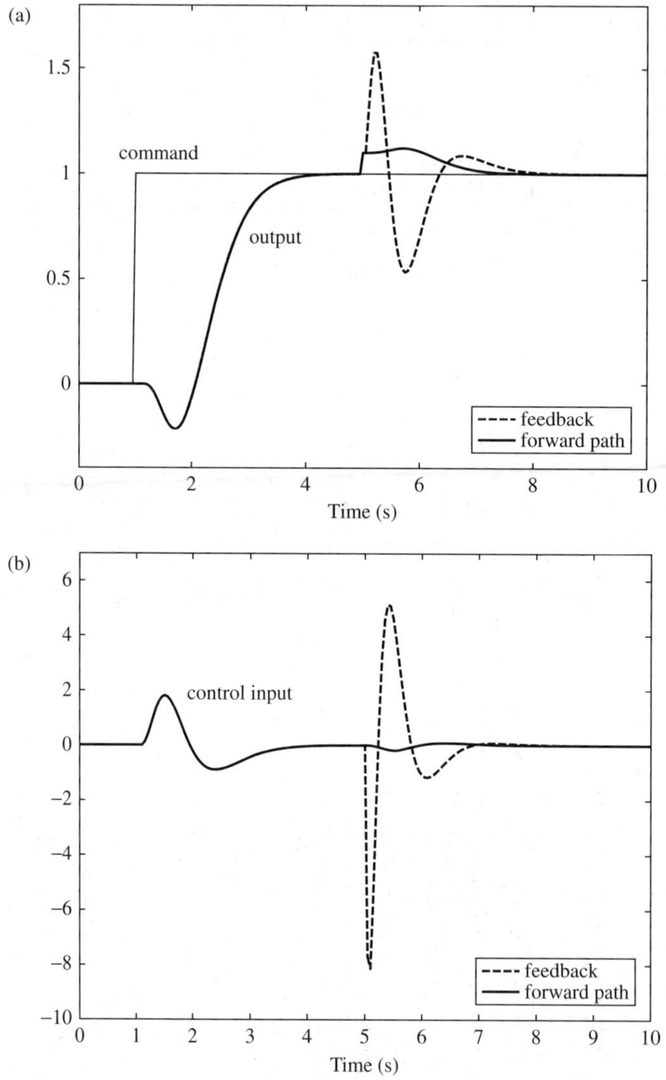

Figure 9.6 Closed-loop response for a unit step in the command level (thin trace) for δ-operator PIP control of non-minimum phase double integrator system in Example 9.1, with the addition of a step load disturbance (magnitude 0.1) to the output signal after 5 s. (a) Output response obtained using the feedback (dashed trace) and forward path (bold solid trace) forms; (b) Control input signals

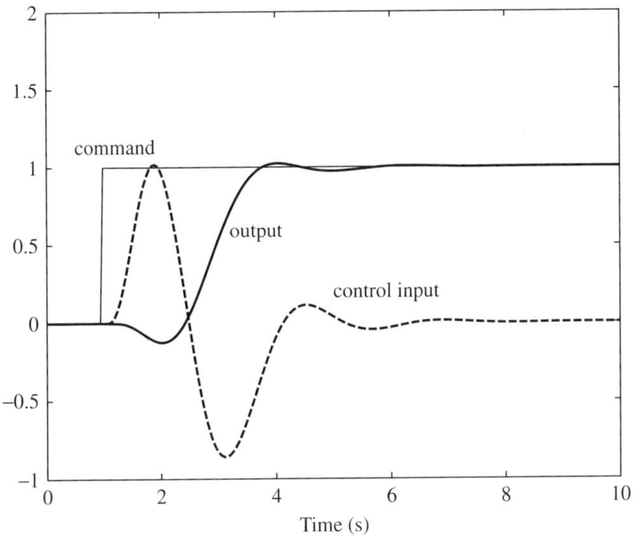

Figure 9.7 Closed-loop response for a unit step in the command level (thin trace) for δ-operator PIP-LQ optimal control of non-minimum phase double integrator system in Example 9.1, showing the output (bold solid trace) and control input (dashed trace)

path pre-compensation implementation of Figure 9.2c yields superior performance: the output is only slightly displaced away from the desired level by the load disturbance and there is little effect on the input signal. In contrast, the feedback form of the controller produces an oscillation in the output that is much larger than the magnitude of the disturbance itself and is accompanied by an even greater oscillation in the control input.

Finally, Figure 9.7 shows the closed-loop responses obtained using the LQ design procedure. Here, the input weighting in the cost function r is set to unity and the $\boldsymbol{Q}$ weighting matrix is chosen to be diagonal, with the first four elements set to unity and the integral of error weighting set to 10^4. The control gains generated by this LQ design are:

$$f_1 = 175.074; \quad f_2 = 169.938; \quad g_1 = 8.631; \quad g_2 = 123.699; \quad k_1 = 80.521 \tag{9.39}$$

These gains produce closed-loop poles at $-1.15 \pm 2.44j$, $-2.27 \pm 1.01j$ and -1.78. This choice of $\boldsymbol{Q}$ and r yields a closed-loop response time similar to that of the pole assignment design, as shown in Figure 9.7 but with somewhat less non-minimum phase undershoot and a small overshoot.

9.2 Time Variable Parameter Estimation

The present section briefly reviews an approach to TVP modelling that can be used for real-time TDC applications, such as self-tuning or adaptive control. More significantly, however, the approach also facilitates the development of SDP techniques for nonlinear PIP control, as discussed in section 9.3. In self-tuning control, it is assumed that the controlled system dynamics are not changing over time, so that the model parameters are constant. In this situation, the variance of the recursive parameter estimates becomes smaller as time progresses and the data size increases, so that the model used for control system design becomes ever more accurate. In many practical situations, however, the assumption of time-invariant parameters will be inappropriate since, at least to some extent, most real systems exhibit dynamic characteristics that change over time. Consequently, it is probably safer to assume that parameter changes may occur and introduce some method of recursive estimation that allows for the estimation of such changes. Amongst the different approaches to such TVP estimation, the following three are deserving of most attention here:

1. **The *Extended* (or re-linearised) *Kalman Filter* (EKF)** (e.g. Kopp and Orford 1963; Ljung 1979). Here, the stochastic state space model of the dynamic system is extended to include simple stochastic models for the TVPs (e.g. the simple random walk model: see later). The resulting model is nonlinear because the original system state variables are multiplied by the state variables arising from the adjoined TVPs. As a result, the state estimates of this nonlinear model are then updated by a KF-type algorithm with the equations linearised in some manner at each recursive update, based on the current recursive parameter estimates. This is widely applicable but the parameter tracking ability is dependent on the application.
2. **Shaping the memory of the algorithm** (e.g. Bray *et al.* 1965; Young 1969; Wellstead and Zarrop 1991). Here, the predecessor of the *Kalman Filter* (KF), the recursive algorithm developed by K.F. Gauss (1826) for estimating the constant parameters in linear regression models (Young 2011), is modified to include a 'forgetting factor' or 'weighting kernel' that shapes the memory of the estimator and so allows for TVP estimation. This is much the most popular approach to TVP estimation but its performance is rather limited when compared with the third approach.
3. **Modelling the parameter variations** (e.g. Mayne 1963; Lee 1964; Young 1969). Here, the roles of the state equations and observation equations are reversed. The model of the system now appears in the observation equation and the state equations are used to model the TVPs appearing in this model, again using simple stochastic models such as the random walk. This is the most sophisticated and flexible approach and represents the current state of the art in TVP estimation.

Until comparatively recently, the main emphasis in all three of these approaches has been the 'online' or 'real-time' estimation of the TVPs. As a result, most algorithms have been of the 'filtering' type, where the estimate $\hat{\boldsymbol{\rho}}(k|k)$ of the TVP vector $\boldsymbol{\rho}(k)$, at any sampling instant k, is a function of all the data up to and including this kth instant. Surprisingly, given its ultimate power, the extension of these methods to the 'offline' analysis situation was not considered very much at first, despite the fact that mechanisms for such 'smoothing' estimation were suggested in the 1960s. In these *Fixed Interval Smoothing* (FIS) algorithms (e.g. Bryson and Ho 1969 and the prior references therein; Norton 1975, 1986; Young 1984, 1999, 2011), the FIS estimate $\hat{\boldsymbol{\rho}}(k|N)$ of $\boldsymbol{\rho}(k)$ is based on *all* of the data available over a 'fixed interval' of N samples, usually the full sample length of the time series data. The later research in the above references placed this approach in an optimal context based on *Maximum Likelihood* (ML) estimation of the associated '*hyper-parameters*' that control the nature of the smoothing operations (see later). MATLAB® Toolboxes that allow for the estimation of such TVP models based on KF/FIS methods, and with ML optimisation of the hyper-parameters, include CAPTAIN (Appendix G; Taylor *et al.* 2007a) and SSpace (Pedregal and Taylor 2012).

9.2.1 Simple Limited Memory Algorithms

Before illustrating the value of a unified, statistical approach to TVP estimation based on modelling the parameter variations, it is instructive to take a brief look at the less sophisticated, deterministic algorithms based on explicitly restricting the memory of the recursive estimation algorithm. For simplicity, let us consider a SISO system (although the extension to multi-input systems is straightforward). In the case of a TVP or *Dynamic*[3] *Transfer Function* (DTF) representation, the model takes the following form:

$$y(k) = \frac{B(z^{-1}, k)}{A(z^{-1}, k)} u(k - \delta) + \xi(k) \qquad t = 1, \ldots, N \tag{9.40}$$

where $\delta \geq 0$ represents the time delay (not to be confused with the δ-operator of the previous subsection) and $\xi(k)$ is the additive noise, while $A(z^{-1}, k)$ and $B(z^{-1}, k)$ are now *time variable coefficient polynomials* in z^{-1} of the following form:

$$\begin{aligned} A(z^{-1}, k) &= 1 + a_1(k)z^{-1} + a_2(k)z^{-2} + \cdots + a_n(k)z^{-n} \\ B(z^{-1}, k) &= b_0(k) + b_1(k)z^{-1} + b_2(k)z^{-2} + \cdots + b_m(k)z^{-m} \end{aligned} \tag{9.41}$$

In the more restricted case of the *Dynamic Auto-Regressive, eXogenous variables* (DARX) model, the additive noise $\xi(k)$ is defined as:

$$\xi(k) = \frac{1}{A(z^{-1}, k)} e(k) \qquad e_k = \mathcal{N}(0, \sigma^2) \tag{9.42}$$

where, as in the constant parameter situation, $e(k)$ is assumed to be is a zero mean, normally distributed, white noise sequence with variance σ^2. This model has the advantage that equation (9.40) can be written in the following alternative vector equation or 'regression' form:

$$y(k) = \boldsymbol{\phi}^T(k)\boldsymbol{\rho}(k) + e(k) \tag{9.43}$$

where

$$\begin{aligned} \boldsymbol{\phi}^T(k) &= [-y(k) \ - y(k-1) \ \cdots \ - y(k-n) \ u(k-\delta) \ \cdots \ u(k-\delta-m)] \\ \boldsymbol{\rho}(k) &= [a_1(k) \ a_2(k) \ \cdots \ a_n(k) \ b_0(k) \ b_1(k) \ \cdots \ b_m(k)]^T = [\rho_1(k) \ \rho_2(k) \ \cdots \ \rho_{n+m+1}(k)]^T \end{aligned} \tag{9.44}$$

Equation (9.40) and equation (9.42), or equivalently equation (9.43), can be compared with the time-invariant model given by equations (8.8).

[3] The term 'dynamic' is used here for historical reasons, primarily because the parameters are defined as evolving in a stochastic, dynamic manner.

If we wish to limit the memory of the estimation algorithm, it is necessary to specify the nature of the memory process. Two main memory functions have been suggested: *Rectangular-Weighting-into-the-Past* (RWP); and *Exponential-Weighting-into-the-Past* (EWP). The latter approach is the most popular and can be introduced into the least squares problem formulation by considering EWP least squares optimisation of the form [cf. equation (8.9)]:

$$\hat{\rho}(N) = \arg\min_{\rho} \mathcal{J}_2^{EWP}(\rho) \qquad \mathcal{J}_2^{EWP}(\rho) = \sum_{k=1}^{N} [y(k) - \boldsymbol{\phi}^T(k)\boldsymbol{\rho}]^2 \alpha(N-k) \tag{9.45}$$

where $0 < \alpha < 1.0$ is a constant related to the time constant T_e of the exponential weighting by the expression $\alpha = \exp(-\Delta t / T_e)$, and Δt is the sampling interval in time units appropriate to the application. Of course, with $\alpha = 1.0$, $\mathcal{J}_2^{EWP}$ becomes the usual, constant parameter, least squares cost function $\mathcal{J}_2$.

The recursive algorithm derived by the minimisation of the EWP cost function (9.45) takes the form:

$$\begin{aligned}\hat{\rho}(k) &= \hat{\rho}(k-1) + \boldsymbol{P}(k-1)\boldsymbol{\phi}(k)[\alpha + \boldsymbol{\phi}^T(k)\boldsymbol{P}(k-1)\boldsymbol{\phi}(k)]^{-1}\{y(k) - \boldsymbol{\phi}^T(k-1)\hat{\rho}(k-1)\}\\ \boldsymbol{P}(k) &= \frac{1}{\alpha}\{\boldsymbol{P}(k-1) - \boldsymbol{P}(k-1)\boldsymbol{\phi}(k)[\alpha + \boldsymbol{\phi}^T(k)\boldsymbol{P}(k-1)\boldsymbol{\phi}(k)]^{-1}\boldsymbol{\phi}^T(k)\boldsymbol{P}(k-1)\}\end{aligned} \tag{9.46}$$

This estimation algorithm is one form of the recursive EWP least squares algorithm, although other forms are possible. These can all be considered in terms of the EWP coefficient α, or 'forgetting factor', as it is often called. Amongst the possibilities (see e.g. Wellstead and Zarrop 1991) are *constant trace algorithms*, the use of *variable or adaptive forgetting factors*, including *start-up forgetting factors*; and *Directional Forgetting* (DF), first suggested by Kulhavy (1987).

In the latter DF algorithm, the $\boldsymbol{P}(k)$ matrix update takes the form:

$$\boldsymbol{P}(k) = \{\boldsymbol{P}(k-1) - \boldsymbol{P}(k-1)\boldsymbol{\phi}(k)[r^{-1}(k-1) + \boldsymbol{\phi}^T(k)\boldsymbol{P}(k-1)\boldsymbol{\phi}(k)]^{-1}\phi^T(k)\boldsymbol{P}(k-1)\} \tag{9.47}$$

where a typical choice for $r(k)$ is:

$$r(k) = \alpha^* + \frac{1-\alpha^*}{\boldsymbol{\phi}^T(k+1)\boldsymbol{P}(k)\boldsymbol{\phi}(k+1)} \tag{9.48}$$

in which α^* plays a similar role to α in the EWP algorithm (9.46).

9.2.2 *Modelling the Parameter Variations*

TVP estimation is best considered as a unified operation that involves both recursive filtering and smoothing, based on modelling the parameter variations in a stochastic, state space manner. Here, the time series data are processed sequentially, first by the 'forward-pass' filtering algorithm that provides the online parametric estimate $\hat{\boldsymbol{\rho}}(k \mid k)$ $k = 1, 2, \ldots N$, as well as any predictions (forecasts) $\hat{\boldsymbol{\rho}}(k + f \mid k)$ into the future, where f is the forecasting horizon. Following this, the 'backward-pass' FIS algorithm updates these filtered estimates to yield the smoothed estimate $\hat{\boldsymbol{\rho}}(k \mid N)$ $k = N, N - 1, \ldots, 1$, as well as any interpolations over gaps in the data; or backcasts $\hat{\boldsymbol{\rho}}(k - b|k)$, where b is the backcasting horizon (normally for $k = 1$ at the beginning of the data set to yield a backcast into the past). Although TVP estimation of this type can be applied most easily to ARX and ARMAX models, we will now consider how it has been extended (Young 2000) to the DTF model (9.40).

Reflecting the statistical setting of the analysis and referring to previous research on this topic, it seems desirable if the temporal variability of the model parameter vector $\boldsymbol{\rho}(k)$ is characterised in some stochastic manner. Normally, when little is known about the nature of the time variability, this model needs to be both simple and flexible. One of the simplest and most generally useful class of stochastic, state space models involves the assumption the ith parameter, $\rho_i(k)$, $i = 1, 2, \ldots, n + m + 1$, in $\boldsymbol{\rho}(k)$ is defined by a two-dimensional stochastic state vector $\boldsymbol{x}_i(k) = [\rho_i(k) \;\; \nabla\rho_i(k)]^T$, where $\rho_i(k)$ and $\nabla\rho_i(k)$ are, respectively, the changing 'level' and 'slope' of the associated TVP. This selection of a two-dimensional state representation of the TVPs is based on practical experience over a number of years. Initial research tended to use a simple scalar random walk model for the parameter variations but subsequent research showed the value of modelling not only the level changes in the TVPs but also their rates of change.

The stochastic evolution of each $\boldsymbol{x}_i(k)$ [and, therefore, of each of the $n + m + 1$ parameters in $\boldsymbol{\rho}(k)$] is assumed to be described by *one* of the *Generalised Random Walk* (GRW: Young and Ng 1989; Young 1999, 2011) family defined in the following state space terms:

$$\boldsymbol{x}_i(k) = \boldsymbol{F}_i\boldsymbol{x}_i(k-1) + \boldsymbol{G}_i\boldsymbol{\eta}_i(k) \qquad i = 1, 2, \ldots, n + m + 1 \tag{9.49}$$

where[4]

$$\boldsymbol{F}_i = \begin{bmatrix} \alpha_i & \beta_i \\ 0 & \gamma_i \end{bmatrix}; \qquad \boldsymbol{G}_i = \begin{bmatrix} \delta_i & 0 \\ 0 & \varepsilon_i \end{bmatrix} \tag{9.50}$$

and $\boldsymbol{\eta}_i(k) = [\eta_{1i}(k)\eta_{2i}(k)]^T$ is a 2×1, zero mean, white noise vector that allows for stochastic variability in the parameters and is assumed to be characterised by a (normally diagonal) covariance matrix $\boldsymbol{Q}_{\eta i}$. Of course, equation (9.49) is a generic model formulated in this manner only to unify various random walk-type models: it is never used in its entirety since it is clearly over-parameterised.

This general model comprises, as special cases, the integrated random walk (IRW: $\alpha_i = \beta_i = \gamma_i = \varepsilon_i = 1; \delta_i = 0$); the scalar random walk [RW: scalar but equivalent to (9.49) if $\beta_i = \gamma_i = \varepsilon_i = 0; \alpha_i = \delta_i = 1$]; the first order auto-regressive AR(1) model (also scalar, with $\beta_i = \gamma_i = \varepsilon_i = 0$; $0 < \alpha_i < 1$; $\delta_i = 1$); the intermediate case of smoothed random walk (SRW: $0 < \alpha_i < 1$; $\beta_i = \gamma_i = \varepsilon_i = 1$; $\delta = 0$); and, finally, both the local linear trend (LLT: $\alpha_i = \beta_i = \gamma_i = \varepsilon_i = 1$; $\delta_i = 1$) and damped trend ($\alpha_i = \beta_i = \delta_i = \varepsilon_i = 1$; $0 < \gamma_i v < 1$). Note that the LLT model can be considered simply as the combination of the simpler

[4] Do not confuse the subscripted α_i here with the α used in the limited memory algorithms.

RW and IRW models. The various, normally constant, parameters in this GRW model (α_i, β_i, γ_i, δ_i, ε_i) and the elements of $\boldsymbol{Q}_{\eta_i}$ are normally referred to as 'hyper-parameters'. This is to differentiate them from the TVPs that are the main object of the estimation analysis. However, the hyper-parameters are also assumed to be unknown *a priori* and need to be estimated from the data, normally under the assumption that they are time-invariant.

In the case of the RW model, i.e.

$$\rho_i(k) = \rho_i(k-1) + \eta_{1i}(k) \tag{9.51}$$

each parameter can be assumed to be time-invariant if the variance of the white noise input $\eta_{1i}(k)$ is set to zero. Then the stochastic TVP setting reverts to the more normal, constant parameter TF model situation. In other words, if RW models with zero variance, white noise inputs are specified for the model parameters, the recursive *Instrumental Variable* (IV) estimation algorithm described below for the general stochastic TVP case will provide recursive estimates that are identical to those obtained with the normal recursive IV estimation algorithm for TF models with constant parameters (see section 8.2).

9.2.3 State Space Model for DTF Estimation

Having introduced the GRW models for the individual parameter variations, an overall state space model can then be constructed straightforwardly by the aggregation of the subsystem matrices defined in (9.50), with the 'observation' equation defined by the model equation (9.43), i.e.

$$\begin{aligned} \text{Observation equation: } y(k) &= \boldsymbol{h}^T(k)\boldsymbol{x}(k) + \mu(k) \quad \text{(i)} \\ \text{State equations: } \boldsymbol{x}(k) &= \boldsymbol{F}\boldsymbol{x}(k-1) + \boldsymbol{G}\boldsymbol{\eta}(k) \quad \text{(ii)} \end{aligned} \tag{9.52}$$

If $n_p = n + m + 1$, then

$$\boldsymbol{x}(k) = \left[\rho_1(k) \quad \nabla\rho_1(k) \quad \rho_2(k) \quad \nabla\rho_2(k) \ \ldots \ \rho_{n_p}(k) \quad \nabla\rho_{n_p}(k)\right]^T \tag{9.53}$$

while $\boldsymbol{F}$ is a $2n_p \times 2n_p$ block diagonal with blocks defined by the $\boldsymbol{F}_i$ matrices in (9.50); $\boldsymbol{G}$ is a $2n_p \times 2n_p$ block diagonal matrix with blocks defined by the corresponding sub–system matrices $\boldsymbol{G}_i$ in (9.50); and $\boldsymbol{\eta}(k)$ is a $2n_p$-dimensional vector containing, in appropriate locations, the white noise input vectors $\eta_i(k)$ ('system disturbances' in normal state space terminology) to each of the GRW models. These white noise inputs, which provide the stochastic degree of freedom to allow for parametric change in the model, are assumed to be independent of the observation noise $e(k)$ and have a block-diagonal covariance matrix $\boldsymbol{Q}$ formed from the combination of the individual covariance matrices $\boldsymbol{Q}_{\eta i}$. Finally, $\boldsymbol{h}^T(k)$ is a $1 \times 2n_p$ vector of the following form:

$$\boldsymbol{h}^T(k) = [-y(k-1)\ 0 \quad -y(k-2)\ 0 \cdots -y(k-n) \ \ 0 \ \ u(k-\delta) \ \ 0 \cdots u(k-\delta-1) \ \ 0] \tag{9.54}$$

that relates the scalar observation $y(k)$ to the state variables defined by (9.52) (ii), so that it represents the DTF model (9.40), with each parameter defined as a GRW process and

$\mu(k) = A(z^{-1}, k)\xi(k)$, which is a complex, non-stationary noise process. In the case of the scalar RW and AR(1) models for parameter variation, the alternate zeros are omitted.

In the unlikely event that the noise variable $\mu(k)$ in (9.52) (i) happens to be zero mean, white noise, then the above TVP model reduces to the simpler DARX model in which the system noise component is defined by equation (9.42). It is well known that this DARX model can be treated as a 'linear regression' relationship and that the standard forms of the KF and FIS algorithms can be used, very successfully, to estimate the TVPs. The problem is, of course, that $\mu(k)$ is not white and Gaussian, even if $\xi(k)$ has these desirable properties. This difference is very important in the DTF context since it can be shown that the TVP estimates obtained from the standard recursive filtering/smoothing algorithm will be inconsistent and so asymptotically biased away from their 'true' values. Note that, strictly, the terms 'consistency' and 'bias', as used here, apply only to constant parameter, stationary models but they are used informally here because similar behaviour is encountered with TVP models.

The level of this asymptotic bias is dependent on the magnitude of the measurement noise and it can be problematic in high noise situations, particularly if the parameters are physically meaningful. For this reason, it is necessary to modify the standard algorithms to avoid biasing problems. This can be achieved by attempting to model the noise $\mu(k)$ in some moving average manner (see e.g. Norton 1975, 1986). However, since $\mu(k)$ is a complex, non-stationary noise process, its complete estimation is not straightforward.

An alternative approach, which does not require modelling $\mu(k)$ provided it is independent of the input $u(k)$, is the recursive-iterative DTF algorithm. This algorithm can be implemented in various ways (Young 2011). The simplest version is outlined below to give some idea of its main features, while the interested reader can use the *dtfm* routine in the CAPTAIN Toolbox (Appendix G; Taylor *et al.* 2007a) for applications. In the first part of the algorithm, a 'symmetric matrix' version of the standard iterative IV algorithm, as described in section 8.2, is used to estimate the time variable parameters. The results obtained at the final iteration are then processed by the recursive FIS algorithm.

1. **Forward-Pass Symmetric IV Equations** (iterative)
   Iterate recursive equation (9.55) and equation (9.56) for $j = 1, 2, \ldots, I_T$, with *Recursive Least Squares* (RLS) estimation used at the first iteration, i.e. $\hat{\boldsymbol{h}}(k) = \boldsymbol{h}(k)$ for $j = 1$:
   Prediction:

$$\begin{aligned}\hat{\boldsymbol{x}}(k \mid k-1) &= \boldsymbol{F}\hat{\boldsymbol{x}}(k-1)\\ \hat{\boldsymbol{P}}(k \mid k-1) &= \boldsymbol{F}\hat{\boldsymbol{P}}(k-1)\boldsymbol{F}^T + \boldsymbol{G}\boldsymbol{Q}_r\boldsymbol{G}^T\end{aligned} \tag{9.55}$$

Correction:

$$\begin{aligned}\hat{\boldsymbol{x}}(k) &= \hat{\boldsymbol{x}}(k \mid k-1) + \hat{\boldsymbol{P}}(k \mid k-1)\hat{\boldsymbol{h}}(k)\left[1 + \hat{\boldsymbol{h}}^T(k)\hat{\boldsymbol{P}}(k \mid k-1)\hat{\boldsymbol{h}}(k)\right]^{-1}\\ &\quad \{\, y(k) - \boldsymbol{h}^T(k)\hat{\boldsymbol{x}}(k \mid k-1)\,\}\\ \hat{\boldsymbol{P}}(k) &= \hat{\boldsymbol{P}}(k \mid k-1) + \hat{\boldsymbol{P}}(k \mid k-1)\hat{\boldsymbol{h}}(k)\left[1 + \hat{\boldsymbol{h}}^T(k)\hat{\boldsymbol{P}}(k \mid k-1)\hat{\boldsymbol{h}}(k)\right]^{-1}\\ &\quad \hat{\boldsymbol{h}}^T(k)\hat{\boldsymbol{P}}(k \mid k-1)\end{aligned} \tag{9.56}$$

where, for $j > 1$,

$$\begin{aligned}\hat{\boldsymbol{h}}^T(k) &= [-\hat{x}(k-1), 0, -\hat{x}(k-2), 0, \ldots, \hat{x}(k-n), 0, \ldots, u(k-\delta),\\ &\quad 0, \ldots, u(k-\delta-m), 0]\\ \hat{x}(k) &= \frac{\hat{B}_{j-1}(z^{-1}, k)}{\hat{A}_{j-1}(z^{-1}, k)} u(k-\delta)\end{aligned} \tag{9.57}$$

The following FIS algorithm is in the form of a backward recursion operating from the end of the sample set to the beginning.

2. **Backward-Pass Fixed Interval Smoothing IV equations** (single pass)

$$\begin{aligned}\hat{\boldsymbol{x}}(k \mid N) &= \boldsymbol{F}^{-1}\left[\hat{\boldsymbol{x}}(k+1 \mid N) + \boldsymbol{G}\boldsymbol{Q}_r\boldsymbol{G}^T\boldsymbol{L}(k)\right]\\ \boldsymbol{L}(k) &= [\boldsymbol{I} - \hat{\boldsymbol{P}}(k+1)\hat{\boldsymbol{h}}(k+1)\hat{\boldsymbol{h}}^T(k+1)]^T\\ &\quad [\boldsymbol{F}^T\boldsymbol{L}(k+1) - \boldsymbol{h}(k+1)\{y(k+1) - \hat{\boldsymbol{h}}^T(k+1)\hat{\boldsymbol{x}}(k+1)\}]\\ \hat{\boldsymbol{P}}(k \mid N) &= \hat{\boldsymbol{P}}(k) + \hat{\boldsymbol{P}}(k)\boldsymbol{F}^T\hat{\boldsymbol{P}}^{-1}(k+1 \mid k)\left[\hat{\boldsymbol{P}}(k+1 \mid N) - \hat{\boldsymbol{P}}(k+1 \mid k)\right]\hat{\boldsymbol{P}}^{-1}(k+1 \mid k)\boldsymbol{F}\hat{\boldsymbol{P}}(k)\end{aligned} \tag{9.58}$$

with $\boldsymbol{L}(N) = 0$. In these recursions, the $n_p \times n_p$ *Noise Variance Ratio* (NVR) matrix $\boldsymbol{Q}_r$ and the $n_p \times n_p$ matrix $\hat{\boldsymbol{P}}(k)$ are defined as follows:

$$\boldsymbol{Q}_r = \frac{\boldsymbol{Q}}{\sigma^2}; \quad \hat{\boldsymbol{P}}(k) = \frac{\boldsymbol{P}(k)}{\sigma^2} \tag{9.59}$$

Here, $\boldsymbol{P}(k) = \sigma^2\hat{\boldsymbol{P}}(k)$ is the estimated parametric covariance matrix in the DARX case, where the square roots of its diagonal elements provide an estimate of the *Standard Errors* (SEs) on the TVP estimates. However, care must be taken with its interpretation in the DTF case, where the estimation provides only an approximation in this regard. Note also that an alternative FIS algorithm is available in which, at each backwards recursion, the estimate $\hat{\boldsymbol{x}}(k \mid N)$ is based on an update of the filtering estimate $\hat{\boldsymbol{x}}(k)$. This can be specified as an alternative to (9.58) in the *dtfm* algorithm of the CAPTAIN Toolbox. The advantage of the smoothing recursions is that they provide 'lag-free', lower variance estimates of the TVPs.

The main difference between the above algorithm and the standard recursive filtering–smoothing algorithms is the introduction of 'hats' on the $\hat{\boldsymbol{h}}(k)$ and the $\hat{\boldsymbol{P}}(k)$ matrix, and the use of an iterative IV solution in the forward-pass algorithm. In the standard algorithm, which applies for the simpler DARX model, $\hat{\boldsymbol{h}}(k)$ is replaced by $\boldsymbol{h}(k)$ in (9.57) and there is no need for iteration in the forward-pass. In equations (9.56) $\hat{\boldsymbol{h}}(k)$ is the IV vector, which is used by the algorithm in the generation of all the $\hat{\boldsymbol{P}}(k)$ terms and is the main vehicle in removing the bias from the TVP estimates, as discussed in Chapter 8. The subscript *j*–1 on $\hat{A}_{j-1}(z^{-1}, k)$ and $\hat{B}_{j-1}(z^{-1}, k)$ indicates that the estimated DTF polynomials in the auxiliary model (9.57), which generates the instrumental variables $\hat{x}(k)$ that appear in the definition of $\hat{\boldsymbol{h}}(k)$, are updated in an iterative manner, starting with the least squares estimates of these polynomials for $j = 1$. In order to ensure the stability of the auxiliary model at each iteration a stability check is applied to the estimated denominator polynomial $\hat{A}(z^{-1}, k)$ and, if necessary, the polynomial is stabilised (using, for example, the MATLAB® *polystab* routine). Iteration is continued until the forward pass IV estimates are no longer changing significantly: normally only 4 or 5 iterations are required.

This recursive-iterative approach exploits the 'symmetric gain' version (see **III**s in section 8.3.4) of the IV algorithm (Young 1970), rather than the more usual asymmetric version. This is necessary in order that the standard recursive FIS algorithm in (9.58) can be used to generate the smoothed estimates of the TVPs. As we have seen previously in Chapter 8, optimal RIV algorithms exploit adaptive prefiltering to induce optimality in a ML sense. It is possible to extend the above algorithm to include such adaptive prefiltering but this requires TVP prefilters, increasing the complexity of the algorithm.

Note that when it is applied to the general TF model, the DTF algorithm is an *offline* algorithm used for investigating the nature of any time variable parameters prior to online, real-time estimation (e.g. in self-tuning or adaptive control applications). The DARX option can be used in real-time but is limited to ARX models. The recently developed *Real-Time Refined IV* algorithm does include adaptive prefiltering (Young 2010) and the *Recursive Prediction Error Minimisation* (RPEM) algorithm in the MATLAB® System Identification Toolbox fulfils a similar role, although it does not use IV estimation or have a facility (such as *dtfmopt* in the CAPTAIN Toolbox) for optimising the hyper-parameters. However, neither of these real-time algorithms have been extended yet to include FIS estimation for offline estimation purposes.

9.2.4 *Optimisation of the Hyper-parameters*

In order to utilise the above TVP estimation algorithm, it is necessary to optimise the values of the hyper-parameters in relation to the data being modelled, including $\boldsymbol{Q}_r$ in equations (9.59). This optimisation can be achieved in various ways but the best known approach is to use ML optimisation based on 'prediction error decomposition' (Schweppe 1965). This derives originally from research that showed how to generate likelihood functions for Gaussian signals using the Kalman Filter (see also Bryson and Ho 1969). Its importance in the present context was probably first recognised by Harvey (1981) and Kitagawa (1981). It has become one of the two standard approaches to the problem [the other being the *Expectation and Minimisation* (EM) algorithm]. Other alternatives are simply to optimise the hyper-parameters so that they minimise the sum of the squares of the innovation errors $y(k) - \hat{\boldsymbol{h}}^T(k)\hat{\boldsymbol{x}}(k \mid k-1)$ in (9.56); or to use frequency domain optimisation. The advantages of an optimal approach to hyper-parameter estimation is that it is relatively objective and removes the need for manual selection of the hyper-parameters that characterises less sophisticated TVP estimation algorithms, such as the EWP forgetting algorithm. This is illustrated for a simulated DARX model in Example 9.3. An example of full DTF estimation applied to simulated data is given by Young (2000); and its application to real data is described in the second example of Young (2002).

Example 9.3 Estimation of a Simulated DARX Model This example is based on an example given in Wellstead and Zarrop (1991) and the results obtained here are similar to those of Young (2011). These results are obtained using the statistical DARX option of the DTF algorithm described above. The simulation model takes the following DARX form:

$$y(k) = \frac{b_0(k)}{1 + a_1 z^{-1} + a_2 z^{-2}} u(k-1) + \frac{1}{1 + a_1 z^{-1} + a_2 z^{-2}} e(k) \tag{9.60}$$

or, in discrete-time equation terms:

$$y(k) = -a_1 y(k-1) - a_2 y(k-2) + b_0(k) u(k-1) + e(k) \tag{9.61}$$

where $e(k)$ is zero mean, white noise with variance 0.16 [10% noise by *Standard Deviation* (SD)]. The b_0 parameter changes from 1.0 to 2.0 at $t = 200$, and then back to 1.0 at $t = 900$; while $a_1 = -1.0$ and $a_2 = 0.25$ are time-invariant. The simulated input–output data are shown in Figure 9.8: the input $u(k)$ changes from a square wave between plus and minus one to a very small amplitude square wave of plus and minus 0.002 at $t = 400$, reverting to the original large square wave at $k = 800$. This choice of input signal induces 'estimator wind-up' in the case of the standard EWP algorithm because the information content in the data during the period of low input activity is not sufficient to ensure good performance from this rather crude TVP estimation algorithm. The DF algorithm, which is designed specifically to limit estimator wind-up produces a distinct improvement over the standard EWP algorithm, with the worst excesses of the wind-up no longer occurring. However, the response to the parametric change is relatively slow and there is considerable interaction between the estimates over the period of input inactivity.

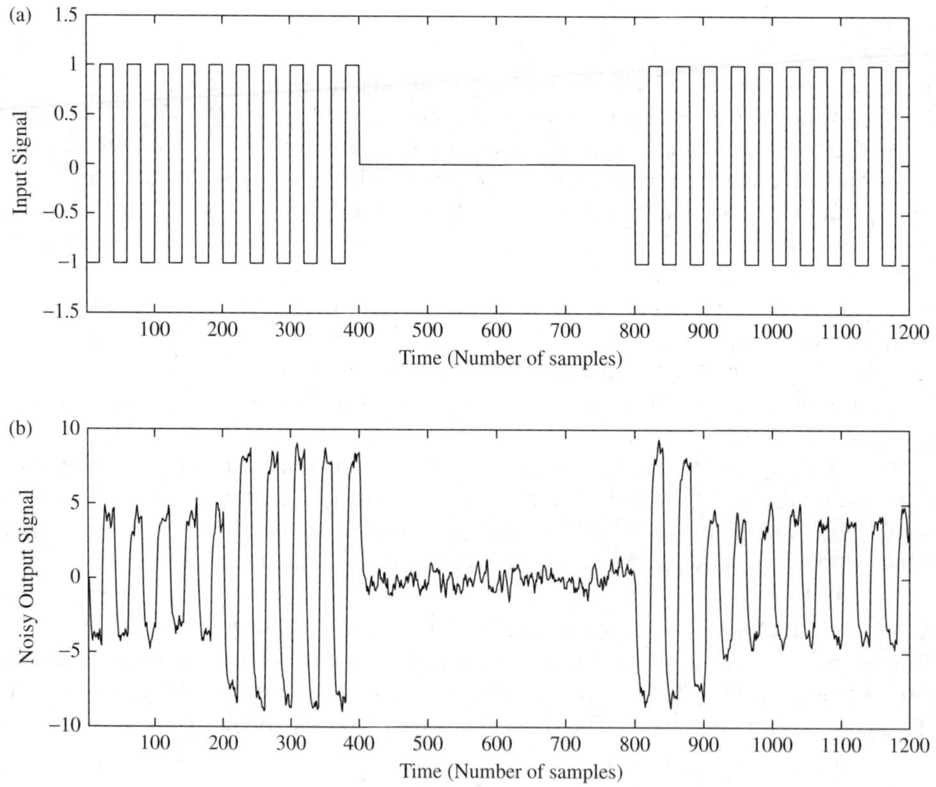

Figure 9.8 Input $u(k)$ (a) and noisy output $y(k)$ (b) for Example 9.3

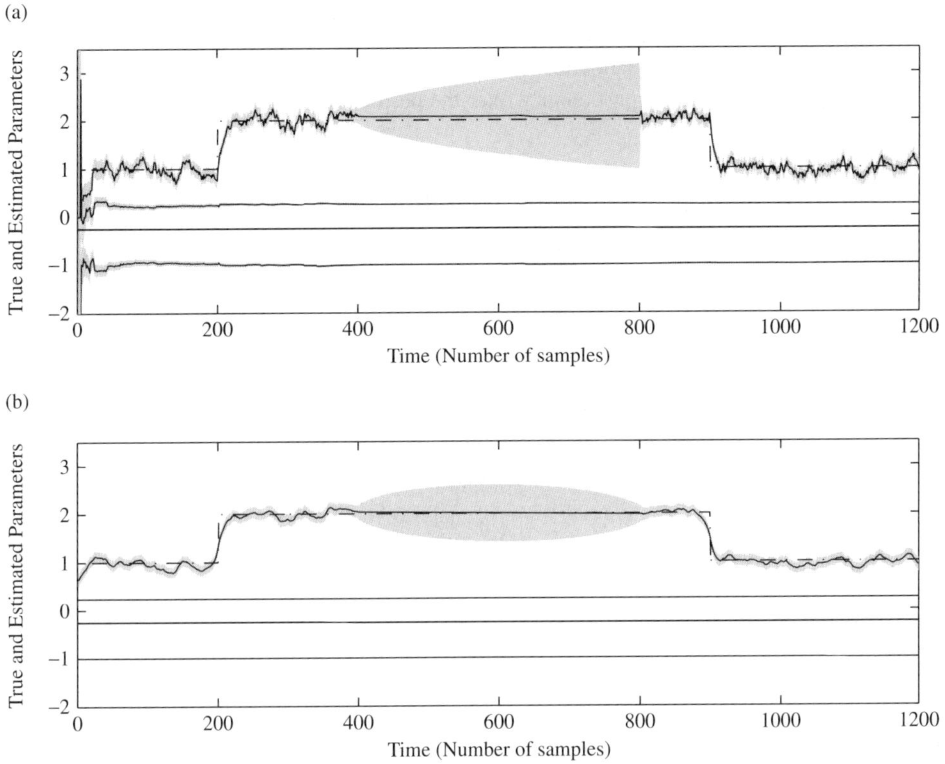

Figure 9.9 Estimation results for Example 9.3 using the ML optimised DARX estimation algorithm: filtering results (a) and smoothing results (b). Estimated SE bounds are shown as the grey area in both cases

By far the best results are those shown in Figure 9.9, as obtained using the *dtfmopt* and *dtfm* routines in CAPTAIN, based on the simpler DARX model form, which is then of similar complexity to the EWP and DF algorithms. Here, however, the elements of the diagonal NVR matrix $\boldsymbol{Q}_r$ are the ML-optimised NVR hyper-parameters:

$$NVR_{a_1} = 1.5 \times 10^{-16}; \quad NVR_{a_2} = 4.6 \times 10^{-20}; \quad NVR_{b_0} = 0.0186. \tag{9.62}$$

We see that the NVR_{a1} and NVR_{a2} are both insignificantly different from zero, illustrating how the ML optimisation has, quite objectively, inferred from the data that the associated a_1 and a_2 parameters are time-invariant. However, NVR_{b_0} is significant and has been optimised at a value that gives good tracking of the step changes in the b_0 parameter.

Figure 9.9b shows the backward-pass FIS estimation results, as obtained from the simplest RW version of the FIS algorithm (9.58). In comparison with the filtering results in Figure 9.9a, it will be noted that the smoothed estimates are more accurate; the step changes are anticipated because the estimates are based on the whole data set; and the estimates of the constant parameters are now themselves constant. Note also that, since both the filtering and smoothing algorithms are statistical in nature, they provide information on the statistical properties of the estimates, as shown here by the SE bounds plotted as the shaded area either side of the estimates. As expected, the SE bounds on the smoothed estimates are visibly smaller than those of the filtered estimates, particularly over the period of input inactivity.

9.3 State-Dependent Parameter Modelling and PIP Control

The previous section has shown that the estimation of time variable parameters can help to model 'non-stationary' systems whose static and dynamic behaviour changes over time. On the other hand, if the changes in the parameters are functions of the state or input variables (i.e. the parameters and/or their changes actually constitute stochastic state variables), then the stochastic system is truly nonlinear and likely to exhibit severe nonlinear behaviour. Normally, this cannot be approximated in a simple TVP manner; in which case, recourse must be made to other approaches, such as the SDP modelling methods of the kind proposed by Young (2000) and Young *et al.* (2001). The use of SDP models in the design of PIP control systems for nonlinear systems was briefly outlined by Young (1996) and McCabe *et al.* (2000). In this chapter, we will introduce the basic SDP control approach and refer the reader to recent research by Taylor *et al.* (2007b, 2009, 2011) on its wider application within a control systems design context.

9.3.1 The SDP-TF Model

While the DARX model (9.40) can produce fairly complex response characteristics, it is only when the parameters are functions of the system variables, and so vary at a rate commensurate with these variables, that the resultant model can behave in a heavily nonlinear or even chaotic manner. In its simplest SISO form, the *State-Dependent Parameter ARX* (SDARX) model can be written most conveniently in the following form:

$$y(k) = \boldsymbol{\phi}^T(k)\boldsymbol{\rho}(k) + e(k); \qquad e(k) = \mathcal{N}(0, \sigma^2) \tag{9.63}$$

where

$$\begin{aligned} \boldsymbol{\phi}^T(k) &= [-y(k-1) - y(k-2) \cdots - y(k-n) \;\; u(k-\delta) \cdots u(k-\delta-m)] \\ \boldsymbol{\rho}(k) &= [a_1\{\boldsymbol{\chi}(k)\} \;\; a_2\{\boldsymbol{\chi}(k)\} \cdots a_n\{\boldsymbol{\chi}(k)\} \, b_0\{\boldsymbol{\chi}(k)\} \cdots \;\; b_m\{\boldsymbol{\chi}(k)\}]^T \end{aligned} \tag{9.64}$$

in which $a_i\{\boldsymbol{\chi}(k)\},\ i = 1, 2, \ldots, n$, and $b_j\{\boldsymbol{\chi}(k)\},\ j = 0, 1, 2, \ldots, m$, are the state-dependent parameters, which are each assumed to be functions of one or more of the variables in a non-minimal state vector $\boldsymbol{\chi}^T(k) = [\boldsymbol{\phi}^T(k)\ \boldsymbol{U}^T(k)]$. Here $\boldsymbol{U}(k) = [U_1(k)\ U_2(k)\ \cdots\ U_r(k)]^T$ is a vector of other variables that may affect the relationship between the two primary input–output variables but do not appear in $\boldsymbol{\phi}(k)$: for example, measured 'air data' variables used in the self-adaptive autostabilisation of airborne vehicles (see e.g. Young 1981, 1984 and Chapter 4 in Young 2011); or, in the case of the joint angle (or velocity) of a robotic manipulator, the angles of *other* joints, representing the present configuration of the device (Taylor and Seward 2010). Finally, δ represents a pure time delay on the input variable and $e(k)$ is a zero mean, white noise input with Gaussian normal amplitude distribution and variance σ^2 (although this assumption is not essential to the practical application of the resulting estimation algorithms).

The SDARX model (9.63) can be written in the alternative pseudo-TF form:

$$y(k) = \frac{B\left\{\chi(k), z^{-1}\right\}}{A\left\{\chi(k), z^{-1}\right\}} u(k-\delta) + \frac{1}{A\left\{\chi(k), z^{-1}\right\}} e(k) \tag{9.65}$$

where

$$\begin{aligned} B\left\{\chi(k), z^{-1}\right\} &= b_0\{\chi(k)\} + b_1\{\chi(k)\} z^{-1} + \cdots + b_m\{\chi(k)\} z^{-m} \\ A\left\{\chi(k), z^{-1}\right\} &= 1 + a_1\{\chi(k)\} z^{-1} + a_2\{\chi(k)\} z^{-2} + \cdots + a_n\{\chi(k)\} z^{-n} \end{aligned} \tag{9.66}$$

The more general SDP-TF model is:

$$y(k) = \frac{B\left\{\chi(k), z^{-1}\right\}}{A\left\{\chi(k), z^{-1}\right\}} u(k-\delta) + \xi(k) \tag{9.67}$$

where $\xi(k)$ is a general additive noise term: e.g. it could be ARMA noise or, indeed, nonlinear noise described by a SDP relationship. Note that multi-state SDP models, where each TF model parameter is a function of several variables, are now being considered (e.g. Sadhegi *et al.* 2010; Tych *et al.* 2012) and offer the possibility of still richer nonlinear models. However, these have not yet been applied to the analysis of real-time series data and their practical utility still has to be evaluated.

Example 9.4 State-Dependent Parameter Representation of the Logistic Growth Equation A typical, simple deterministic example of equation (9.63) is a SDARX model in the form of the following nonlinear 'forced logistic growth' equation:

$$y(k) = \alpha_1 y(k-1) - \alpha_2 y(k-1)^2 + u(k) \tag{9.68}$$

or

$$y(k) = -a_1\{y(k-1)\}\, y(k-1) + b_0\{u(k)\}\, u(k) \tag{9.69}$$

where, in this example, b_0 is not state-dependent, i.e.

$$a_1\{y(k-1)\} = -\alpha_1 + \alpha_2 y(k-1) \;\; \text{and} \;\; b_0\{u(k)\} = 1.0 \quad \forall k \tag{9.70}$$

Although it is simple, this model can exhibit rich behavioural patterns: from simple to chaotic responses, depending on α_1 and α_2 (Young 2000). For example, Figure 9.10 shows the typical chaotic response of the model obtained when $\alpha_1 = \alpha_2 = 4$ and the initial condition $y(0) = 0.1$. The noisy response shown in Figure 9.10 relates to a *Monte Carlo Simulation* (MSC) study in Example 9.5.

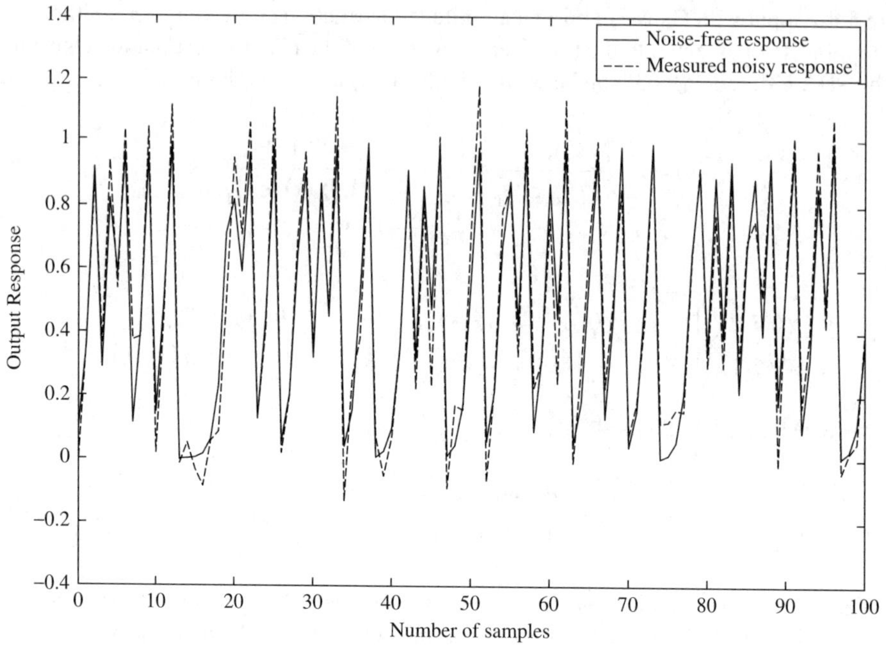

Figure 9.10 Typical chaotic response of the logistic growth equation (9.68)

9.3.2 State-Dependent Parameter Model Identification and Estimation

The SDP modelling procedure is briefly summarised as follows:

1. If feasible linear models can be obtained (e.g. for small perturbations close to an operating condition), then the underlying model structure and potential state variables are first identified by statistical estimation of discrete- or continuous-time, linear TF models, using the methods discussed in Chapter 8. Otherwise, the modeller can move straight to stage 2 but, in this case, further investigations into possible model structures are required.
2. The second stage of the analysis is based on the estimation of stochastic state space models with TVPs, using the recursive KF and FIS algorithms discussed in section 9.2. However, in order to address systems that exhibit severe nonlinear or chaotic behaviour, these are embedded within an iterative 'back-fitting' routine (i.e. the *sdp* function in the CAPTAIN Toolbox) that involves re-ordering of the time series data into an SDP space which is able to reveal the nature of the state-dependency using standard TVP estimation. Here, each parameter in the model has a different value at each sample and can only be viewed in state-dependent form as a graph, so that the model is effectively 'non-parametric'.
3. In the final optimisation stage, the non-parametrically defined nonlinearities obtained initially by SDP estimation are now parameterised in some manner, in terms of their associated dependent variables. This can be achieved by defining an appropriate parametric model, such as polynomial functions, trigonometric functions, wavelets, radial basis functions, etc., estimated directly from the input–output data using some method of dynamic model optimisation, such as *fminsearch* in MATLAB® (see e.g. Taylor *et al.* 2007b; Beven *et al.*

2012). Initial conditions for this optimisation are based on the SDP estimates obtained from the KF/FIS stage of the analysis, which helps to avoid the potential problem of finding a locally optimal set of parameters.

It should be noted that SDP models can be linked with what are referred to as *Non-Linear Parameter Varying* (NLPV) models in the control and systems literature (see e.g. Previdi and Lovera 2004); a link that is demonstrated in Young (2005). This name derives from the earlier LPV models, while the state on which the parameters are assumed to depend is called the 'scheduling' variable because of the link with scheduled gain control systems. The main difference between the SDP and NLPV modelling approaches is that the former involves an initial non-parametric SDP identification stage to identify the location and nature (in graphical or look-up table terms) of any significant nonlinearities in the model. This stage is normally omitted in NLPV (and LPV) modelling, where the location of the SDPs and their *parametric* form appear to be selected either by assuming that all parameters are state-dependent and modelling them using parametric functions of various kinds (neural/neuro-fuzzy networks, wavelets, etc.), or by selecting them based on prior knowledge of the physical system being modelled. Although such parametric modelling is also used in SDP estimation, the selection of the parametric functions is guided by the shapes of the non-parametric functions identified in stage 2 above.

SDP estimation is not discussed further here because it is dealt with in detail elsewhere (e.g. Pedregal *et al.* 2007; Young 2011). Nonetheless, as a pointer to future TDC design studies, we will briefly consider the control of such SDP identified systems.

9.3.3 *Proportional-Integral-Plus Control of SDP Modelled Systems*

The SDP-NMSS representation of the deterministic SDP model (9.67) with $\delta > 0$ and $\xi(k) = 0$ is as follows:

$$\boldsymbol{x}(k+1) = \boldsymbol{F}\left\{\chi(k)\right\}\boldsymbol{x}(k) + \boldsymbol{g}\left\{\chi(k)\right\}u(k) + \boldsymbol{d}\,y_d(k); \quad y(k) = \boldsymbol{h}\boldsymbol{x}(k) \tag{9.71}$$

where $\boldsymbol{x}(k)$ is the usual servomechanism non-minimal state vector, i.e. equation (5.14), and the various matrices are defined in a similar manner to before. In this case, however, $\boldsymbol{F}\left\{\chi(k)\right\}$ and $\boldsymbol{g}\left\{\chi(k)\right\}$ are expressed in terms of the SDPs, i.e. the model coefficients a_i $(i = 1 \ldots n)$ and b_i $(i = 1 \ldots m)$ in equation (5.15) and equation (5.16) are straightforwardly replaced by $a_i\left\{\chi(k)\right\}$ and $b_i\left\{\chi(k)\right\}$: see Example 9.5.

The SDP-PIP control law takes the following state variable feedback form:

$$u(k) = -\boldsymbol{k}^T\left\{\chi(k)\right\}\boldsymbol{x}(k) \tag{9.72}$$

where, at an arbitrary *k*th sampling instant, the state dependent control gain vector:

$$\boldsymbol{k}^T\left\{\chi(k)\right\} = \left[f_0\left\{\chi(k)\right\}, \ldots, f_{n-1}\left\{\chi(k)\right\}, g_1\left\{\chi(k)\right\}, \ldots g_{m-1}\left\{\chi(k)\right\}, -k_I\left\{\chi(k)\right\}\right] \tag{9.73}$$

could be selected using any of the methods discussed for PIP control in previous chapters, i.e. based on the 'snapshot' of the SDP model at this same *k*th sampling instant. The block diagram form of the controller takes a similar form to Figure 5.3, here with the SDP model and state dependent control gains replacing the linear TF models and time-invariant control filters of the standard controller. The SDP-PIP algorithm could also be implemented using the alternative control structures discussed in Chapter 6.

Consider the following three approaches for determining the control gains:

1. **Scheduled LQ design.** The standard LQ control problem (see section 5.4) can be solved online at each sampling instant, on an assumption of point-wise controllability. Since the functional representations of each parameter are estimated offline, this pragmatic approach yields scheduled rather than adaptive control systems (although this depends on the viewpoint of the designer). Unfortunately, the optimality of the design is determined by the choice of SDP model, with suboptimal solutions obtained in the general case (Taylor *et al.* 2007b). Furthermore, while some theoretical advances have been made regarding the asymptotic stability of this type of 'state-dependent Riccati equation' approach, the conditions obtained can be difficult to fulfil in practical applications (Banks *et al.* 2007).
2. **Scheduled pole assignment.** Another approach is to solve a version of the standard linear pole assignment problem at each sampling interval, i.e. by equating the closed-loop characteristic equation at sample k with the elements of a desired (user-defined) characteristic polynomial. Section 5.3 develops an algorithmic pole assignment method that is ideal for an online implementation of the resulting SDP-PIP controller. Alternatively, an algebraic solution can be determined for the system under study, which can simplify the practical implementation (see Taylor *et al.* 2009 and Example 9.5). For time-invariant systems, the associated eigenvalues are, of course, equivalent to the poles of the closed-loop system and hence determine the dynamic behaviour of the control system. Unfortunately, it is well known that, for time-varying systems as here, these eigenvalues do not completely determine the transient response and stability of the closed-loop system.
3. **Stabilising pole assignment for all-pole SDP models.** The limitations noted above for standard pole assignment have motivated the development of a new algorithm for SDP-NMSS systems, as discussed in recent articles by Taylor *et al.* (2009, 2011). The initially developed approach is constrained to 'all-pole' systems in which there are no open-loop zeros. The required pole assignability conditions are particularly transparent in this case and can be identified offline. Furthermore, using this formulation, the closed-loop system reduces to a linear TF with the specified (design) poles. Hence, assuming pole assignability at each sample and no model mismatch, stability of the nonlinear system is guaranteed for stable design poles. Present research is evaluating the robustness and performance properties of the algorithm when there is model mismatch, as there always will be in practical applications.

Notwithstanding the various caveats above, practical examples show that SDP-PIP control system design can yield robust and practically useful control systems for nonlinear problems. For example, the scheduled LQ design, scheduled pole assignment and stabilising pole assignment methods have been utilised for the hydraulic manipulator of a vibro-lance ground compaction system (Taylor *et al.* 2007b), ventilation rate in a micro-climate test chamber (Stables and Taylor 2006) and for a mobile nuclear decommissioning robot (e.g. Taylor and Seward 2010; Robertson and Taylor 2012), respectively. However, for tutorial purposes, the final example of this book returns to the simulated logistic growth equation introduced above.

Example 9.5 SDP-PIP Control of the Logistic Growth Equation This example is concerned with the logistic growth equation model (9.68) with $\alpha_1 = \alpha_2 = 4$, where the natural open-loop, chaotic response is shown in Figure 9.10. The reader should verify from the above equations that the nonlinear NMSS form of this SDP model is as follows:

$$\begin{bmatrix} y(k) \\ z(k) \end{bmatrix} = \begin{bmatrix} 4-4y(k-1) & 0 \\ -4+4y(k-1) & 1 \end{bmatrix} \begin{bmatrix} y(k-1) \\ z(k-1) \end{bmatrix} + \begin{bmatrix} 1 \\ -1 \end{bmatrix} u(k-1) + \begin{bmatrix} 0 \\ 1 \end{bmatrix} y_d(k) \quad (9.74)$$

Here, $z(k) = y_d(k) - y(k)$ is the integral-of-error state as usual; $y_d(k)$ is the command input signal; and the control input $u(k)$ is generated by the following SVF control law:

$$u(k) = -f_0(k)y(k) + k_I(k)z(k) \quad (9.75)$$

where the control gains $f_0(k)$ and $k_I(k)$ at an arbitrary kth sampling instant are selected using one of the methods discussed above. Equation (9.74) and equation (9.75) can be compared with the time-invariant equivalents (for a similar model structure) given by equation (5.6) and equation (5.10). Using the scheduled pole assignment approach, let the desired polynomial [the general form of which is given by equation (5.57)] have roots at 0.5 and zero: i.e. $D(z^{-1}) = 1 - 0.5z^{-1}$, so that $d_1 = -0.5$ and $d_2 = 0$. At the kth sampling instant, the parameters of the SDP model are $a_1(k) = 4 - 4y(k-1)$ and $b_1 = 1$, hence the linear PIP pole assignment computation algorithm (5.67) yields[5]:

$$f_0(k) = 4 - 4y(k-1) - d_2 = 4 - 4y(k-1) \quad \text{and} \quad k_I = b_0 + d_1 + d_2 = 1 - 0.5 = 0.5 \quad (9.76)$$

In this case, $f_0(k)$ varies with time because it is state dependent but k_I is constant. Consequently, the controller is of the standard PIP form but with the control gain $f_0(k)$ being changed at each sample to reflect the nonlinearity introduced by the SDP.

It might be assumed that this SDP-PIP controller would be sensitive to uncertainty in the model parameters but this is not necessarily the case, as shown in Figure 9.11, which gives plots of the results obtained from a MCS exercise. Here, the uncertainty was quantified by estimating the model from 110 samples of data generated by the uncontrolled (chaotic) system, in which $u(k)$ is not present and white noise is added to the measurement $y(k)$ to yield a noise–signal ratio of 0.27 by SD: these noisy open-loop data are shown in Figure 9.10. The resulting *Linear Least Squares* (LLS) estimates and their estimated covariance matrix $\boldsymbol{P}$, are as follows:

$$\hat{\alpha}_1 = 3.89\,(0.07); \quad \hat{\alpha}_2 = 3.89\,(0.08); \quad \boldsymbol{P} = \begin{bmatrix} 0.0054 & -0.0061 \\ -0.0061 & 0.0072 \end{bmatrix} \quad (9.77)$$

[5] For this low order example, it is alternatively straightforward to obtain (9.76) directly by equating the closed-loop characteristic polynomial with the desired polynomial; see Example 2.7.

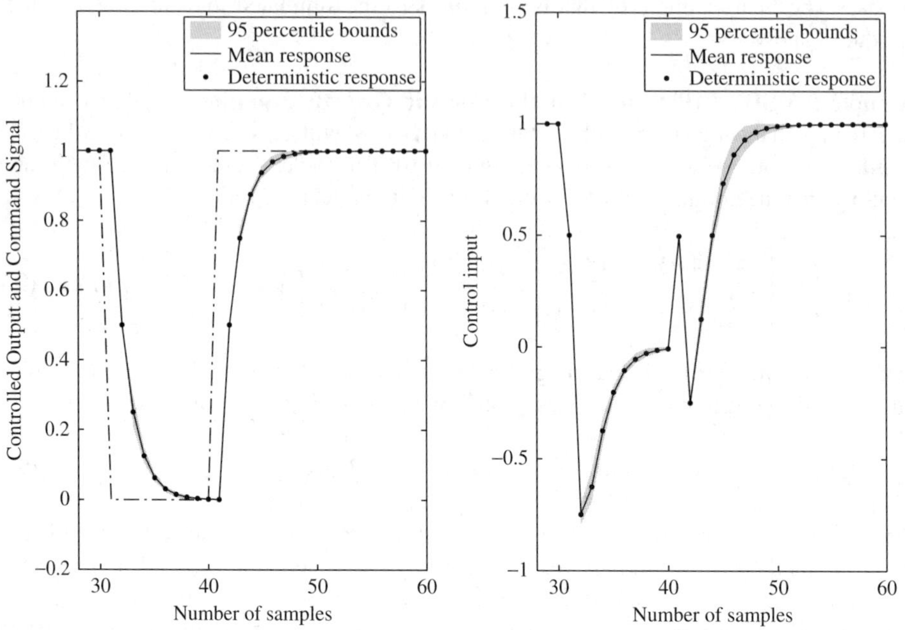

Figure 9.11 MCS response of SDP-PIP controlled logistic growth equation in Example 9.5 to a sequence of step changes in the command input, showing the relative insensitivity of the controller to uncertainty in the model parameters

where the estimated parameters are asymptotically biased away from their true values because of the crude least squares estimation. These estimates were used in the usual MCS manner to generate 1000 realisations of the SDP-PIP controlled system (this time without the additive noise in order to more clearly illustrate the closed-loop dynamic behaviour), as shown in Figure 9.11. The associated changes in the SDP gain $f_0(k)$ are plotted in Figure 9.12.

It is clear that the controlled system is robust to this level of uncertainty: indeed all the realisations remain stable even if the uncertainty in the parameters was increased 10-fold. One might expect that the performance would be degraded if the controller was designed for much more rapid 'dead-beat' operation (i.e. $d_1 = d_2 = 0$) but, once again, this is not the case, as shown in Figure 9.13, where the 95 percentile bounds remain relatively small.

Unfortunately, it is not possible to guarantee the robustness and apparently strong stability shown in this example to all SDP-PIP implementations. Research is proceeding to investigate what conditions are required to guarantee stability, at least in the deterministic (no model mismatch) case. The most recent research in this regard (Taylor *et al.* 2009, 2011) has shown that such a guarantee is possible in the 'all-pole' system situation, i.e. when the numerator polynomial $B\{\chi(k), z^{-1}\}$ in the SDP model (9.67) is limited to:

$$B\{\chi(k), z^{-1}\} = b_0\{\chi(k)\} \tag{9.78}$$

Furthermore, for unity time-delay systems that are also constrained by (9.78), the basic scheduled pole assignment and new stabilising pole assignment algorithms mentioned above

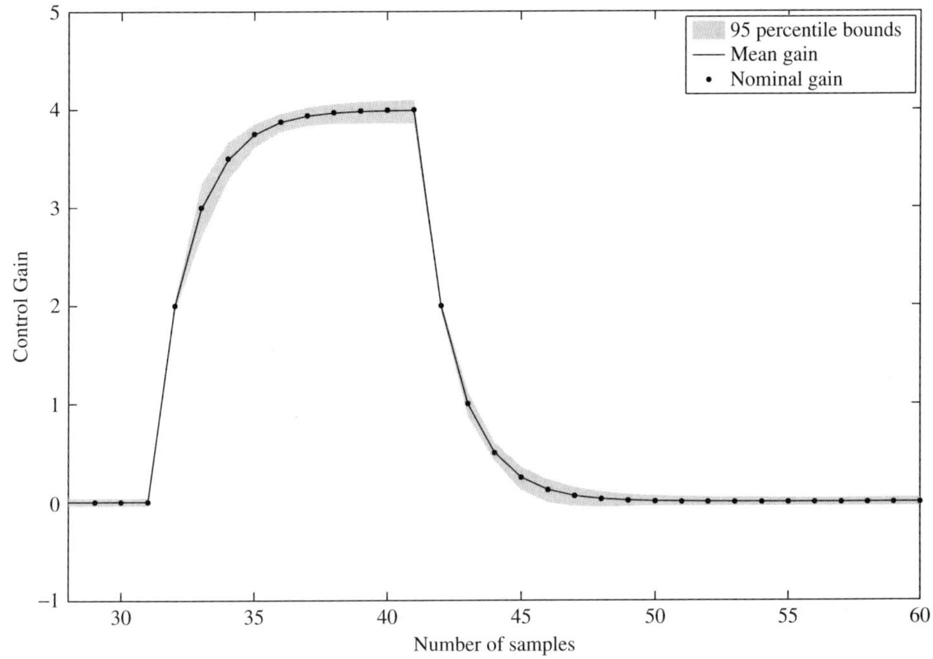

Figure 9.12 Changes in the SDP-PIP gain $f_0(k)$ that led to the closed-loop responses in Figure 9.11

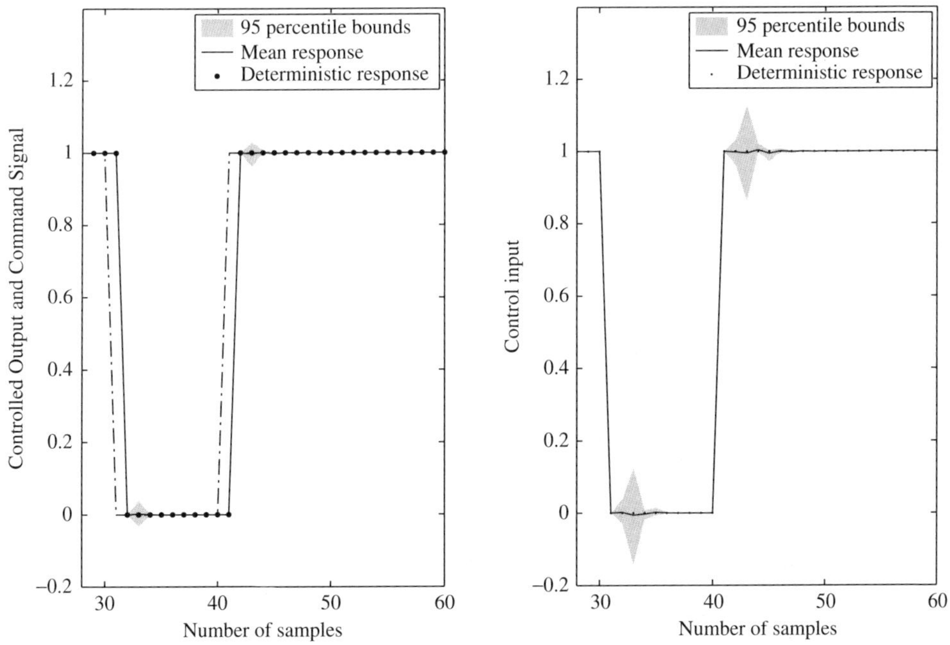

Figure 9.13 MCS response of SDP-PIP dead-beat controlled logistic growth equation in Example 9.5 to a sequence of step changes in the command input

are equivalent: this is why the expected design responses, such as dead-beat, are successfully obtained for the present example. Although limited, the *time-delay* SDP model (9.67) with (9.78), still encompasses a wide range of nonlinear structures and has proven particularly useful for the control of hydraulically operated robotic manipulators (e.g. Taylor *et al.* 2010; Taylor and Seward 2010; Robertson and Taylor 2012).

9.4 Concluding Remarks

This final chapter has considered two additional topics that illustrate the wide applicability of the TDC concepts and methods discussed in previous chapters of this book. We believe that the PIP control of δ-operator systems has important practical implications in those situations where rapid sampling is essential and we are surprised that, so far, it has not been exploited more in practical applications. Similarly, adaptive TVP and scheduled SDP control methods are important because they open the door to the PIP control of nonlinear stochastic systems. Although the SDP-PIP pole assignment approach demonstrated above has initially been limited in general terms to systems that can be described by all-pole SDP models, without model mismatch, this does not mean that systems that are described by the more general SDP-TF model (9.67), with model mismatch, cannot be controlled at all; it simply means that a general proof of closed-loop stability has not been established so far in this situation. Indeed, it seems likely that SDP-PIP control can be applied to such systems within some defined region of the SDP model parameter space and research is proceeding in this direction. Clearly, therefore, SDP modelling and nonlinear SDP-PIP control are both subjects that have great potential for practical application, and are an attractive topic for future theoretical and practical research.

References

Banks, H.T., Lewis, B.M. and Tran, H.T. (2007) Nonlinear feedback controllers and compensators: a state-dependent Riccati equation approach, *Computational Optimization and Applications*, **37**, pp. 177–218.

Beven, K.J., Leedal, D.T., Smith, P.J. and Young, P.C. (2012) Identification and representation of state dependent non-linearities in flood forecasting using the DBM methodology. In L. Wang and H. Garnier (Eds), *System Identification, Environmetric Modelling and Control*, Springer-Verlag, Berlin, pp. 341–366.

Bray, J.W., High, R.J., McCann, A.D. and Jemmeson, H. (1965) On-line model making for chemical plant, *Transactions of the Society of Instrument Technology*, **17**, pp. 1–9.

Bryson, A.E. and Ho, Y.C. (1969) *Applied Optimal Control*, Blaisdell, Waltham, MA.

Chotai, A., Young, P.C., McKenna, P.G. and Tych, W. (1998) Proportional-Integral-Plus (PIP) design for delta operator systems: Part 2, MIMO systems, *International Journal of Control*, **70**, pp. 149–168.

Gauss, K. (1826) Theoria combinationis observationum erroribus minimis obnoxiae. *Werke*, **4** (Parts 1, 2 and Supplement), pp. 1–108.

Harvey, A.C. (1981) *Time Series Models*, Phillip Allen, Oxford.

Kitagawa, G. (1981) A non-stationary time series model and its fitting by a recursive filter, *Journal of Time Series Analysis*, **2**, pp. 103–116.

Kopp, R.E. and Orford, R.J. (1963) Linear regression applied to system identification for adaptive control systems, *American Institution of Aeronautics and Astronautics Journal*, **1**, pp. 2300–236.

Kulhavy, R. (1987) Restricted exponential forgetting in real time identification. *Automatica*, **23**, pp. 589–600.

Lee, R.C.K. (1964) *Optimal Identification, Estimation and Control*, MIT Press, Cambridge, MA.

Ljung, L. (1979) Asymptotic behaviour of the extended Kalman filter as a parameter estimator for linear systems, *IEEE Transactions on Automatic Control*, **24**, pp. 36–50.

Mayne, D.Q. (1963) Optimal nonstationary estimation of the parameters of a dynamic system, *Journal of Electronics and Control*, **14**, pp. 101–112.

McCabe, A.P., Young, P., Chotai, A. and Taylor, C.J. (2000) Proportional-Integral-Plus (PIP) control of non-linear systems, *Systems Science* (Wrocław), **26**, pp. 25–46.

McKenna, P.G. (1997) *Delta Operator Modelling, Forecasting and Control*, PhD thesis, Environmental Science Department, Lancaster University.

Middleton, R.H. and Goodwin, G.C. (1990) *Digital Control and Estimation – A Unified Approach*, Prentice Hall, Englewood Cliffs, NJ.

Norton, J.P. (1975) Optimal smoothing in the identification of linear time-varying systems, *Proceedings Institute Electrical Engineers*, **122**, pp. 663–668.

Norton, J.P. (1986) *An Introduction to Identification*, Academic Press, New York.

Pedregal, D.J. and Taylor, C.J. (2012) SSpace: A flexible and general state space toolbox for MATLAB. In L. Wang and H. Garnier (Eds), *System Identification, Environmental Modelling and Control System Design*, Springer-Verlag, Berlin, pp. 615–635.

Pedregal, D.J., Taylor, C.J. and Young, P.C. (2007) *System Identification, Time Series Analysis and Forecasting: The Captain Toolbox Handbook*, Lancaster University, Lancaster.

Previdi, F. and Lovera, M. (2004) Identification on non-linear parametrically varying models using separable least squares, *International Journal of Control*, **77**, pp. 1382–1392.

Robertson, D. and Taylor, C.J (2012) State-dependent control of a hydraulically-actuated nuclear decommissioning robot, *UKACC International Conference Control*, September, Cardiff, UK, Paper 48.

Sadeghi, J., Tych, W., Chotai, A. and Young, P.C. (2010) Multi-state dependent parameter model identification and estimation for control system design, *Electronics Letters*, **46**, pp. 1265–1266.

Schweppe, F. (1965) Evaluation of likelihood functions for gaussian signals, *IEEE Transactions on Information Theory*, **11**, pp. 61–70.

Stables, M.A. and Taylor, C.J. (2006) Nonlinear control of ventilation rate using state dependent parameter models, *Biosystems Engineering*, **95**, pp. 7–18.

Taylor, C.J., Chotai, A. and Burnham, K.J. (2011) Controllable forms for stabilising pole assignment design of generalised bilinear systems, *Electronics Letters*, **47**, pp. 437–439.

Taylor, C.J., Chotai, A. and Robertson, D. (2010) State dependent control of a robotic manipulator used for nuclear decommissioning activities, *IEEE RSJ International Conference on Intelligent Robots and Systems* (IROS-10), October, Taipei, Taiwan, pp. 2413–2418.

Taylor, C.J., Chotai, A. and Young, P.C. (1998) Proportional-Integral-Plus (PIP) control of time-delay systems, *IMECHE Proceedings: Systems and Control Engineering*, **212**, pp. 37–48.

Taylor, C.J., Chotai, A. and Young, P.C. (2009) Nonlinear control by input–output state variable feedback pole assignment, *International Journal of Control*, **82**, pp. 1029–1044.

Taylor, C.J., Pedregal, D.J., Young, P.C. and Tych, W. (2007a) Environmental time series analysis and forecasting with the Captain toolbox, *Environmental Modelling and Software*, **22**, pp. 797–814.

Taylor, J. and Seward, D. (2010) Control of a dual-arm robotic manipulator, *Nuclear Engineering International*, **55**, pp. 24–26.

Taylor, C.J., Shaban, E.M., Stables, M.A. and Ako, S. (2007b) Proportional-Integral-Plus (PIP) control applications of state dependent parameter models, *IMECHE Proceedings: Systems and Control Engineering*, **221**, pp. 1019–1032.

Tych, W., Sadeghi, J., Smith, P., Chotai, A. and Taylor, C.J. (2012) Multi-state dependent parameter model identification and estimation. In L. Wang and H. Garnier (Eds), *System Identification, Environmetric Modelling and Control*, Springer-Verlag, Berlin, pp. 191–210.

Wellstead, P.E. and Zarrop, M. B. (1991) *Self-Tuning Systems: Control and Signal Processing*, John Wiley & Sons, Ltd, New York.

Young, P.C. (1969) Applying parameter estimation to dynamic systems: Part I – theory. *Control Engineering*, **16**, pp. 119–125.

Young, P.C. (1970) An instrumental variable method for real-time identification of a noisy process. *Automatica*, **6**, pp. 271–287.

Young, P.C. (1981) A second generation adaptive autostabilization system for airborne vehicles, *Automatica*, **17**, pp. 459–470.

Young, P.C. (1984) *Recursive Estimation and Time-Series Analysis: An Introduction*, Springer-Verlag, Berlin.

Young, P.C. (1996) A general approach to identification, estimation and control for a class of nonlinear dynamic systems. In M.I. Friswell and J.E. Mottershead (Eds), *Identification in Engineering Systems*, University of Wales, Swansea, pp. 436–445.

Young, P.C. (1999) Nonstationary time series analysis and forecasting. *Progress in Environmental Science*, **1**, pp. 3–48.

Young, P.C. (2000) Stochastic, dynamic modelling and signal processing: time variable and state dependent parameter estimation. In W.J. Fitzgerald, A. Walden, R. Smith and P.C. Young (Eds), *Nonlinear and Nonstationary Signal Processing*, Cambridge University Press, Cambridge, pp. 74–114.

Young, P.C. (2002) Identification of time varying systems. In H. Unbehauen (Ed.), *Encyclopedia of Life Support Systems (EOLSS)*, volume 6.43: Control Systems, Robotics and Automation, UNESCO, EOLSS Publishers, Oxford.

Young, P.C. (2005) Comments on 'identification on non-linear parametrically varying models using separable least squares' by F. Previdi and M. Lovera: black-box or open box? *International Journal of Control*, **78**, pp. 122–127.

Young, P.C. (2010) Gauss, Kalman and advances in recursive parameter estimation. *Journal of Forecasting (special issue celebrating 50 years of the Kalman Filter)*, **30**, pp. 104–146

Young, P.C. (2011) *Recursive Estimation and Time-Series Analysis: An Introduction for the Student and Practitioner*, Springer-Verlag, Berlin.

Young, P.C., Chotai, A., McKenna, P.G. and Tych, W. (1998) Proportional-Integral-Plus (PIP) design for delta operator systems: Part 1, SISO systems. *International Journal of Control*, **70**, pp. 123–147.

Young, P.C., Chotai, A. and Tych, W. (1991) Identification, estimation and control of continuous–time systems described by delta operator models. In N. Sinha and G. Rao (Eds), *Identification of Continuous-Time Systems*, Kluwer, Dordrecht, pp. 363–418.

Young, P.C., McKenna, P.G. and Bruun, J. (2001) Identification of nonlinear stochastic systems by state dependent parameter estimation, *International Journal of Control*, **74**, pp. 1837–1857.

Young, P.C. and Ng, C.N. (1989) Variance intervention, *Journal of Forecasting*, **8**, pp. 399–416.

A

Matrices and Matrix Algebra

A.1 Matrices

A matrix is defined as a rectangular array of elements arranged in rows and columns; in this book it is denoted by a bold italics capital letter, e.g.

$$\boldsymbol{A} = \begin{bmatrix} a_{11} & a_{12} & \cdots & a_{1n} \\ a_{21} & a_{22} & \cdots & a_{2n} \\ \cdots & \cdots & \cdots & \cdots \\ a_{m1} & a_{m2} & \cdots & a_{mn} \end{bmatrix} \tag{A.1}$$

Often $\boldsymbol{A}$ is alternatively denoted by $[a_{ij}]$ to indicate that it is characterised by elements a_{ij},

$$i = 1, 2, \ldots, m; \quad j = 1, 2, \ldots, n$$

If it has mn elements arranged in m rows and n columns, then it is said to be of order m by n, usually written $m \times n$.

The following should be noted in relation to matrices:

(i) A *null* matrix has all of its elements set to zero, i.e. $a_{ij} = 0$ for all i, j.
(ii) A *symmetric* matrix is a square matrix in which $a_{ij} = a_{ji}$; i.e. it is symmetric about the diagonal elements.
(iii) The *trace* of a square $n \times n$ matrix, denoted by *Tr.*, is the sum of its diagonal elements, i.e. $Tr. \boldsymbol{A} = a_{11} + a_{22} + \cdots + a_{nn}$.
(iv) A *diagonal* matrix is a square matrix with all its elements *except those on the diagonal* set to zero, i.e.

$$\boldsymbol{A} = \begin{bmatrix} a_{11} & 0 & \cdots & 0 \\ 0 & a_{22} & \cdots & 0 \\ \cdots & \cdots & \cdots & \cdots \\ 0 & 0 & \cdots & a_{nn} \end{bmatrix}$$

True Digital Control: Statistical Modelling and Non-Minimal State Space Design, First Edition.
C. James Taylor, Peter C. Young and Arun Chotai.

(v) An $n \times n$ diagonal matrix with elements set to *unity* is denoted by I_n and termed the *identity* (or unit) matrix of order n, e.g. for a 3×3 identity matrix:

$$\boldsymbol{I}_3 = \begin{bmatrix} 1 & 0 & 0 \\ 0 & 1 & 0 \\ 0 & 0 & 1 \end{bmatrix}.$$

Sometimes the subscript is omitted if the order is obvious.

(vi) An *idempotent* matrix is a square matrix such that $\boldsymbol{A}^2 = \boldsymbol{A}\boldsymbol{A} = \boldsymbol{A}$, i.e. it remains unchanged when multiplied by itself.

A.2 Vectors

A matrix of order $m \times 1$ contains a single column of m elements and is termed a *column vector* (or sometimes just a *vector*); in this book, it is denoted by a bold italics lower case letter, i.e. for a vector $\boldsymbol{b}$:

$$\boldsymbol{b} = \begin{bmatrix} b_1 \\ b_2 \\ \vdots \\ b_n \end{bmatrix} \tag{A.2}$$

A.3 Matrix Addition (or Subtraction)

If two matrices $\boldsymbol{A}$ and $\boldsymbol{B}$ are of the same order then we define $\boldsymbol{A} + \boldsymbol{B}$ to be a new matrix $\boldsymbol{C}$, where

$$c_{ij} = a_{ij} + b_{ij}$$

In other words, the addition of the matrices is accomplished by adding *corresponding* elements, with $\boldsymbol{A} - \boldsymbol{B}$ defined in an analogous manner.

A.4 Matrix or Vector Transpose

The *transpose* of a matrix $\boldsymbol{A}$ is obtained from $\boldsymbol{A}$ by interchanging the rows and columns; in this book, it is denoted by a superscript capital T, e.g. for $\boldsymbol{A}$ defined in (A.1):

$$\boldsymbol{A}^T = \begin{bmatrix} a_{11} & a_{21} & \cdots & a_{m1} \\ a_{12} & a_{22} & \cdots & a_{m2} \\ \cdots & \cdots & \cdots & \cdots \\ a_{1n} & a_{2n} & \cdots & a_{mn} \end{bmatrix}$$

The transpose of a column vector $\boldsymbol{b}$, denoted by $\boldsymbol{b}^T$ is termed a *row vector*, e.g. for $\boldsymbol{b}$ in (A.2), $\boldsymbol{b}^T = [b_1\ b_2\ \ldots\ b_n]$.

Note that:

(i) in the case of a symmetric matrix $\boldsymbol{A}^T = \boldsymbol{A}$;
(ii) $[\boldsymbol{A}^T]^T = \boldsymbol{A}$;
(iii) $[\boldsymbol{A} + \boldsymbol{B}]^T = \boldsymbol{A}^T + \boldsymbol{B}^T$.

A.5 Matrix Multiplication

If $\boldsymbol{A}$ is of order $m \times n$ and $\boldsymbol{B}$ is of order $n \times p$ then the product $\boldsymbol{AB}$ is defined to be a matrix of order $m \times p$ whose (*ij*)th element c_{ij} is given by:

$$c_{ij} = \sum_{k=1}^{n} a_{ik} b_{kj}$$

i.e. the (*ij*)th element is obtained by, in turn, multiplying the elements of the *i*th row of the matrix $\boldsymbol{A}$ by the *j*th column of the matrix $\boldsymbol{B}$ and summing over all terms. Therefore, the number of elements (n) in each row of $\boldsymbol{A}$ must be equal to the number of elements in each column of $\boldsymbol{B}$. Note that, in general, the commutative law of multiplication which applies for scalars does *not* apply for matrices, i.e.

$$\boldsymbol{AB} \neq \boldsymbol{BA}$$

so that *pre-multiplication* of $\boldsymbol{B}$ by $\boldsymbol{A}$ does not, in general, yield the same as *post-multiplication* of $\boldsymbol{B}$ by$\boldsymbol{A}$. However, pre-multiplying or post-multiplying by the identity matrix leaves the matrix unchanged, i.e.

$$\boldsymbol{AI}_n = \boldsymbol{I}_n\boldsymbol{A} = \boldsymbol{A}$$

Note also that for $\boldsymbol{A}$ of order $m \times n$, $\boldsymbol{B}$ of order $n \times p$ and $\boldsymbol{C}$ of order $p \times q$ the following results apply:

(i) $(\boldsymbol{AB})\boldsymbol{C} = \boldsymbol{A}(\boldsymbol{BC})$;
(ii) $\boldsymbol{A}(\boldsymbol{B} + \boldsymbol{C}) = \boldsymbol{AB} + \boldsymbol{AC}$ with orders m, n, p and q chosen appropriately;
(iii) $(\boldsymbol{B} + \boldsymbol{C})\boldsymbol{A} = \boldsymbol{BA} + \boldsymbol{CA}$;
(iv) for $\boldsymbol{A}$, $\boldsymbol{B}$ and $\boldsymbol{C}$, the multiplication by a scalar λ yields a corresponding matrix with all its elements multiplied by λ, i.e .$\lambda\boldsymbol{A} = [\lambda a_{ij}]$;
(v) $[\boldsymbol{AB}]^T = \boldsymbol{B}^T\boldsymbol{A}^T$;
(vi) $[\boldsymbol{ABC}]^T = \boldsymbol{C}^T\boldsymbol{B}^T\boldsymbol{A}^T$, since $[\boldsymbol{ABC}]^T = [(\boldsymbol{AB})\boldsymbol{C}]^T = \boldsymbol{C}^T[\boldsymbol{AB}]^T = \boldsymbol{C}^T\boldsymbol{B}^T\boldsymbol{A}^T$ from (v).

Finally, it should be observed that, for a vector $\boldsymbol{x} = [x_1\ x_2 \cdots x_n]^T$, the *inner product* $\boldsymbol{x}^T\boldsymbol{x}$ yields a scalar quantity which is the sum of the squares of the elements of $\boldsymbol{x}$, i.e.

$$[x_1\ x_2\ \cdots\ x_n]\begin{bmatrix} x_1 \\ x_2 \\ \vdots \\ x_n \end{bmatrix} = x_1^2 + x_2^2 + \cdots + x_n^2$$

The product $\boldsymbol{x}\boldsymbol{x}^T$, on the other hand yields a symmetric square matrix of order $n \times n$, whose elements are the squares (on the diagonal) and cross products (elsewhere) of the $\boldsymbol{x}$ elements, i.e.

$$\begin{bmatrix} x_1 \\ x_2 \\ \cdots \\ x_n \end{bmatrix}[x_1\ x_2\ \cdots\ x_n] = \begin{bmatrix} x_1^2 & x_1x_2 & \cdots & x_1x_n \\ x_2x_1 & x_2^2 & \cdots & x_2x_n \\ \cdots & \cdots & \cdots & \cdots \\ x_nx_1 & x_nx_2 & \cdots & x_n^2 \end{bmatrix}$$

Both products are of importance in the present text.

A.6 Determinant of a Matrix

The *determinant* of a square $n \times n$ matrix $\boldsymbol{A}$ is a scalar quantity, denoted by $|\boldsymbol{A}|$ or det.$[\boldsymbol{A}]$, obtained by performing certain systematic operations on the matrix elements. In particular, if the *cofactors* c_{ij} of $\boldsymbol{A}$ are defined as follows:

$$c_{ij} = (-1)^{i+j}|\boldsymbol{A}_{ij}| \tag{A.3}$$

where $|\boldsymbol{A}_{ij}|$ is the determinant of the submatrix obtained when the ith row and jth column are deleted from $\boldsymbol{A}$, then the determinant of $\boldsymbol{A}$ can be defined as follows in terms of the elements of the ith row or their cofactors:

$$|\boldsymbol{A}| = a_{i1}c_{i1} + a_{i2}c_{i2} + \cdots + a_{in}c_{in} \tag{A.4}$$

$|\boldsymbol{A}|$ may be similarly expanded in terms of the elements of any row or column.

Note that, for a matrix of order greater than 2, it is necessary to nest the operation (A.3) and operation (A.4) and apply them repeatedly until $\boldsymbol{A}_{ij}$ is reduced to a scalar, in which case the determinant is equal to the scalar. The following example demonstrates this process:

$$\boldsymbol{A} = \begin{bmatrix} a_{11} & a_{12} & a_{13} \\ a_{21} & a_{22} & a_{23} \\ a_{31} & a_{32} & a_{33} \end{bmatrix}$$

then

$$|\boldsymbol{A}| = a_{11}\begin{vmatrix} a_{22} & a_{23} \\ a_{32} & a_{33} \end{vmatrix} - a_{12}\begin{vmatrix} a_{21} & a_{23} \\ a_{31} & a_{33} \end{vmatrix} + a_{13}\begin{vmatrix} a_{21} & a_{22} \\ a_{31} & a_{32} \end{vmatrix}$$

so that, applying (A.3) and (A.4) again to the subdeterminants, we obtain:

$$|\boldsymbol{A}| = a_{11}(a_{22}a_{33} - a_{32}a_{23}) - a_{12}(a_{21}a_{33} - a_{31}a_{23}) + a_{13}(a_{21}a_{32} - a_{31}a_{22})$$

For further discussion on determinants see e.g. Johnston (1963).

A.7 Partitioned Matrices

Since a matrix is a rectangular array of elements, we may divide it up by means of horizontal and vertical dotted lines into smaller rectangular arrays of submatrices, e.g.

$$\boldsymbol{A} = \left[\begin{array}{ccc:c} a_{11} & a_{12} & a_{13} & a_{14} \\ a_{21} & a_{22} & a_{23} & a_{24} \\ \hdashline a_{31} & a_{32} & a_{33} & a_{34} \end{array}\right]$$

has been divided in this manner into four submatrices:

$$\boldsymbol{A}_{11} = \begin{bmatrix} a_{11} & a_{12} & a_{13} \\ a_{21} & a_{22} & a_{23} \end{bmatrix};\ \boldsymbol{A}_{12} = \begin{bmatrix} a_{14} \\ a_{24} \end{bmatrix};\ \boldsymbol{A}_{21} = [a_{31}\ a_{32}\ a_{33}];\ \boldsymbol{A}_{22} = a_{34}$$

So that $\boldsymbol{A}_{11}$ is a 2×3 submatrix, $\boldsymbol{A}_{12}$ is a 2×1 column vector, $\boldsymbol{A}_{21}$ is a 1×3 row vector, and $\boldsymbol{A}_{22}$ is a scalar. As a result $\boldsymbol{A}$ can be denoted by:

$$\boldsymbol{A} = \left[\begin{array}{c:c} \boldsymbol{A}_{11} & \boldsymbol{A}_{12} \\ \hdashline \boldsymbol{A}_{21} & \boldsymbol{A}_{22} \end{array}\right]$$

The basic operations for addition, multiplication and transposition apply for partitioned matrices but the matrices must be partitioned conformably to allow for such operations. A multiplicative example is:

$$\boldsymbol{AB} = \left[\begin{array}{c:c} \boldsymbol{A}_{11} & \boldsymbol{A}_{12} \\ \hdashline \boldsymbol{A}_{21} & \boldsymbol{A}_{22} \end{array}\right]\begin{bmatrix} \boldsymbol{B}_{11} \\ \boldsymbol{B}_{21} \end{bmatrix} = \begin{bmatrix} \boldsymbol{A}_{11}\boldsymbol{B}_{11} + \boldsymbol{A}_{12}\boldsymbol{B}_{21} \\ \boldsymbol{A}_{21}\boldsymbol{B}_{11} + \boldsymbol{A}_{22}\boldsymbol{B}_{21} \end{bmatrix}$$

The results of such operations will be the same as would be obtained by multiplying the unpartitioned matrices element by element (as in section A.5) but the partitioning approach may be extremely useful in simplifying the analysis.

One theorem for partitioned matrices that is sometimes useful in the context of this book concerns the determinant of a partitioned matrix $\boldsymbol{A}$, where

$$\boldsymbol{A} = \left[\begin{array}{c|c} \boldsymbol{A}_{11} & \boldsymbol{A}_{12} \\ \hline \boldsymbol{A}_{21} & \boldsymbol{A}_{22} \end{array}\right]$$

It can be shown (e.g. Gantmacher 1960; Dhrymes 1970) that:

$$|\boldsymbol{A}| = |\boldsymbol{A}_{22}|.|\boldsymbol{A}_{11} - \boldsymbol{A}_{12}\boldsymbol{A}_{22}^{-1}\boldsymbol{A}_{21}|$$

or, alternatively,

$$|\boldsymbol{A}| = |\boldsymbol{A}_{11}|.|\boldsymbol{A}_{22} - \boldsymbol{A}_{21}\boldsymbol{A}_{11}^{-1}\boldsymbol{A}_{12}|$$

where $\boldsymbol{A}_{11}^{-1}$ and $\boldsymbol{A}_{22}^{-1}$ are, respectively, the *inverses* of the matrices $\boldsymbol{A}_{11}$ and $\boldsymbol{A}_{22}$, respectively, as defined in section A.8.

A.8 Inverse of a Matrix

If a matrix $\boldsymbol{A}^{-1}$ exists such that:

$$\boldsymbol{A}\boldsymbol{A}^{-1} = \boldsymbol{A}^{-1}\boldsymbol{A} = \boldsymbol{I}$$

where $\boldsymbol{I}$ is an appropriately ordered identity matrix, then $\boldsymbol{A}^{-1}$ is termed the *inverse* (or reciprocal) of $\boldsymbol{A}$ by analogy with the scalar situation.

The inverse of a square matrix $\boldsymbol{A}$ of order $n \times n$ is obtained from $\boldsymbol{A}$ by means of the formula:

$$\boldsymbol{A}^{-1} = \frac{1}{|\boldsymbol{A}|}[\text{Adj}.\boldsymbol{A}] = \begin{bmatrix} \frac{c_{11}}{|\boldsymbol{A}|} & \frac{c_{21}}{|\boldsymbol{A}|} & \cdots & \frac{c_{n1}}{|\boldsymbol{A}|} \\ \cdots & \cdots & \cdots & \cdots \\ \frac{c_{1n}}{|\boldsymbol{A}|} & \frac{c_{2n}}{|\boldsymbol{A}|} & \cdots & \frac{c_{nn}}{|\boldsymbol{A}|} \end{bmatrix}$$

where Adj.$\boldsymbol{A}$ denotes the *adjoint* of the matrix $\boldsymbol{A}$ and is obtained as the *transpose* of an $n \times n$ matrix $\boldsymbol{C}$ with elements c_{ij} which are the *cofactors* of $\boldsymbol{A}$ as defined by (A.3), i.e.

$$c_{11} = \left|\boldsymbol{A}_{11}^{-1}\right|;\ c_{12} = -\left|\boldsymbol{A}_{12}^{-1}\right|;\ c_{22} = \left|\boldsymbol{A}_{22}^{-1}\right|,\ \text{etc.}$$

Note that, by definition, the inverse will only exist if $|\boldsymbol{A}| \neq 0$; otherwise the matrix is non-invertible or *singular*. A non-singular matrix is, therefore, invertible.

Several theorems on inverse matrices are useful, e.g.

(i) $[\boldsymbol{A}\boldsymbol{B}]^{-1} = \boldsymbol{B}^{-1}\boldsymbol{A}^{-1}$;
(ii) $[\boldsymbol{A}\boldsymbol{B}][\boldsymbol{B}^{-1}\boldsymbol{A}^{-1}] = \boldsymbol{A}[\boldsymbol{B}\boldsymbol{B}^{-1}]\boldsymbol{A}^{-1} = \boldsymbol{A}\boldsymbol{I}\boldsymbol{A}^{-1} = \boldsymbol{A}\boldsymbol{A}^{-1} = \boldsymbol{I}$;
(iii) $[\boldsymbol{A}\boldsymbol{B}\boldsymbol{C}]^{-1} = \boldsymbol{C}^{-1}\boldsymbol{B}^{-1}\boldsymbol{A}^{-1}$;

(iv) $[\boldsymbol{A}^T]^{-1} = [\boldsymbol{A}^{-1}]^T$;
(v) $|\boldsymbol{A}^{-1}| = 1/|\boldsymbol{A}|$.

One of the most common uses of the inverse matrix is in solving a set of algebraic, simultaneous equations such as:

$$\boldsymbol{X}\boldsymbol{a} = \boldsymbol{b} \tag{A.5}$$

where $\boldsymbol{X}$ is a known $n \times n$ matrix, $\boldsymbol{a}$ is an $n \times 1$ vector of *unknowns*, and $\boldsymbol{b}$ is a known $n \times 1$ vector. The reader can easily verify that this represents a set of simultaneous equations in the elements of $\boldsymbol{a}$, where $\boldsymbol{a} = [a_1\ a_2\ \ldots\ a_n]^T$, by defining $\boldsymbol{X} = [x_{ij}]$ and $\boldsymbol{b} = [b_1\ b_2\ \ldots\ b_n]^T$. Premultiplying both sides of (A.5) by $\boldsymbol{X}^{-1}$ we obtain:

$$\boldsymbol{X}^{-1}\boldsymbol{X}\boldsymbol{a} = \boldsymbol{X}^{-1}\boldsymbol{b} \quad \text{or} \quad \boldsymbol{I}\boldsymbol{a} = \boldsymbol{X}^{-1}\boldsymbol{b}$$

so that

$$\boldsymbol{a} = \boldsymbol{X}^{-1}\boldsymbol{b} = \frac{1}{|\boldsymbol{X}|}[adj.\boldsymbol{X}]\boldsymbol{b}$$

which is the required solution for $\boldsymbol{a}$ and is an alternative to other methods of solution such as pivotal elimination. For further discussion on matrix inverses see e.g. Johnston (1963).

A.9 Quadratic Forms

A *quadratic form* in a vector $\boldsymbol{e} = [e_1\ e_2\ \ldots\ e_n]^T$ is defined as $\boldsymbol{e}^T\boldsymbol{Q}\boldsymbol{e}$ where $\boldsymbol{Q}$ is a *symmetric* matrix of order $n \times n$. The reader can verify that, for $\boldsymbol{Q} = [q_{ij}]$ with off-diagonal elements $q_{ij} = q_{ji}$, $\boldsymbol{e}^T\boldsymbol{Q}\boldsymbol{e}$ is a scalar given by:

$$\boldsymbol{e}^T\,\boldsymbol{Q}\boldsymbol{e} = q_{11}e_1^2 + 2q_{12}e_1e_2 + \cdots + 2q_{1n}e_1e_n + q_{22}e_2^2 + \cdots + 2q_{2n}e_2e_n + q_{nn}e_n^2 \tag{A.6}$$

Note that if $\boldsymbol{Q}$ is *diagonal*, then this reduces to (cf. inner product):

$$\boldsymbol{e}^T\,\boldsymbol{Q}\boldsymbol{e} = q_{11}e_1^2 + q_{22}e_2^2 + \cdots + q_{nn}e_n^2$$

A quadratic form such as (A.6) is sometimes termed the *weighted Euclidian Squared Norm* of the vector $\boldsymbol{e}$ and is denoted by:

$$||\boldsymbol{e}||_Q^2 \tag{A.7}$$

As we see, it represents a very general or weighted (by the elements of $\boldsymbol{Q}$) 'sum of squares' type operation on the elements of $\boldsymbol{e}$. It proves particularly useful as a cost (or criterion function) if $\boldsymbol{e}$ represents a vector of errors (e.g. a lack of fit) associated with some model (see Chapter 8) or relates to the states of a control system (see Chapter 1, Chapter 5, Chapter 6 and Chapter 7). For instance, the cost function associated with the *Linear Quadratic* (LQ) criterion for a discrete-time state space model, as utilised extensively in this book, is based on (A.6): see e.g. equation (5.75).

A.10 Positive Definite or Semi-Definite Matrices

A symmetric matrix $\boldsymbol{A}$ is said to be *positive definite* (p.d.) if $\boldsymbol{x}^T \boldsymbol{Q} x > 0$ where $\boldsymbol{x}$ is any non-null vector. It is termed positive semi-definite (p.s.d.) if $\boldsymbol{x}^T \boldsymbol{Q} x \geq 0$.

For an $n \times n$ p.d. matrix $\boldsymbol{A}$, $a_{ii} > 0, i = 1, 2, \ldots, n$; for a p.s.d. matrix $a_{ii} \geq 0, i = 1, 2, \ldots, n$.

Note that if $\boldsymbol{A}$ is p.d. then $\boldsymbol{A}$ is *non-singular* and can be inverted; if $\boldsymbol{A}$ is p.s.d. (but not p.d.) then A is singular (Dhrymes 1970).

A.11 The Rank of a Matrix

The *rank* of a matrix is the order of its largest submatrix that is non-singular and so has a non-zero determinant. Thus for a square *nxn* matrix the rank must be *n* (i.e. the matrix must be *full rank*) for the matrix to be non-singular and invertible. For further discussion on the rank of a matrix see e.g. Johnston (1963). The rank of a matrix is particularly useful in this book in relation to the rank test for controllability (e.g. section 3.4).

A.12 Differentiation of Vectors and Matrices

The differentiation of vectors and matrices is most important in optimisation and statistical analysis. The main result concerns the differentiation of an inner product of two vectors with respect to the elements of one of the vectors.

Consider the inner product of two $(n \times 1)$ vectors $\boldsymbol{x}$ and $\boldsymbol{a}$, i.e.

$$\boldsymbol{x}^T \boldsymbol{a} = [x_1 \, x_2 \, \ldots \, x_n] \begin{bmatrix} a_1 \\ a_2 \\ \vdots \\ a_n \end{bmatrix}$$

It is clear that for all i, $i = 1, 2, \ldots, n$, the partial differentials with respect to a_i are given by:

$$\frac{\partial \, (\boldsymbol{x}^T \boldsymbol{a})}{\partial \, a_i} = \boldsymbol{x}_i$$

As a result, if the partial differentials are arranged in order of their subscripts as a vector, then this vector is simply $\boldsymbol{x}$. Thus it is convenient to refer to the process of *vector differentiation* in shorthand as:

$$\frac{\partial \, (\boldsymbol{x}^T \boldsymbol{a})}{\partial \, \boldsymbol{a}} = \boldsymbol{x} \quad \text{or} \quad \frac{\partial \, (\boldsymbol{x}^T \boldsymbol{a})}{\partial \, \boldsymbol{a}^T} = \boldsymbol{x}^T$$

The analogy with scalar differentiation is apparent from the above result. A particularly important example of vector differentiation, which occurs in Chapter 8 of this book, is concerned with the differentiation of a least squares cost function J_2 which, in its simplest form, is defined as:

$$\sum_{i=1}^{k} e_i^2$$

where $e_i = \boldsymbol{x}_i^T \hat{\boldsymbol{a}} - y_i$ is an error measure based on a vector of estimated coefficients or parameters $\hat{\boldsymbol{a}}$. In order to obtain the estimate $\hat{\boldsymbol{a}}$, it is necessary to differentiate J_2 with respect to all of the elements $\hat{a}_i$, $i = 1, 2, \ldots n$, of $\hat{\boldsymbol{a}}$. Using the above results, we see that since

$$J_2 = \sum_{i=1}^{k} [\boldsymbol{x}^T(i)\,\hat{\boldsymbol{a}})^2 - 2\boldsymbol{x}(i)^T\,\hat{\boldsymbol{a}}\,y(i) + y(i)^2]$$

then

$$\frac{\partial J_2}{\partial \hat{\boldsymbol{a}}} = \sum_{i=1}^{k}[2\boldsymbol{x}(i)\boldsymbol{x}(i)^T\,\hat{\boldsymbol{a}} - 2\boldsymbol{x}(i)y(i)] = 2\sum_{i=1}^{k} \boldsymbol{x}(i)\boldsymbol{x}(i)^T\,\hat{\boldsymbol{a}} - \boldsymbol{x}(i)y(i) \tag{A.8}$$

which, when set to zero in the usual manner, constitutes a set of n simultaneous equations in the n unknowns $\hat{a}_i$, $i = 1, 2, \ldots, n$; the normal equations.

Alternatively, we can proceed by forming the $k \times n$ matrix $\boldsymbol{X}$ with rows defined by $\boldsymbol{x}^T(i)$, $i = 1,2,\ldots, k$. The reader can then verify that the vector $\boldsymbol{e} = [e_1\ e_2 \ldots e_n]^T$ is defined by:

$$\boldsymbol{e} = \boldsymbol{X}\hat{\boldsymbol{a}} - \boldsymbol{y}$$

where $\boldsymbol{e} = [y_1\ y_2 \ldots y_k]^T$, and so

$$J_2 = [\boldsymbol{X}\hat{\boldsymbol{a}} - \boldsymbol{y}]^T[\boldsymbol{X}\hat{\boldsymbol{a}} - \boldsymbol{y}] = \hat{\boldsymbol{a}}^T\boldsymbol{X}^T\boldsymbol{X}\hat{\boldsymbol{a}} - 2\boldsymbol{a}^T\boldsymbol{X}^T\boldsymbol{y} + \boldsymbol{y}^T\boldsymbol{y}$$

since $\boldsymbol{y}^T\boldsymbol{X}\hat{\boldsymbol{a}}$ is a scalar and so equal to its transpose $\boldsymbol{a}^T\boldsymbol{X}^T\boldsymbol{y}$. It now follows straightforwardly that

$$\frac{\partial J_2}{\partial \hat{\boldsymbol{a}}} = 2\boldsymbol{X}^T\boldsymbol{X}\hat{\boldsymbol{a}} - 2\boldsymbol{X}^T\boldsymbol{y}$$

which will be seen as identical to (A.8) by substituting for $\boldsymbol{X}$ in terms of $\boldsymbol{x}(i)$.

If J_2 is replaced by the more general weighted least squares cost function, i.e.

$$J_2 = [\boldsymbol{X}\hat{\boldsymbol{a}} - \boldsymbol{y}]^T[\boldsymbol{Q}\boldsymbol{X}\hat{\boldsymbol{a}} - \boldsymbol{y}] = ||\boldsymbol{X}\hat{\boldsymbol{a}} - \boldsymbol{y}||_Q^2 \tag{A.9}$$

where $\boldsymbol{Q}$ is a symmetric p.d. weighting matrix, then it is straightforward to show that

$$\frac{\partial J_2}{\partial \hat{\boldsymbol{a}}} = 2X^T \boldsymbol{Q} X \hat{\boldsymbol{a}} - 2X^T \boldsymbol{Q} y$$

References

Dhrymes, P.J. (1970) *Econometrics: Statistical Foundations and Applications*, Harper and Row, New York.
Gantmacher, F.R. (1960) *Matrix Theory*, volume 1, Chelsea, New York.
Johnston, J. (1963) *Econometric Methods*, McGraw-Hill, New York.

B

The Time Constant

The continuous-time concept of a *time constant* is referred to occasionally in this book. In brief, the time constant relates to the analytical solution for the unit step response of a first order differential equation, and is the time taken for the output to reach 63% of the steady-state value; see e.g. Franklin *et al.* (2006, p. 107). The unit step response of a first order differential equation

$$T\frac{dy(t)}{dt} + y(t) = Gu(t) \tag{B.1}$$

is given by:

$$y(t) = G(1 - e^{-t/T}) \tag{B.2}$$

Hence, $y(t \to \infty) = G$ and $y(T) = 0.632G$. The TF form of equation (B.1) is:

$$\frac{G}{Ts + 1} \tag{B.3}$$

where G is the steady-state gain and T is the time constant. For the first order discrete-time system defined by equation (2.4), with sampling interval Δt, the equivalent time constant is given by:

$$-\frac{\Delta t}{\ln(-a_1)} \tag{B.4}$$

in which ln is the natural logarithm.

Reference

Franklin, G.F., Powell, J.D. and Emami-Naeini, A. (2006) *Feedback Control of Dynamic Systems*, Fifth Edition, Pearson Prentice Hall, Upper Saddle River, NJ.

True Digital Control: Statistical Modelling and Non-Minimal State Space Design, First Edition.
C. James Taylor, Peter C. Young and Arun Chotai.

C

Proof of Theorem 4.1

Theorem 4.1 Controllability of the Non-Minimal State Space (NMSS) Representation is stated in section 4.2. The proof is based on Wang and Young (1988). According to the *Popov, Belevitch and Hautus* (PBH) test (Kailath 1980), the state spacc system (4.4) defined by $[\boldsymbol{F}, \boldsymbol{g}]$ is uncontrollable (or unreachable) if and only if there exists a non-zero row vector $\boldsymbol{q}$ (a left eigenvector of $\boldsymbol{F}$), which is orthogonal to $\boldsymbol{g}$, i.e. we need to find a vector:

$$\boldsymbol{q} = \begin{bmatrix} q_1 & q_2 & \cdots & q_{n+m-1} \end{bmatrix} \neq \boldsymbol{0} \tag{C.1}$$

such that the following two equations hold simultaneously:

$$\boldsymbol{q}\boldsymbol{F} = \lambda \boldsymbol{q} \tag{C.2}$$

$$\boldsymbol{q}\,\boldsymbol{g} = 0 \tag{C.3}$$

where λ is an eigenvalue of $\boldsymbol{F}$. If there exists a row vector $\boldsymbol{q}$ such that equation (C.2) and (C.3) are fulfilled, then expanding these yields the following scalar equations:

$$\left.\begin{array}{ll}
-q_1 a_1 + q_2 = \lambda q_1 & (1) \\
-q_1 a_2 + q_3 = \lambda q_2 & (2) \\
\vdots & \vdots \\
-q_1 a_{n-1} + q_n = \lambda q_{n-1} & (n-1) \\
-q_1 a_n = \lambda q_n & (n) \\
q_1 b_2 + q_{n+2} = \lambda q_{n+1} & (n+1) \\
\vdots & \vdots \\
q_1 b_{m-1} + q_{n+m-1} = \lambda q_{n+m-2} & (n+m-2) \\
q_1 b_m = \lambda q_{n+m-1} & (n+m-1) \\
q_1 b_1 = -q_{n+1} & (n+m)
\end{array}\right\} \tag{C.4}$$

True Digital Control: Statistical Modelling and Non-Minimal State Space Design, First Edition.
C. James Taylor, Peter C. Young and Arun Chotai.

Let us consider the implications of the above scalar equations under the conditions $\lambda = 0$ and $\lambda \neq 0$ separately.

(i) $\lambda = 0$. In the first instance, we note that since $a_n \neq 0$, then $\lambda = 0$ is not a root of $A^*(\lambda)$ in equation (4.13). Also, from the nth equation in (C.4), we know that $q_1 = 0$ which, from the other equations of (C.4), implies that the row vector $\boldsymbol{q}$ is a zero vector. As a result, the modes $\lambda = 0$ are completely controllable.
(ii) $\lambda = \lambda_1 \neq 0$ and $A^*(\lambda_1) = 0$.

In order to consider this case, we need to manipulate equations (C.4). By multiplying equation (1) to equation $(n-1)$ of (C.4) by $\lambda^{n-1}, \lambda^{n-2}, \ldots, \lambda$, respectively, and subsequently adding all of them to equation n, we obtain:

$$-q_1(a_1\lambda^{n-1} + \cdots + a_n) = \lambda^n q_1 \tag{C.5}$$

which can be written as:

$$q_1(\lambda^n + a_1\lambda^{n-1} + \cdots + a_n) = q_1 A^*(\lambda) = 0 \tag{C.6}$$

Similarly, by multiplying equation $(n+m)$ of (C.4) by λ^{n-1} and equation $(n+1)$ to equation $(n+m-2)$ by $\lambda^{m-2}, \lambda^{m-3}, \ldots, \lambda$, respectively, before finally adding all of them to equation $(n+m-1)$, we obtain:

$$q_1 B^*(\lambda) = 0 \tag{C.7}$$

where

$$B^*(\lambda) = b_1\lambda^{m-1} + b_2\lambda^{m-2} + \cdots + b_m \tag{C.8}$$

Now assume that $A^*(\lambda)$ and $B^*(\lambda)$ are *not* coprime, i.e. that there is at least one common factor between $A^*(\lambda)$ and $B^*(\lambda)$. In this case, there is a $\lambda = \lambda_1$, such that $A^*(\lambda_1) = 0$ and $B^*(\lambda_1) = 0$. Therefore, we can always find a non-zero vector $\boldsymbol{q}$, so that equations (C.4) all hold. Conversely, if $A^*(\lambda)$ and $B^*(\lambda)$ are coprime, then $B^*(\lambda_1) \neq 0$, which implies from equation (C.7) that $q_1 = 0$. It follows from equations (C.4) that the whole vector $\boldsymbol{q}$ is a zero vector, so that the modes specified by the roots of $A^*(\lambda) = 0$ are controllable if and only if $A^*(\lambda)$ and $B^*(\lambda)$ are coprime. Finally, in this scalar case, coprime $A^*(\lambda)$ and $B^*(\lambda)$ is equivalent to coprime $A(z^{-1})$ and $B(z^{-1})$, i.e. the controllability conditions stated by Theorem 4.1.

References

Kailath, T. (1980) *Linear Systems*, Prentice Hall, Englewood Cliffs, NJ.

Wang, C.L. and Young, P.C. (1988) Direct digital control by input–output, state variable feedback: theoretical background, *International Journal of Control*, **47**, pp. 97–109.

D

Derivative Action Form of the Controller

In order to demonstrate that *Proportional-Integral-Plus* (PIP) control contains derivative action, as noted in section 5.2, we need to find the polynomials $\mathcal{F}(\Delta)$ and $\mathcal{G}(\Delta)$ using the difference operator $\Delta = 1 - z^{-1}$, such that:

$$\mathcal{F}(\Delta) \equiv F(z^{-1}) \quad \text{and} \quad \mathcal{G}(\Delta) \equiv G(z^{-1}) \tag{D.1}$$

where $F(z^{-1})$ and $G(z^{-1})$ are defined by equations (5.34).

Let

$$\mathcal{F}(\Delta) = k_0 + k_1\Delta + k_2\Delta^2 + \cdots + k_{n-1}\Delta^{n-1} \tag{D.2}$$

and

$$\mathcal{G}(\Delta) = l_0 + l_1\Delta + l_2\Delta^2 + \cdots + l_{m-1}\Delta^{m-1} \tag{D.3}$$

After substituting for $\Delta = 1 - z^{-1}$, the expression for $\mathcal{F}(\Delta)$ is expanded and the coefficients for like powers of z^i equated to those of the polynomial $F(z^{-1})$. This results in the following relationships between the coefficients of both polynomials:

$$f_j = \sum_{i=j}^{n-1} {}_iC_j \, k_i \tag{D.4}$$

where

$${}_iC_j = \frac{(-1)^j \, i\,!}{(i-j)!\, j\,!} \tag{D.5}$$

True Digital Control: Statistical Modelling and Non-Minimal State Space Design, First Edition.
C. James Taylor, Peter C. Young and Arun Chotai.

are binomial coefficients and $j = 0, 1, \ldots, n-1$. Here $i!$ is the i factorial and $0! = 1$ by definition. These relationships are expressed in vector matrix form as follows:

$$\begin{bmatrix} f_0 \\ f_1 \\ f_2 \\ \vdots \\ f_{n-1} \end{bmatrix} = \boldsymbol{T} \begin{bmatrix} k_0 \\ k_1 \\ k_2 \\ \vdots \\ k_{n-1} \end{bmatrix} = \begin{bmatrix} 1 & 1 & 1 & \cdots & 1 \\ 0 & {}_1C_1 & {}_2C_1 & \cdots & {}_{n-1}C_1 \\ 0 & 0 & {}_2C_2 & \cdots & {}_{n-1}C_2 \\ \vdots & \vdots & & \ddots & \vdots \\ 0 & 0 & 0 & 0 & {}_{n-1}C_{n-1} \end{bmatrix} \begin{bmatrix} k_0 \\ k_1 \\ k_2 \\ \vdots \\ k_{n-1} \end{bmatrix} \tag{D.6}$$

Similarly, we can show that the coefficients $g_i (i = 0, 1, \ldots, m-1)$ are related to the derivative action coefficients $l_j (j = 0, 1, \ldots, m-1)$ by the following vector-matrix equations, with $g_0 = 1$:

$$\begin{bmatrix} g_0 \\ g_1 \\ g_2 \\ \vdots \\ g_{m-1} \end{bmatrix} = \boldsymbol{T} \begin{bmatrix} l_0 \\ l_1 \\ l_2 \\ \vdots \\ l_{m-1} \end{bmatrix} = \begin{bmatrix} 1 & 1 & 1 & \cdots & 1 \\ 0 & {}_1C_1 & {}_2C_1 & \cdots & {}_{m-1}C_1 \\ 0 & 0 & {}_2C_2 & \cdots & {}_{m-1}C_2 \\ \vdots & \vdots & & \ddots & \vdots \\ 0 & 0 & 0 & 0 & {}_{m-1}C_{m-1} \end{bmatrix} \begin{bmatrix} l_0 \\ l_1 \\ l_2 \\ \vdots \\ l_{m-1} \end{bmatrix} \tag{D.7}$$

Finally, it should be noted that the transformation matrices $\mathbf{T}$ in equation (D.6) and equation (D.7) are both upper triangular matrices, such that $\boldsymbol{T} = \boldsymbol{T}^{-1}$.

Hence, equation (D.6) can be written alternatively:

$$\begin{bmatrix} k_0 \\ k_1 \\ k_2 \\ \vdots \\ k_{n-1} \end{bmatrix} = \begin{bmatrix} 1 & 1 & 1 & \cdots & 1 \\ 0 & {}_1C_1 & {}_2C_1 & \cdots & {}_{n-1}C_1 \\ 0 & 0 & {}_2C_2 & \cdots & {}_{n-1}C_2 \\ \vdots & \vdots & & \ddots & \vdots \\ 0 & 0 & 0 & 0 & {}_{n-1}C_{n-1} \end{bmatrix} \begin{bmatrix} f_0 \\ f_1 \\ f_2 \\ \vdots \\ f_{n-1} \end{bmatrix} \tag{D.8}$$

E

Block Diagram Derivation of PIP Pole Placement Algorithm

One approach for developing a general pole assignment algorithm for *Proportional-Integral-Plus* (PIP) control utilises polynomial algebra based on Figure 5.4 and the closed-loop characteristic polynomial (5.36). The closed-loop characteristic polynomial is equated to the desired characteristic polynomial as follows:

$$\Delta \left(G(z^{-1})A(z^{-1}) + F(z^{-1})B(z^{-1})\right) + k_I B(z^{-1}) = D(z^{-1}) \tag{E.1}$$

where $\Delta = 1 - z^{-1}$ is the difference operator and

$$D(z^{-1}) = 1 + d_1 z^{-1} + d_2 z^{-2} + \cdots + d_{n+m} z^{-(n+m)} \tag{E.2}$$

in which d_i are the user-specified coefficients. The left-hand side of equation (E.1) is written in the form of the following Diophantine equation:

$$A'(z^{-1})G(z^{-1}) + B(z^{-1})F'(z^{-1}) \tag{E.3}$$

where

$$\left.\begin{aligned} A'(z^{-1}) &= \Delta\, A(z^{-1}) \\ F'(z^{-1}) &= \Delta\, F(z^{-1}) + k_I \end{aligned}\right\} \tag{E.4}$$

If the system is controllable, the Diophantine equation has a unique solution. This solution can be obtained straightforwardly by equating the coefficients of like powers of z^{-i} in equation (E.1) to yield a set of simultaneous algebraic equations in the $(n + m)$ unknown pole assignment control gains: $f_0, f_1, \ldots f_{n-1}, g_1, \ldots g_{m-1}, k_I$. These equations are, of course, identical with those obtained from the state space analysis in the main text, except that the controllability requirement is not so transparent.

True Digital Control: Statistical Modelling and Non-Minimal State Space Design, First Edition.
C. James Taylor, Peter C. Young and Arun Chotai.

For computational purposes, the pole assignment equations obtained in the above manner are conveniently written in the following general form:

$$\boldsymbol{\Sigma}\,\boldsymbol{k} = \boldsymbol{d} - \boldsymbol{p} \tag{E.5}$$

where $\boldsymbol{d}$ and $\boldsymbol{p}$ are defined by equation (5.64) and equation (5.65), respectively, while $\boldsymbol{\Sigma}$ is a $(n+m)$ by $(n+m)$ matrix:

$$\boldsymbol{\Sigma} = \left[\,\boldsymbol{\Sigma}_1 \,|\, \boldsymbol{\Sigma}_2 \,|\, \boldsymbol{\Sigma}_3\,\right] \tag{E.6}$$

in which the component matrices are as follows:

$$\boldsymbol{\Sigma}_1 = \begin{bmatrix} b_1 & 0 & \cdots & 0 & 0 \\ b_2 - b_1 & b_1 & \cdots & 0 & 0 \\ b_3 - b_2 & b_2 - b_1 & \cdots & 0 & 0 \\ \vdots & \vdots & \cdots & \vdots & \vdots \\ b_m - b_{m-1} & b_{m-1} - b_{m-2} & \cdots & b_2 - b_1 & b_1 \\ -b_m & b_m - b_{m-1} & \cdots & b_3 - b_2 & b_2 - b_1 \\ 0 & -b_m & \cdots & b_4 - b_3 & b_3 - b_2 \\ \vdots & 0 & \cdots & \vdots & \vdots \\ \vdots & \vdots & \cdots & \vdots & \vdots \\ 0 & 0 & \cdots & -b_m & b_m - b_{m-1} \\ 0 & 0 & \cdots & 0 & -b_m \end{bmatrix} \tag{E.7}$$

$$\boldsymbol{\Sigma}_2 = \begin{bmatrix} 1 & 0 & \cdots & 0 & 0 \\ a_1 - 1 & 1 & \cdots & 0 & 0 \\ a_2 - a_1 & a_1 - 1 & \cdots & 0 & 0 \\ \vdots & \vdots & \cdots & \vdots & \vdots \\ \vdots & \vdots & \cdots & a_2 - a_1 & a_1 - 1 \\ a_n - a_{n-1} & a_{n-1} - a_{n-2} & \cdots & a_3 - a_2 & a_2 - a_1 \\ -a_n & a_n - a_{n-1} & \cdots & a_4 - a_3 & a_3 - a_2 \\ 0 & -a_n & \cdots & \vdots & \vdots \\ \vdots & \vdots & \cdots & \vdots & \vdots \\ 0 & 0 & \cdots & -a_n & a_n - a_{n-1} \\ 0 & 0 & \cdots & 0 & -a_n \end{bmatrix} \tag{E.8}$$

$$\boldsymbol{\Sigma}_3 = \left[b_1 \quad b_2 \quad b_3 \quad \cdots \quad b_m \quad 0 \quad 0 \quad \cdots \quad \cdots \quad 0 \quad 0\right]^T \tag{E.9}$$

Comparing (E.6) and (5.66) in the main text, it is clear that $\boldsymbol{\Sigma} = \boldsymbol{M}\boldsymbol{S}_1^T$.

We can see from these equations that the PIP control gain vector $\boldsymbol{k}$ can only be computed if the matrix $\boldsymbol{\Sigma} = \boldsymbol{M}\boldsymbol{S}_1^T$ is non-singular. This 'pole assignability' condition (a practical requirement of the above approach) and the controllability conditions developed in Chapter 5 are equivalent (see Theorem 5.1 and Theorem 5.2).

F

Proof of Theorem 6.1

Theorem 6.1 Relationship between Proportional-Integral-Plus (PIP) and Smith Predictor (SP)-PIP Control Gains is stated in section 6.3. The proof is based on Taylor *et al.* (1998). With the conditions of Theorem 6.1, the closed-loop systems (6.3) and (6.29) in Chapter 6 have exactly the same denominator polynomials. Since both systems have a gain of unity, by virtue of the inherent integral action, then the closed-loop numerator polynomials must also be equal *at steady state*. Moreover, the scalar integral gains are equal, since $B_s(z^{-1})z^{-\tau+1} = B(z^{-1})$ by definition. In this case, the numerator polynomials of the nominal PIP and SP-PIP closed loop systems are always equal.

To derive equation (6.30), consider the PIP pole assignment algorithm (E.5) developed in Appendix E, i.e.

$$\boldsymbol{\Sigma k} = \boldsymbol{d} - \boldsymbol{p}$$

where $\boldsymbol{d}$ and $\boldsymbol{p}$ are the vectors of coefficients of the desired closed-loop characteristic polynomial of the nominal PIP system and the open-loop characteristic polynomial of the NMSS model, respectively, modified here to explicitly show the time delay [see equation (6.24)], i.e.

$$\mathbf{d}^T = \begin{bmatrix} d'_1 & d'_2 & d'_3 & \cdots & d'_n & d'_{n+1} & \cdots & d'_{n+m+\tau-2} & d'_{n+m+\tau-1} \end{bmatrix} \quad \text{(F.1)}$$

$$\boldsymbol{p}^T = \begin{bmatrix} a_1 - 1 & a_2 - a_1 & a_3 - a_2 & \cdots & a_n - a_{n-1} & -a_n & \cdots & 0 & 0 \end{bmatrix} \quad \text{(F.2)}$$

The equivalent result for the SP-PIP case is:

$$\boldsymbol{\Sigma}_s \cdot \boldsymbol{k}_s = \boldsymbol{d}_s - \boldsymbol{p}_s \quad \text{(F.3)}$$

where

$$\begin{aligned} \boldsymbol{d}_s &= \begin{bmatrix} d''_1 & d''_2 & \cdots & d''_{n+m-1} & d''_{n+m} \end{bmatrix}^T \\ \boldsymbol{p}_s &= \begin{bmatrix} a_1 - 1 & a_2 - a_1 & \cdots & a_n - a_{n-1} & -a_n & 0 & \cdots & 0 \end{bmatrix}^T \end{aligned} \quad \text{(F.4)}$$

True Digital Control: Statistical Modelling and Non-Minimal State Space Design, First Edition.
C. James Taylor, Peter C. Young and Arun Chotai.

Using the conditions of Theorem 6.1:

$$d'_1 = d''_1, \;\; d'_2 = d''_2, \;\; \ldots, \;\; d'_{n+m-1} = d''_{n+m-1}, \;\; d'_{n+m} = d''_{n+m} \tag{F.5}$$

and

$$d'_{n+m+1} = d'_{n+m+2} = \cdots = d'_{n+m+\tau-2} = d'_{n+m+\tau-1} = 0 \tag{F.6}$$

If the SP-PIP $\boldsymbol{\Sigma}_s$ matrix and the vector of gains $\boldsymbol{k}_s$ are padded with an appropriate number of zeros, in order to ensure that the matrices and vectors are all of the same order:

$$\boldsymbol{\Sigma} \cdot \boldsymbol{k} = \boldsymbol{\Sigma}_s \cdot \boldsymbol{k}_s \tag{F.7}$$

Note that the $\boldsymbol{\Sigma}$ matrix is always invertible if the system satisfies the controllability conditions, i.e. the pole assignability conditions of Theorem 5.2, hence $\boldsymbol{k} = \boldsymbol{\Sigma}^{-1}\boldsymbol{\Sigma}_s \cdot \boldsymbol{k}_s$.

Reference

Taylor, C.J., Chotai, A. and Young, P.C. (1998) Proportional-Integral-Plus (PIP) control of time-delay systems, *IMECHE Proceedings: Systems and Control Engineering*, **212**, pp. 37–48.

G

The CAPTAIN Toolbox

The *Computer-Aided Program for Time series Analysis and Identification of Noisy systems* (CAPTAIN) Toolbox provides access to novel algorithms for various important aspects of identification, estimation, non-stationary time series analysis, signal processing, adaptive forecasting and automatic control system design. These have been developed between 1981 and the present at Lancaster University, UK. Much of the underlying research into the modelling tools in CAPTAIN was carried out by the second author at the University of Cambridge, UK (1965–1975) and the Australian National University, Canberra, Australia (1975–1981). Although it has its origins in the CAPTAIN and micro-CAPTAIN (Young and Benner 1991) packages, the CAPTAIN Toolbox (Taylor *et al.* 2007) developed for the MATLAB®[1] software environment is much more flexible, and provides access to many more algorithms that have been added over the 1990s and continue to be developed up to the present day, most of them only available in this Toolbox.

Essentially, the CAPTAIN Toolbox is a collection of routines (MATLAB® m-file scripts) for: the identification and estimation of the various *Transfer Function* (TF) model types discussed in Chapter 8, together with numerous other model types considered elsewhere (e.g. Pedregal *et al.* 2007; Young 2011), including the *Time Varying Parameter* (TVP) and *State-Dependent Parameter* (SDP) models defined in Chapter 9. These are organised around a core of the *Recursive Least Squares* (RLS), *Kalman Filter* (KF), *Fixed Interval Smoothing* (FIS) and *Refined Instrumental Variable* (RIV) algorithms. However, in order to allow for straightforward user access, CAPTAIN consists of numerous 'shells', i.e. top level functions that automatically generate the necessary model structures, with default options based on the experience of the developers (but user-adjustable to override these). In this regard, the main areas of functionality related to the present book are listed below.

G.1 Transfer Functions and Control System Design

The functional pairs **rivid/riv** (for discrete-time models) and **rivcid/rivc** (for continuous-time models estimated from discrete-time sampled data) are provided for order/structure

[1] MATLAB®, The MathWorks Inc., Natick, MA, USA.

True Digital Control: Statistical Modelling and Non-Minimal State Space Design, First Edition.
C. James Taylor, Peter C. Young and Arun Chotai.

identification and parameter estimation in the case of constant parameter, linear TF models. These routines include, as options, both recursive and *en bloc* versions of the optimal RIV and *Simplified Refined Instrumental Variable* (SRIV) algorithms, in addition to conventional least squares based approaches. More recent enhancements are the introduction of the **rivbj/rivcbj** and **rivbjid/rivcbjid** routines for full *Box–Jenkins* (BJ) models with *Auto-Regressive Moving-Average eXogenous variables* (ARMA) additive noise.

In all of the order/structure identification routines (the above routines ending in 'id'), the model listing, for the selection of models chosen, can be reported in the order of various statistics, such as the coefficient of determination R_T^2 and various identification statistics, including the *Akaike Information Criterion* (AIC), *Bayesian Information Criterion* (BIC) and *Young Information Criterion* (YIC), as discussed in Chapter 8. The **ivarmaid/ivarma** routines are for the identification and estimation of ARMA noise models; **ivarma** is part of the **rivbj/rivcbj** routines but is made available separately for univariate noise model estimation.

The *Proportional-Integral-Plus* (PIP) control system design routines in CAPTAIN include the **pip** algorithm for pole assignment and **pipopt** for *Linear Quadratic* (LQ) design, together with other required support routines. These are integrated into SIMULINK[2] objects that can be used in simulation studies and, potentially, for online use. Generalised PIP control routines and multivariable versions of the algorithms are also available from the first author.

G.2 Other Routines

Various other model structures are unified in terms of the unobserved components model. Here, the output time series is assumed to be composed of an additive or multiplicative combination of different components that have defined statistical characteristics but which cannot be observed directly. Such components may include a trend or low frequency component, a seasonal component (e.g. annual), additional sustained cyclical or quasi-cyclical components, stochastic perturbations, and a component that captures the influence of exogenous input signals. If the system is non-stationary, then the analysis typically utilises statistical estimation methods that are based on the identification and estimation of stochastic TVP models, as discussed in Chapter 9. An important example that is used for nonlinear PIP control is the SDP model considered in section 9.3. The modelling approach is based around the **sdp** routine which yields non-parametric (graphical) estimates of SDPs. If required, the user can parameterise these graphically defined nonlinearities using specified nonlinear functions (e.g. exponential, power law, radial basis functions, etc.) that can then be optimised using standard MATLAB® functions.

Various conventional models, identification tools and auxiliary functions, too numerous to list individually here are included. Of these, the largest is the **kalmanfis** routine, which provides a shell to the KF/FIS algorithms for general state space filtering, smoothing and forecasting purposes.

System identification is inherent to the modelling approach utilised by most of the functions already discussed. Other routines include: **acf** to determine the sample and partial autocorrelation function; **ccf** for the sample cross-correlation; **period** to estimate the periodogram; and **statist** for some sample descriptive statistics. Additional statistical diagnostics include:

[2] SIMULINK™, The MathWorks Inc., Natick, MA, USA.

boxcox (optimal Box–Cox transformation for homoscedasticity); **cusum** (cusum recursive test for time varying mean) and **cusumsq** (recursive test for time varying variance); and **histon** (histogram over normal distribution and Bera–Jarque statistical test); while useful general routines are: **del** for generating a matrix of delayed variables; **irwsm** for smoothing, decimation or for fitting a simple trend to a time series; **prepz** to prepare data for TF modelling (e.g. baseline removal and input scaling); **stand** to standardise or de-standardise a matrix by columns; and **reconst** to reconstruct a time series by removing any dramatic jumps in the trend.

Note that almost all of the recursive estimation and smoothing procedures outlined above will automatically handle missing data in the time series, represented in MATLAB® by Not-a-Number (**NaN**) variables. Indeed, by appending or pre-pending such variables to the data set using the **fcast** function, the routines will forecast, interpolate or backcast as appropriate, without requiring further user intervention.

Finally, the toolbox is supported by online help and numerous demonstration examples. These demonstrations, including several for control system design, are invoked by typing the instruction **captdemo** in the MATLAB® command line. The various command line demos can also be opened and edited, as conventional scripts.

G.3 Download

http://www.lancs.ac.uk/staff/taylorcj/tdc/

References

Pedregal, D.J., Taylor, C.J. and Young, P.C. (2007) *System Identification, Time Series Analysis and Forecasting: The Captain Toolbox Handbook*, Lancaster University, Lancaster.

Taylor, C.J., Pedregal, D.J., Young, P.C. and Tych, W. (2007) Environmental time series analysis and forecasting with the Captain toolbox, *Environmental Modelling and Software*, **22**, pp. 797–814.

Young, P.C. (2011) *Recursive Estimation and Time-Series Analysis: An Introduction for the Student and Practitioner*, Springer-Verlag, Berlin.

Young, P.C. and Benner, S. (1991) *microCAPTAIN 2 User Handbook*, Lancaster University, Lancaster.

H

The Theorem of D.A. Pierce (1972)

The theoretical justification for the *Refined Instrumental Variable* (RIV) algorithm developed in Chapter 8 is twofold. First, the development of the normal equations (8.60) for the estimation of the parameters in the Box–Jenkins model (8.38) when it is considered in pseudo-linear form; and secondly, the following theorem of D.A. Pierce (1972) that establishes the statistical properties of the *Maximum Likelihood* (ML) estimates obtained as the solution to these normal equations and shows that the ML estimates $\hat{\boldsymbol{\rho}}$ and $\hat{\boldsymbol{\eta}}$ of the system and noise parameter vectors, respectively, are asymptotically independent, so justifying an important aspect of the iterative RIV algorithm.

If, in the model (8.38) of Chapter 8:

1. *the $e(k)$ are independent and identically distributed with zero mean, variance σ^2 and skewness and kurtosis κ_1 and κ_2;*
2. *the parameter values are admissible (i.e. the model is stable and identifiable); and*
3. *the $u(k)$ are persistently exciting*[1]

then the ML estimates $\hat{\boldsymbol{\rho}}(N)$, $\hat{\boldsymbol{\eta}}(N)$ and $\hat{\sigma}^2$ obtained from a data set of N samples, possess a limiting normal distribution, such that:

1. *The asymptotic covariance matrix of the estimation errors associated with the estimate $\hat{\boldsymbol{\rho}}(N)$ is of the form:*

$$\boldsymbol{P}^* = \frac{\sigma^2}{N}\left[p\lim \frac{1}{N}\sum \hat{\boldsymbol{\phi}}_f(k)\hat{\boldsymbol{\phi}}_f^T(k)\right]^{-1} \tag{H.1}$$

[1] Identifiability and persistent excitation are important concepts discussed in most books on transfer function identification and estimation (e.g. Ljung 1999; Young 2011).

True Digital Control: Statistical Modelling and Non-Minimal State Space Design, First Edition.
C. James Taylor, Peter C. Young and Arun Chotai.

2. *The estimate* $\hat{\boldsymbol{\eta}}(N)$ *is asymptotically independent of* $\hat{\boldsymbol{\rho}}(N)$ *and has an error covariance matrix of the form*:

$$\boldsymbol{P}^*_\eta = \frac{\sigma^2}{N}[E\{\hat{\boldsymbol{\psi}}(k)\hat{\boldsymbol{\psi}}^T(k)\}]^{-1} \tag{H.2}$$

3. *The estimate* $\hat{\sigma}^2$ *has asymptotic variance* $(2\sigma^4/T)(1 + 0.5\kappa_2)$ *and, if* $\kappa_1 = 0$*, is independent of the above estimates.*

References

Ljung, L. (1999) *System Identification. Theory for the User*, Second Edition, Prentice Hall, Upper Saddle River, NJ.
Pierce, D.A. (1972) Least squares estimation in dynamic disturbance time-series models. *Biometrika*, **5**, pp. 73–78.
Young, P.C. (2011) *Recursive Estimation and Time-Series Analysis: An Introduction for the Student and Practitioner*, Springer-Verlag, Berlin.

Index

True Digital Control: Statistical Modelling and Non-Minimal State Space Design, First Edition.
C. James Taylor, Peter C. Young and Arun Chotai.